Using R for Introductory Statistics

Second Edition

Chapman & Hall/CRC
The R Series

Series Editors

John M. Chambers
Department of Statistics
Stanford University
Stanford, California, USA

Torsten Hothorn
Division of Biostatistics
University of Zurich
Switzerland

Duncan Temple Lang
Department of Statistics
University of California, Davis
Davis, California, USA

Hadley Wickham
RStudio
Boston, Massachusetts, USA

Aims and Scope

This book series reflects the recent rapid growth in the development and application of R, the programming language and software environment for statistical computing and graphics. R is now widely used in academic research, education, and industry. It is constantly growing, with new versions of the core software released regularly and more than 5,000 packages available. It is difficult for the documentation to keep pace with the expansion of the software, and this vital book series provides a forum for the publication of books covering many aspects of the development and application of R.

The scope of the series is wide, covering three main threads:

- Applications of R to specific disciplines such as biology, epidemiology, genetics, engineering, finance, and the social sciences.
- Using R for the study of topics of statistical methodology, such as linear and mixed modeling, time series, Bayesian methods, and missing data.
- The development of R, including programming, building packages, and graphics.

The books will appeal to programmers and developers of R software, as well as applied statisticians and data analysts in many fields. The books will feature detailed worked examples and R code fully integrated into the text, ensuring their usefulness to researchers, practitioners and students.

Published Titles

Using R for Numerical Analysis in Science and Engineering, *Victor A. Bloomfield*

Event History Analysis with R, *Göran Broström*

Computational Actuarial Science with R, *Arthur Charpentier*

Statistical Computing in C++ and R, *Randall L. Eubank and Ana Kupresanin*

Reproducible Research with R and RStudio, *Christopher Gandrud*

Introduction to Scientific Programming and Simulation Using R, Second Edition, *Owen Jones, Robert Maillardet, and Andrew Robinson*

Displaying Time Series, Spatial, and Space-Time Data with R, *Oscar Perpiñán Lamigueiro*

Programming Graphical User Interfaces with R, *Michael F. Lawrence and John Verzani*

Analyzing Baseball Data with R, *Max Marchi and Jim Albert*

Growth Curve Analysis and Visualization Using R, *Daniel Mirman*

R Graphics, Second Edition, *Paul Murrell*

Multiple Factor Analysis by Example Using R, *Jérôme Pagès*

Customer and Business Analytics: Applied Data Mining for Business Decision Making Using R, *Daniel S. Putler and Robert E. Krider*

Implementing Reproducible Research, *Victoria Stodden, Friedrich Leisch, and Roger D. Peng*

Using R for Introductory Statistics, Second Edition, *John Verzani*

Dynamic Documents with R and knitr, *Yihui Xie*

Using R for Introductory Statistics

Second Edition

John Verzani

CUNY/College of Staten Island

New York, USA

CRC Press
Taylor & Francis Group
Boca Raton London New York

CRC Press is an imprint of the
Taylor & Francis Group, an **informa** business

A CHAPMAN & HALL BOOK

First published 2014 by Chapman & Hall

Published 2019 by CRC Press
Taylor & Francis Group
6000 Broken Sound Parkway NW, Suite 300
Boca Raton, FL 33487-2742

ISBN 13: 978-1-4665-9073-1 (hbk)

Library of Congress Cataloging-in-Publication Data

Verzani, John.
  Using R for introductory statistics / John Verzani. -- Second edition.
      pages cm. -- (Chapman & Hall/CRC the R series)
  "A CRC title."
  Includes bibliographical references and index.
      ISBN 978-1-4665-9073-1 (hardcover : alk. paper) 1. Statistics--Data processing. 2. R
  (Computer program language) I. Title.

QA276.4.V47 2014
519.50285'5133--dc23   2014018934

Visit the Taylor & Francis Web site at
http://www.taylorandfrancis.com

and the CRC Press Web site at
http://www.crcpress.com

Contents

Preface

About this book

This is a second edition of a book that introduces R alongside the introductory statistics curriculum. The first edition found its niche with individuals looking to get started with both areas outside of a classroom environment. It is the hope, that this second edition will be even more useful for that task.

The book was first published in 2004, when R was at version 2.0.0. Now, as of writing, R is past version 3.0.0 (3.1.0 and climbing). In that time so much has changed. For example:

- The number of R users has grown enormously. A recent survey ranked R the 15th most used programming language.

- The number of add-on packages for R has grown four- or five-fold to over 5,000. The depth and range of applications has grown considerably.

- The number of books including material on R has grown at least ten-fold.[1]

- The internet has developed many additional R communities beyond the initial mailing list. Two key additions are the question and answer site stackoverflow.com which has nearly 50,000 questions tagged with "r" and the blog aggregator r-bloggers.com which has over 13,000 blog entries related to R.

Basically, the amount of material out there related to learning and using R is now enormous. This book doesn't try to canvas even a sliver, rather it tries to guide the reader through the learning of the basics of R so that it is possible to take advantage of the contributions made by the R community. Though R—like other programming languages—has a reputation of having

---

[1]For example, there are many other texts introducing R, as this one does, that can be chosen to learn from. For example, [15], [64], [13], [14], [36], [12], [56], and http://www.openintro.org/stat/.

a steep learning curve, we try to break this down into small, task-oriented steps.

In this edition we place a greater emphasis on more idiomatic R. For a small example, despite the greater familiarity of using = for the assignment operator, we now use the <- operator. Another example comes in Chapter 4, where we resist the temptation to illustrate some data manipulations with the widely used plyr package and instead utilize similar functions from base R. For our limited demands, the corner cases that led to the desire for a plyr-type approach are not present, and we have the belief that it is good to start with a grounding in the functionality provided by base R.

We also try to avoid as many of the pitfalls as possible for new R users by encouraging the use of RStudio, a feature-rich, cross-platform development environment for interacting with R. RStudio has very good integration with R's help system and its administrative tools; it has an integrated debugger, a powerful editor, and much more. Though relatively new to the R community, the company has already made an enormous contribution.

This book was written using the excellent knitr package for R. This package allows one to embed R code into a document with ease. The formatting of code blocks follows a convention championed by the knitr author. We think it makes the code much easier to read, and hence, reason about. It also encourages thinking of interacting with R using a script, rather than the command line directly. This style of usage is facilitated by RStudio.

In addition to changes with R, the teaching of introductory statistics (by which we mean a non-calculus approach to inferential statistics) has changed in the last decade, or so. For example, primarily due to the widespread availability of computational resources but also for pedagogical reasons, there have been pushes to include resampling approaches, permutation methods, and Bayesian analysis into the first-year course. The topics of this text hew closely to the traditional ones, be we have added a bit on these computer-intensive approaches, in particular to motivate the more traditional approach. We continue with an emphasis on realistic data and examples (which required updating some now not-so-topical examples) and we rely on visualization techniques to gather insight. Fortunately, the R language makes such inclusion quite easy.

Organization The text has three main parts. The first five chapters introduce the basics of exploratory data analysis and data manipulation in R. The approach is a little slower than it need be. We postpone until Chapter 4 the details of using R's data frames. These are the primary means to store multivariate data in R, and in Chapters 4 and 5 we demonstrate many tools that can act with data frames to make data investigation very convenient. However, most of these techniques are a bit more abstract, so in the first chapters we emphasize a more direct, easier to learn approach, albeit sometimes more tedious. Most all of this material was rewritten for the second edition.

Chapters 6 through 10 cover the core of statistical inference. We added the material in Chapter 7 to introduce the major themes of inference using computation, rather than probability calculations, to give insight into questions on inference.

Chapters 11 through 13 introduce the topic of analyzing statistical models with R, covering the regression model and its specialization to analysis of variance, before ending with a brief introduction to the logistic model and non-linear models. The goal is to cover the main introduction to this topic, and to show that the basic interface R provides extends naturally to cover a wide variety of models.

The appendix on programming discusses some of the details of writing programs in the R language. In the main part of the text, user-written functions are fairly straightforward, so this material is just supplemental.

The UsingR **package**   The book has an accompanying package, UsingR. This package is available from CRAN, R's repository of user-contributed packages. Installation should be painless. The package contains the data sets mentioned in the text (data(package="UsingR")), answers to selected problems (answers()), a few demonstrations (demo()), the errata (errata()), and sample code from the text.

Thanks   The author would like to thank Chapman & Hall/CRC Press. Not just the editors who have pushed for this new edition, but the company as a whole for its work on numerous titles on R-related topics. In a similar manner, the author would like to thank statistics.com. They offer a variety of R-related courses, including one that features this text. The feedback from the students of that course has been important guidance in the redrafting of parts of this text. Finally and most importantly, the author would like to warmly acknowledge the continued support he has received from his family on this and other projects.

John Verzani
February, 2014

Getting started

1.1 What is data?

Data and their statistical summaries and interpretations are ubiquitous. For example, we found these four articles during a typical day reading the paper:

● **Example 1.1: To compile evidence to establish cause and effect**
In an opinion piece, Joe Nocera [46] discusses the prevalence of guns in the movies (in anticipation of yet another "Die Hard" movie). He quotes a spokesperson from the Motion Picture Association of America as

> "There is a predominance of findings that show there is no consistent or convincing evidence that exposure [to gun violence in movies] causes people to be more violent."

However, Nocera immediately refutes this quoting a professor from the University of Wisconsin: "There is tons of research on this."

Clearly the collection and interpretation of data is crucial when making policy decisions. This isn't an easy task, of course. A casual reader may think the above differences of opinion are a matter of political motivation, but this need not be the case. Relationships between variables can exist, even if there is not a cause and effect relationship. Trying to find convincing evidence in data often requires a careful collection of data in order for conclusions to be made. ●●

● **Example 1.2: Price of a hip replacement**
In a news piece, Elisabeth Rosenthal [51] describes the research of Jaime Rosenthal who called more than 100 hospitals, covering every state in the summer of 2012 seeking the price of a hip replacement for a hypothetical, uninsured, 62-year-old female. The results were surprising:

1. Only about half the institutions could provide an estimate

2. Of those that could, the range of prices went from $11,000 to $125,798

Commentary in the article urges people to place the price data in the context of many other factors such as infection rates and unexpected deaths. However, the article summarizes the primary researcher's belief that there is little consistent correlation between higher prices and better quality in American health care.

Even in what is perhaps the most data-driven industry, there is clear need for data and context to place this data within. Further, this example hints at some other difficulties in data collection: e.g., the question of what to do with missing data, as it is often the case that some values will be unavailable. As well, the issue that the actual mechanism for computing this value at a given hospital may vary from that of another.   ••

• **Example 1.3: Safety of the airline industry**
In a front page article titled "Airline Industry at Its Safest Since the Dawn of the Jet Age," authors Jad Mouawad and Christopher Drew [43] summarize the data collected by the Aviation Safety Network pointing out that 2012 had only 23 deadly accidents and 475 fatalities. This may sound high, but putting it into a rate helps give context: this is a risk of one death per 45 million flights. That is, a person could fly daily for an average of 123,000 years before being in a fatal plane crash.

The improvements in safety are not limited to advanced technologies, as the industry (regulators, pilots, and airlines) have created a culture of sharing data about flying hazards with the goal of preventing accidents.

This example shows how a focus on understanding the many factors that can contribute to a given statistic can help improve an area. It wasn't enough that the airline kept statistics, but rather that they used their findings to address shortcomings.   ••

• **Example 1.4: Networking**
On the business page Andrew Sorkin [53] reports on a data base containing names of over two-million deal makers, power brokers and business executives, *and* in many cases the name of spouses, children, associates, political donations, charity work, and more. This information held by a company called Relations Science is compiled by more than 800 people.

The goal of course is to sell this information to people who plan to leverage the network of relationships. Of course, other companies, such as Facebook and LinkedIn have such information on their users, and the NSA seeming has all the data it could ever need, but in this case the information is scraped from web sites—a person need not be a member of a social network or have a security clearance.

How such large data bases get mined and what this means for personal privacy will likely continue to be a major topic of conversation for years to come. Though the statistical techniques of working with so-called "big data" are outside the scope of this text, many of the computational skills will be developed. ●●

In this sampling of articles, we see the analysis of data used in many different ways:

- Under the name "studies," data is used to make a case about social policy (in two different ways!).

- To investigate variability in prices and transparency, data is collected and summarized.

- In an industry, data demonstrates that forward looking practices can have a substantial effect.

- Data and the information it contains is mined to establish a financial advantage.

Data and its analysis is a very wide topic, so wide we couldn't begin to describe it all. In this text we narrow our focus, looking at data with an eye towards *statistical inference*. This is the process of drawing conclusions about populations based on data collected from these populations. To do this, we will use the language of probability. This will give us the flexibility to describe concrete things using data subject to random variation. Exactly how this will be used will require us to make models for our data. This text is roughly organized into three areas: the first to develop techniques for exploring data, the second the basics of statistical inference, and the third area covers the beginnings of modeling with data.

The rest of this chapter is focused on getting started with using R. We save more statistically oriented examples for Chapters 2 and beyond.

1.2 Getting started with R

This section covers the basics of getting started with R, beginning with some notes on installation and continuing with the basics of interacting with R through the command line.

Installing R

Before beginning with R, it must be installed for usage. R is available as source code from CRAN, http://cran.r-project.org/. However, most users probably will install R from a distributed binary. These are also available from CRAN. For example, the Microsoft Windows binary is distributed as a self-extracting .exe file. Simply download the file then install it as any other download. For Microsoft Windows users, the standard installation will

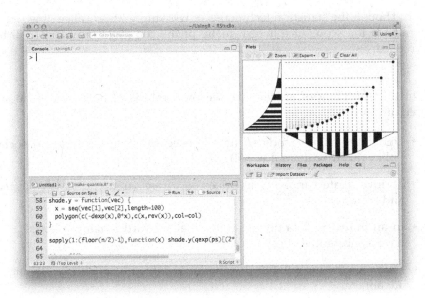

Figure 1.1: The RSTUDIO development environment for R. Visible are the console, the source code editor, the plot pane, and the workspace pane.

create a desktop icon and start menu item for opening R. If started this way, R will open to its standard Microsoft Windows GUI, but we suggest using RSTUDIO®, as described next.

Sometimes installation is a bit more difficult than described. For example, user permissions can be an issue. The "R for Windows FAQ" document, also from CRAN, can be consulted for remedies for the more common issues.

Installing RStudio

In this book we will assume the reader has installed the RSTUDIO IDE. This open-source, integrated desktop environment makes it possible for all R users to have a common R interface, which is greatly enhanced over the R's basic command line interface. Figure 1.1 shows a sample screenshot.

Installation is straightforward in most cases. The RSTUDIO web site http://www.rstudio.com has links to the necessary files to download. If there are issues, the support forum (http://support.rstudio.org/) is available for assistance. When RSTUDIO is started, it starts R with it. Starting RSTUDIOis done in a manner consistent with other applications for your operating system. For example, the Microsoft Windows installation will add an entry to the "Start Menu" to load the program.

Figure 1.2: RStudio's console showing the issuing of the command "2 + 2" and R's response of 4.

R's command line

There are several ways to interact with R, but for us the primary one will be through the *command line*, also known as the console. The command line in RStudio is in the console pane (Figure 1.2). The command line is common to all of R's interactive interfaces. The name comes from it being the place where one types in *commands*.

In the figure we typed the command "2 + 2" then pressed the return key to send the command to R's interpreter. It responded with the answer of 4, prefixed with a [1], which will make sense when we talk about data vectors in Chapter 2.

In this text, rather than show screenshots of the RStudio console, we typeset the command line. The "2 + 2" command would look like:

```
2 + 2

## [1] 4
```

Whereas, the average of five numbers might look like:

```
(1 + 3 + 2 + 12 + 8)/5

## [1] 5.2
```

The output is prefaced with R's comment character # to distinguish it from the input. Any text after a comment character is ignored by R's parser. Placing comments in front of output is not the convention with most R consoles, including RStudio, but is used here for the typesetting of R code used with this text, as we prefer not to include the prompt and need a visual clue to separate input code from output.[1]

R uses standard conventions for mathematical operations: +, -, *, /, and ˆ. Here we find the distance between two points $(1,3)$ and $(2,1)$:

---

[1]This style also is how one would interact with the R process when typing commands into a "script file" and executing these through R's source function or RStudio's "run" features. Using a script makes it much easier to reconstruct one's work in a subsequent session.

```
( (2 - 1)^2 + (1 - 3)^2 )^(1/2)

## [1] 2.236
```

R uses parentheses for grouping, as is done in math texts. Parentheses are also used when calling functions, as described shortly. Square brackets are used to extract and assign values to objects that can contain more than one. Examples will start in Chapter 2, where we discuss a container for a set of data.

Combining commands   We can place more than one command on the command line at once. We use a semicolon, ;, to separate them.

The prompt   The command line has two states, one being it is ready for input, the other expecting a continuation of the currently inputed line. It marks these states with a *prompt*. By default this will be > for a ready state and + for a continuation state.[2] These are not typeset in the text, as they can be distracting while reading. But be warned, the + prompt is indicating the previous command was not complete. If you thought it was, likely you are missing a closing parentheses.

Errors   Of course, there are times where we type in a command that does not make sense to R's interpreter. This can happen, for example, when we misspell a command name or make some syntax error. Here we have two ^ symbols, one too many for R's taste:

```
2 ^^ 2

Error: unexpected '^' in "2 ^^"
```

The error messages generated by R are usually quite informative, though may seem cryptically written to the new R user. The above one is pretty clear. R may also generate warnings, which are similar to an error, but do not stop the flow of a function.[3]

Command history   After one issues a command, it is recorded in R's history. Most command lines allow for scrolling through the previous commands using the up- and down-arrow keys. This can be used to edit and re-execute a previous command.

RStudio has a history pane (Figure 1.3) showing the past commands. One can double click on a command to send it back to the command line. Selecting more than one and then pressing the "To Console" toolbar item will

---

[2]These can be changed through the prompt and continue options, cf. ?options.

[3]If "sourcing" in commands from a script in RStudio, the error message will conveniently contain a line number.

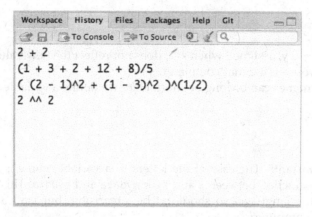

Figure 1.3: RSTUDIO's history pane showing its recording of previously issued commands.

send the collection of commands back to the console. As the history stack can grow quite large, the search panel in the history pane allows one to search through past commands. When the desired one is located, it can be viewed in its context by clicking on the small arrow on the right.

Variables

R can be used like a calculator, as above. But it really is an environment for statistical computing and graphics. The power of R goes well beyond that of a graphing calculator. One immediate difference is the ability to assign names to values.[4] In R this is done with an *assignment operator*. We use the left arrow for assignment. In RSTUDIO, there is a keyboard shortcut to insert the two-character <-, which for Windows users is alt + -.[5]

For example, here we assign a value to x and then refer to x in the subsequent command:

```
x <- 2
y <- x^2 - 2*x + 1
y                                    # assignment does not print output

## [1] 1
```

R is a dynamic language, which means we can redefine and retype values:

---

[4]Assignment basically gives an object a name in such a manner that R can look it up when asked. This process of lookup follows a procedure that defines R's scoping rules. The *scope* of a variable is the context in which the bound variable can be found. Some knowledge of this becomes important when programming new functions.

[5]Alternately, an equals sign may be used for assignment. This is more traditional with programming languages, but we stick with the R community's preferred convention.

```
x <- "two"                                  # x has a new value
```

The value of y, assigned when x=2, does not reflect the new value assigned to x unless you reissue that command.

Variable names can be long or short. Here we define a variable some_data:

```
some_data <- 9.8
```

Case is important   The case of the letters in a variable name is important. There is a distinction between x and X, or mydata and myData. This is the case with everyday language, so shouldn't be surprising, but isn't always true when using computers.

Valid names   The R documentation states that a syntactically valid name consists of letters, numbers and the dot or underline characters and starts with a letter or the dot not followed by a number. While longer names can be more descriptive, shorter ones are, of course, easier to type, but harder to remember what they represent.[6]

Tab completion   R command lines generally have *tab completion*. When a command is partially entered and the tab key is pressed, a list of possible completions for the current token are presented, or, if there is a unique completion, this token is filled in. This can make it much easier to use longer variable names, as one rarely needs to type the entire name. Figure 1.4 shows the options for completion of the token boxpl.

Built-in variables   R has very few built-in variables. One is pi referring to the value π. Another is the variable T referring to the logical TRUE value. These names may have new values bound to them.

Functions

The R language is comprised of numerous built-in functions, providing a rich set of actions. Several of these functions are for the familiar mathematical operations:

```
x <- pi
sin(x)                                  # floating-point inaccuracy
```

---

[6]There are many conventions used for making longer variable names more readable. Here are some alternatives to our use of some_data: some.data, someData, SomeData. The use of a period to separate words is common, but we reserve that for programming S3 functions. The latter two examples are camel case and upper camel case. Both are widely used. We use an underscore, as it seems easier to read, but there is no consensus in the R community on this topic.

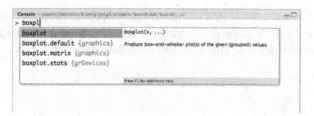

Figure 1.4: Tab completion in RSTUDIO presents the possible choices for completion, if there is more than one. When completion options are shown for a function name, a summary from the help page of each possible function is presented along with one-key access to the full page. Argument completion also shows a description of the argument.

```
## [1] 1.225e-16

sqrt(x)

## [1] 1.772
```

Functions are *called* by their name followed by a pair of parentheses. If there is more than one argument, which is often the case, these are separated by commas. An example of this would be the logarithm function which has an optional argument for the base:

```
log(x)                              # log base e = exp(1)

## [1] 1.145

log(x, 10)                          # log base 10

## [1] 0.4971
```

One of the more commonly used functions in R has the short name c. This function is used to combine values together. Here we combine several numbers and assign them to the variable x:

```
x <- c(74, 122, 235, 111, 292)
```

A typical use of this is to create a data set, of which we discuss much more in Chapter 2. There is a range of statistical functions defined for such objects. For example, we can take the average (or mean) value:

```
mean(x)

## [1] 166.8
```

The mean function in this example takes several numbers and summarizes them with 1. It does so by adding the numbers and dividing by the number of values added. This can also be achieved with:

```
sum(x)/length(x)
```

```
## [1] 166.8
```

There are also many functions for manipulating container objects like x. For example, head and tail which return the first (last) n elements, where by default n=6.

Vectorized functions R has several functions which do not summarize a collection of values with a single number, but rather do the same thing for each number. Such functions are called *vectorized*. Some examples are the standard mathematical functions:

```
x + x
```

```
## [1] 148 244 470 222 584
```

```
sqrt(x)
```

```
## [1]  8.602 11.045 15.330 10.536 17.088
```

This is very natural with statistical use. Here we subtract a single value from each value of x:

```
x - mean(x)
```

```
## [1] -92.8 -44.8  68.2 -55.8 125.2
```

In this last example the sizes of x and mean(x) did not match. R will recycle values from the smaller one to create a new matching-sized object, then do the vectorized subtraction.

Default arguments, named arguments As mentioned, R functions can have one or more arguments. This is a good thing, as it allows the user to customize a call to a function without needing to remember many different function names. To make it much easier to use functions with many arguments, the author can provide reasonable defaults for as many arguments as they see fit. This allows the user to specify relatively few values for common cases, and adjust values as desired for other, less common cases. For example, the mean function has an argument to trim the data before finding the average. This is specified with a value between 0 and 0.5, with a default of 0. With this default, we've seen the familiar average is found. When we specify the other extreme value, 0.5, we actually get the median, or middle value:

```
mean(x, trim=0.5)

## [1] 122

median(x)

## [1] 122
```

The above used a *named argument*, in this case `trim=0.5`. Additional arguments can be matched by position or by keyword. In this example either could have been used. We tend to be explicit by using keywords for additional arguments, as it is easier to see what is being specified.

Generic functions Not only can R programmers create different arguments to give functions extra flexibility, R programmers can also create entirely different function definitions based on the type of these arguments. That is, the same name may refer to different function implementations. Functions for which this is implemented are termed *generic functions*. In most cases, the exact choice of definition to dispatch depends on the *class* of the first argument. We will discuss this feature at more length in Chapter 2 and further in Appendix A. For now we illustrate with an example, using R's summary function:

```
x <- c(74, 122, 235, 111, 292)        # numeric
y <- c(TRUE, FALSE, TRUE, TRUE)        # logical
summary(x)

##    Min. 1st Qu.  Median    Mean 3rd Qu.    Max.
##      74     111     122     167     235     292

summary(y)

##    Mode   FALSE    TRUE    NA's
## logical       1       3       0
```

As seen, the summary of a collection of numbers is a statistical summary. For a collection of logical values, it is a count. The R user needs only be mindful that the function summary presents a reasonable summary of an object, and not worry about what specifically that summary will be.

Though this feature can cause confusion at first, it has a significant advantage in that far fewer function names need be remembered, as similarly behaved functions can be given the same name.[7]

---

[7]In describing functions which are generic in the text, if not noted, the most typically used implementation is described.

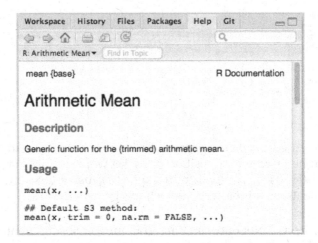

Figure 1.5: The help pane in RSTUDIOdisplaying the help page for the mean function from base R. Along the top the selector on the left is used to select previously displayed topics, the middle search box searches through page, and the rightmost search box searches the help system.

Help   R is comprised of a fairly small set of base functionality and is extended by adding additional packages to one's workspace. For the most part, the data sets and functions that are available in base R and its add-on packages are documented. R's help system allows one to access these help pages. The most basic access is provided by the help function, which has a shortcut ?, as in ?mean.

In RSTUDIO, the help pane provides an interface. Figure 1.5 shows the output from issuing the ?mean command. This command pulled up the help page for the mean function from base R. One can see a description and various ways it can be used. The mean function is a generic function, and the second usage shows what is available by default, when there is no other special implementation for the given arguments.

In Figure 1.4 we see that tab completion in RSTUDIO for a function provides access to the help page through the f1 function key.

R provides several layers of help. Table 1.1 lists a quick summary of what various commands produce, when issued from any R console:[8]

The workspace

After interacting with R one typically has created several objects and perhaps functions. Without doing anything special, R will maintain these objects in a global *Workspace*.[9] When R searches for an object at the command line, this is the first place on its path that it will look.

---

[8]There is also the add-on package SOS to search over contributed packages.
[9]This is kept in an environment returned by globalenv.

Command	Description
apropos("mean")	List objects whose names match 'mean'.
help("mean")	Find help on the mean function. Alias is ?mean.
example("mean")	Run examples found in help page for mean.
help.search("mean")	Search help data base for terms matching 'mean', searching over names, title, alias, keywords, etc. Alias is ??mean.
help(package="MASS")	List general information on the specific package.
vignette()	List all vignettes, supply topic and/or package to narrow.

Table 1.1: Example usage of various commands to access the built-in documentation.

RStudio lists most items in the global workspace along with a short summary in the Workspace pane (Figure 1.6). Clicking an item brings up an editor or viewer, depending on the object.

From the command line, the ls function can be used to list the objects in the global workspace (or other environments). When used at the console, it will list the data sets and functions a user has defined.

For example, the following lists the currently defined objects in the global workspace:

```
ls()
## [1] "a"    "d"    "out" "x"    "y"
```

To get a short summary of an object, the summary function can be used. The str function can give a longer, more cryptic, summary of the structure of an object.

For example, we've seen the summary of the following produces a statistical summary. Here we see what the structure is:

```
x <- c(74, 122, 235, 111, 292)
str(x)
##  num [1:5] 74 122 235 111 292
```

You may wish to remove objects from the workspace. This can be done through the rm function. The following shows how to remove a single object, and how to remove all the objects in the workspace.

```
rm(x)                          # single object
rm(list=ls())                  # all objects
```

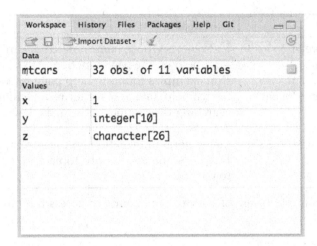

Figure 1.6: The Workspace pane offers a listing of the objects in a user's global workspace by type. Clicking an item opens an appropriate editor or viewer.

The latter, uses ls to return the names of the current objects. As these are character data, the list argument is employed. In RStudio this last action is initiated by the "broom" toolbar icon on the Workspace pane.

Sessions The global workspace and history file contain your currently defined objects and the steps for how they were created. Both are useful to keep, and R can do so from session to session. When one quits R, a prompt to "Save workspace image" is given. The default choice will write the contents of the workspace to a file to be read back in when R is started again.[10] This means that your objects are persistent from session to session.

Projects RStudio users have more options than just keeping track of a history file and global environment from session to session. The project framework allows an RStudio user to specify a directory and its files and subfolders as part of a project. In addition to providing a means to store the session information, projects make it very easy to search over all accompanying files and allows these files to easily be put under version control. Both of these are quite useful when programming with R, though we don't make use of them in this text.

---

[10]R provides the functions save and load to write (and read) representations of R objects to a file. The saved workspace is written to a file .RData in the current working directory. When R is restarted in that directory this file is loaded in as part of the usual startup process. The help page ?Startup documents the startup process.

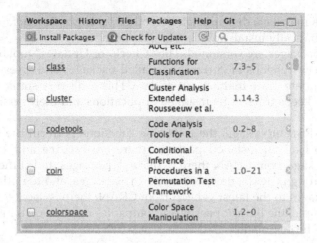

Figure 1.7: RSTUDIO package interface allows one to easily load or unload an installed package, as well one can install packages from CRAN and other sources.

External packages

As mentioned, R can be extended through external *packages* which one can install into a local R environment. There are literally thousands of such packages available.

Packages are primarily available through CRAN, R's worldwide repository of packages and R source. Several packages are also available through the BioConductor project http://www.bioconductor.org, r-forge https://r-forge.r-project.org/, GitHub https://github.com/languages/R, Google Code Page https://code.google.com/, and other sites.

RSTUDIO provides a Package pane for interacting with packages (Figure 1.7). From here one can load and unload currently installed packages by toggling the checkboxes on the left of the package name. Once loaded the (exported) functions and data sets of the package are available for use.

Packages can also be installed onto a user's system. The interface for this requires three pieces of information:

- The package name. As there are so many add-on packages, this is provided through an entry box with autocompletion.

- The repository to install the package from. The default is one of CRAN's repositories. It could also be used to indicate a locally downloaded file or another repository.

- The library of packages to install the package into. When loading an installed package, R searches over available package libraries. Often this can be left to the default, but if there are permission issues or other complications, this may need to be set. For details see ?.libPaths.

Packages may have dependencies on other packages. The default settings are to automatically install any dependent packages.

Like R, packages are versioned. The "Check for Updates" tool button will search for new versions of currently installed packages and gives the user a chance to update those that are out of date. This is all very similar to how a smartphone keeps track of its installed applications and their versions.

For non-RStudio usage, the following functions perform the core functionality: to load an installed package, there are require and library; to install a package from CRAN there is install.packages; to list the packages available through CRAN, there is available.packages; and to update any installed packages to the latest version from CRAN, there is update.packages;

For example, the UsingR package accompanies this book. To install it one could issue the command:

```
install.packages("UsingR")                    # done once
```

If one is not already set, a query, as to which CRAN repository to use for downloading files, will be made. The UsingR package has several dependencies.[11] The defaults for the above call will download and install those not currently installed into the user's package library at the same time.

Once downloaded, the function require (or alternatively, library) is used to attach the package to the workspace:

```
require("UsingR")                            # done each session
```

Data sets

Many packages include accompanying data sets. The UsingR package has several that we will see utilized in the text. This package also calls in, among others, the HistData package that provides data sets from the history of statistics and data visualization. In addition, base R has a datasets package that is loaded automatically, unless one requests something different.

For the most part, the data sets in a package are available in the user's search path, though they don't appear in the Workspace pane by default. For example, the rivers data set is part of the datasets package. Here we show the first 6 values:

```
head(rivers)                              # head displays first 6 only
## [1] 735 320 325 392 524 450
```

---

[11]The package depends on the MASS [57], ggplot [61], lubridate [27], Hmisc [34], coin [31] [32], aplpack [63], vcd [41], LearnEDA [4], quantreg [38], and HistData [24] external packages.

The data function The rivers object cannot be edited directly, any edits will produce a copy in the user's workspace. (This copy will then also display in the Workspace pane of RStudio.) A copy will also be made if one brings the data set into the workspace with the data function:

```
data(rivers)                                  # create local copy of data
```

The data function can also be used to search a package for available data sets, e.g., data(package="UsingR").

The Cavendish (HistData)[12] data set contains data from a series of experiments carried out by Cavendish in 1798 to estimate the gravitational constant, G. We can look at its first 6 values with:

```
require("HistData")                           # only needs to be done once

## Loading required package: HistData

head(Cavendish)

##    density density2 density3
## 1    5.50     5.50      NA
## 2    5.61     5.61      NA
## 3    4.88     5.88      NA
## 4    5.07     5.07      NA
## 5    5.26     5.26      NA
## 6    5.55     5.55      NA
```

Data frames The output above is different from what we have seen so far. This data set is stored as a *data frame*:

```
str(Cavendish)

## 'data.frame': 29 obs. of  3 variables:
##  $ density : num  5.5 5.61 4.88 5.07 5.26 5.55 5.36 5.29 5.58 5.65 ...
##  $ density2: num  5.5 5.61 5.88 5.07 5.26 5.55 5.36 5.29 5.58 5.65 ...
##  $ density3: num  NA NA NA NA NA NA 5.36 5.29 5.58 5.65 ...
```

A data frame is R's way of organizing several related variables into one object. Data frames are rectangular sets of data with each column being a variable and each row representing a case. We discuss much more about data frames in Chapter 4. For now, we just want to indicate how one accesses a variable from a data frame.

---

[12]We use this typesetting convention to refer to data sets in packages that are not loaded by base R.

The output of str(Cavendish) shows there are three variables in this data frame: density, density2, and density3. We can reference the values in, say, density2 through the syntax dataframe_name$variable_name, as in:

```
head(Cavendish$density2)

## [1] 5.50 5.61 5.88 5.07 5.26 5.55
```

Later, we will see other ways to do this task and why we use a dollar sign here, but this is perhaps the most common. For now, we see that we can treat this data just like a data set we may have typed in:

```
summary(Cavendish$density2)

##    Min. 1st Qu.  Median   Mean 3rd Qu.    Max.
##    5.07    5.34    5.47    5.48    5.62    5.88
```

Problems

1.1 Use R as you would a calculator to find numeric answers to the following:

1. $1 + 2(3 + 4)$

2. $4^3 + 3^{2+1}$

3. $\sqrt{(4+3)(2+1)}$

4. $\left(\dfrac{1+2}{3+4}\right)^2$

1.2 Rewrite these R expressions as math expressions, using parentheses to show the order in which R performs the computations:

1. 2 + 3 - 4

2. 2 + 3 * 4

3. 2/3/4

4. 2^3^4

1.3 Use R to compute the following

$$\frac{1 + 2 \cdot 3^4}{5/6 - 7}.$$

1.4 Use R to compute the following

$$\frac{0.25 - 0.2}{\sqrt{0.2 \cdot (1 - 0.2)/100}}.$$

1.5 Assign the numbers 2 through 5 to different variables, then use the variables to multiply all the values.

1.6 The rivers data set is loaded when R is. View the data by typing its name and then the return key. What is the last value listed?

1.7 The exec.pay (UsingR) data set is available from the command line after loading the package UsingR. Load the package, and inspect the data set. Scan the values to find the largest one.

1.8 For the exec.pay (UsingR) data set, apply the functions mean, min, and max. What are the values found?

1.9 The basic mean function has an additional argument trim. When given, the specified proportion of the data is trimmed from the sorted data before the mean is taken. Compare the difference between mean(exec.pay) and mean(exec.pay, trim=0.10).

1.10 The Orange data set is stored as a data frame with three variables. What are the three variables?

1.11 Compute the average age of the trees in the Orange data set using mean.

1.12 Compute the largest circumference of the trees in the Orange data set.

Univariate data

We discuss in this chapter single variable (*univariate*) data sets and various summaries for such data. Univariate data are the building blocks for multivariate data sets, but we resist the temptation to start there, preferring to take our time in the development.

First, what do we mean by a data set? Let's think about it in terms of a data collection process. We may wish to understand measurement or characteristics of several different cases.

A *case* is one of several different possible items of interest. A typical example would be the individuals in some population (a classroom, likely voters). In some texts [42] this is how cases are defined, but we prefer a more generic term to avoid confusion with examples such as hospitals in a state or country, or gas stations in the country.

A *variable* is some measurement or characteristic of a case. For example, with students in a classroom the last test grade; for likely voters their party affiliation, if any; and for gas stations, their current price per gallon.

A *univariate data set* is then a set of measurements for some variable from a collection of cases. We use the subscript notation to represent such a data set:

$$x_1, x_2, \ldots, x_n.$$

The subscript gives an implicit order to the data, which is basically a way to keep track of which case the measurement is for.

Levels of measurement In 1946, Stanley Smith Stevens [54] posited an influential description of various types of data. His ordering consisted of data being:

nominal Such data is qualitative or descriptive, but not numeric. An example might be the name of a person or the town they are from, or the number on a bib a runner wears in a race.

ordinal Ordinal data is data with some order, so that we can sort the data from largest to smallest, say. An example might be the place a runner takes in a race.

interval Interval data is ordinal data where the difference between two values has some interpretation. The clock time a person finishes might be an example. If we know runner A finishes at noon, and runner B at 1PM, then we know that runner B took longer. Since we haven't specified when they started, we don't know what percent longer though.

ratio Ratio data has a meaningful 0. If we record not the time of finishing, but the time since starting, then 0 has a meaning and we can take a ratio of the total time for runner A and B to compare the two.

Though this taxonomy has permeated many textbooks, especially in the social sciences, we prefer to characterize data to match how we work with it on the computer. We have:

factor data When we look at many variables, some may simply record categories used to group the data. In R we will use *factors* to store these variables. An example might be the browser a user has used to view a web site, as gleaned from a web site log. Such information is important to web programmers.

character data Some categorical data are factors, but others are really just identifiers, and are not used for grouping. An example might be a user's IP address. This is basically a unique code identifying a computer, like an address. The distinction between factor and character data can be thought of as the distinction between *categorizing* a case or *characterizing* a case. While both factor and categorical data are "nominal" we keep the distinction as we will interact with such data in R differently.

discrete data Discrete data comes from measurements where there are essentially only distinct and separate possible values that can be counted. For example, the number of visits a person makes to our web site will always be integer data, as will other counting data.

continuous data Continuous data is that which could conceivably come from a continuum of values. The recording of the time in milliseconds of a vist to a web site might be such data. A useful distinction is that for discrete data we expect that cases will share values, whereas for continuous data this will be impossible, or at least very unlikely. There is no fine line though. We can always turn continuous data into discrete data just by truncating (e.g., recording the minute, not the millisecond of a visit) or by binning. Rather than draw distinctions for numeric data between ordinal, interval or ratio, it is more important for statistical theory—in finding a model for the recorded data—to know if the data is discrete or continuous.

date and time data Though we just saw that time data can be considered continuous or discrete depending on resolution, for computers there are often separate ways entirely to handle date and time data. Issues that complicate matters are as familiar as leap days and time zones, but there are even more subtleties. For example, scale. People in finance want millisecond data, but over long time ranges this recording can literally run out of numbers on a computer. Astronomers need precise measurements for durations down to leap seconds. R has several ways to work with such data, that go beyond just storing the values as simple numbers.

hierarchical data The traditional idea of a data set being several measurements for different cases is widely established. Such data nicely fits into a spreadsheet in a rectangular manner, and in tables of data bases. However, this structure doesn't fit well for every data set of interest. For example, data on networks. We don't discuss such examples in the text, but include it here to point out that this list is far from comprehensive.

This chapter will begin with some R basics for working with data sets and then proceed to look at various summaries—numeric and graphical—of different types of data.

2.1 Data vectors

Our notation for a data set, $x_1, x_2, \ldots, x_n$ suggests already a few things: there are n items, and, by using a common name, they are all measurements of the same type. R's basic data structure is perfectly suited for this. The c function can be used to bundle our data set together.

Suppose the number of whale beachings in Texas during the 1990s was

74 122 235 111 292 111 211 133 156 79

We can combine these values into a data set with R through:

```
whale <- c(74, 122, 235, 111, 292, 111, 211, 133, 156, 79)
```

We just separate values with a comma. We refer to the whale object as a *data vector*.[1] It has the same properties as a data set: a size and a common type for the values.

The size of the data set is the "n", and is retrieved in R with the length function:[2]

```
length(whale)
```

```
## [1] 10
```

---

[1]Technically, what we call "data vectors" are more commonly referred to as just "vectors," a more common mathematical usage. They are generalized by adding dimensions to create matrices and arrays.

[2]Both c and length have a much wider usage. Here we describe their specialization to data vectors.

The number of elements is of interest, and R shows a count when it prints a vector. This is the [1] that appears before the output. As even single numbers (scalars) are treated as vectors of length 1, these too are prefixed, as above. When the output spans more than one row, each row is prefixed by the index of the first member of the row.

There are, of course, many other useful functions in R to extract information from a data vector. For example, as seen, we can get the total number using sum:

```
sum(whale)
```

```
## [1] 1524
```

The average number can be found by combining these two:

```
sum(whale)/length(whale)
```

```
## [1] 152.4
```

Or, more commonly, this is done directly through the mean function:

```
mean(whale)
```

```
## [1] 152.4
```

Vectorization   As mentioned, the arithmetic operations and the mathematical functions are vectorized, in that they will be called for each element in a data vector. Where sum, length, and mean are reductions, in that when they are called with a vector they reduce it down to a number, vectorized functions maintain the size of the vectors involved, possibly after recycling to reach a consistent length. Above, the division by length(whale) used vectorization.

Here are more examples, all returning a vector of length 10:

```
whale - mean(whale)                      # mean(whale) recycled
```

```
## [1] -78.4 -30.4  82.6 -41.4 139.6 -41.4  58.6 -19.4   3.6 -73.4
```

```
whale^2 / length(whale)
```

```
## [1]  547.6 1488.4 5522.5 1232.1 8526.4 1232.1 4452.1 1768.9
## [9] 2433.6  624.1
```

```
sqrt(whale)
```

```
## [1]  8.602 11.045 15.330 10.536 17.088 10.536 14.526 11.533
## [9] 12.490  8.888
```

Missing values, NA   Some data sets are not complete. In the initial examples, we discussed a *New York Times* article on a study [33] to look at the price of hip replacements. Simulated data from 15 hospitals is given here:

```
10,500 45,000 74,100 NA     83,500 86,000 38,200 NA
44,300 12,500 55,700 43,900 71,900 NA     62,000
```

Here we use, as the authors did, the value "NA" (not available) for data that was not available. In this case, the surveyed hospitals could not provide the information on the cost of a total hip replacement.

Such situations are quite common. R provides the special variable NA to represent values that are missing. Using this, to enter the above data set into R could be done with:

```
hip_cost <- c(10500, 45000, 74100, NA, 83500, 86000, 38200, NA,
              44300, 12500, 55700, 43900, 71900, NA, 62000)
```

The value NA is interpreted as the value is missing, but could possibly be there. As such, it would be incorrect to assume it has no value—it is just unknown. This effects how the data is used. For example, we could try and take the total costs, as perhaps we work for an insurer and are attempting to estimate a total exposure:

```
sum(hip_cost)
```

```
## [1] NA
```

We see then this sum is also not available. (It can be said that NA values poison subsequent operations.) Many R functions have an argument to specify what to do with missing data. For sum this argument is na.rm (remove NA values). The default is on the side of caution, but we can specify TRUE to change that:

```
sum(hip_cost, na.rm = TRUE)
```

```
## [1] 627600
```

The mean function has the same argument. A consumer may want to know if a quoted charge is out of line. The following shows the average costs for those hospitals that reported a value is just over $52,000:

```
mean(hip_cost, na.rm=TRUE)
```

```
## [1] 52300
```

For multivariate data sets, and the functions that interact with them, we will see there are generally more options for dealing with NA values.

NULL   Somewhat related to NA is the value NULL. NULL is a reserved value usually indicating that some requested action is undefined or unavailable.

NaN, Inf   The value NaN looks like NA, but is different. This value arises from arithmetic operations that are undefined, such as 0/0, or unrepresentable. The value Inf stands for infinity, and comes from evaluations such as 1/0. Inf can be positive or negative and can be used as expected with R's ordering operators, like <. (Comparisons with NaN produce NA.) These values are part of the specification for floating point numbers implemented by R.

Attributes: names   Data vectors may have attributes. The "names" attribute allows the association of a name with each case. For example, the built-in precip data set lists the average rainfall in inches for 70 cities in the United States. Here we list the first 6 using the head function:[3]

```
head(precip)

##        Mobile       Juneau      Phoenix Little Rock Los Angeles
##          67.0         54.7          7.0        48.5        14.0
##    Sacramento
##          17.2
```

The city names are printed above the values. Clearly this is useful information when looking at the data, as otherwise we'd need to be familiar with how the index matches up with the city. In fact, operations may make tracking the index mentally quite challenging. Here we sort the data biggest to smallest to find the top 6 rainiest cities:

```
head( sort(precip, decreasing=TRUE) )

##        Mobile        Miami     San Juan  New Orleans      Juneau
##          67.0         59.8         59.2         56.8        54.7
## Jacksonville
##          54.5
```

The names function returns the names associated with a data vector. Here we show the first 6 names—without the associated data:

```
head(names(precip))

## [1] "Mobile"      "Juneau"       "Phoenix"      "Little Rock"
## [5] "Los Angeles" "Sacramento"
```

---

[3]The head function is used to show the first k elements, with $k = 6$ by default; the tail function is used to show the last k elements, again with $k = 6$ by default. The headtail function is provided by UsingR to show the first and last k elements with ellipses separating them with a default of $k = 3$.

Creating named components   The setting of names is so common, that there
are several means to do so. For example, we can specify the values through
c using what is essentially a key=value syntax:

```
test_scores <- c(Alice = 87, Bob = 72, Shirley = 99)
```

Quoting the key values is optional, though necessary if a name contains
spaces.

Alternatively, the setNames function can be used to set the names. This
is more commonly used with programming, as you'll see the syntax is a bit
more verbose:

```
test_scores <- setNames(c(87, 82, 99), c("Alice", "Bob", "Shirley"))
```

Assignment functions   In R there are several functions which take on two
forms, one to "get" and one to "set" values. The names of these functions
come in pairs, for example names and names<-. The latter is used for assignment. Though they look similar when used, the setting operator appears on
the left-hand side of the assignment, as below.[4]

```
test_scores <- c(87, 782, 99)
names(test_scores) <- c("Alice", "Bob", "Shirley")
test_scores

##   Alice     Bob Shirley
##      87     782      99
```

We used a character vector for the names. The assignment functions are
typically called through fun(x) <- value to set the value for x. This notation
can be a bit confusing at first, as it doesn't fall into the typical x <- value.

Coercion   As mentioned a univariate data set, being comprised of similar
measurements, should have values of the same basic type. This need not be
the case when entered into a data vector:

```
x <- c(1, "two", "III")
x

## [1] "1"    "two" "III"
```

Note how x is printed: all the values are in quotes. What happened? Data
vectors too must all be of the same type when stored in R.[5] Here R silently

---

[4]The assignment functions mutate the values assigned to a variable, unlike most other functions in R which do not.

[5]In R, a list is a generalized vector where each component may have a different type.

coerced, or converted, the numeric value 1 into the string "1". Basically, if we try to combine different types of objects through c, R will promote each element to a common type, which often will be a character type.

One common mistake is to use the character string "NA" instead of the NA variable itself when specifying missing values. Such a mistake will silently convert all values to character.

The coercion can also be done through the "as" functions. Here, we show how to coerce back and forth between numbers and characters:

```
as.numeric("1")

## [1] 1

as.character(1)

## [1] "1"
```

There are several other specific "as" functions and one simply named as to do general coercions.

scan   There are many alternatives to c for entering data into R. We wait until Chapter 4 to discuss most of these, as they often result in a data frame, which are discussed therein. However, one way to read data in from the command line or a file that returns a data vector is the scan function. This function has many arguments, we mention just its first which allows us to indicate if the function should scan the keyboard input or scan the contents of a file. If left empty, it will scan the keyboard until an empty line is specified. With scan we can separate values by spaces:

```
scan()
1: 74 122 235 111 292 111 211 133 156 79
11:

Read 10 items
 [1]   74 122 235 111 292 111 211 133 156   79
```

In the above, we copied and pasted in the values and then added a new line to indicate we were done. This output the 10 values we specified, but did not save them, as we did not assign the values to a name.

Alternatively, we can scan values from a file. For example, suppose the file whale.txt contained the same data. This command will read in the file and store its scanned values in the whale variable.

```
whale <- scan("whale.txt")                    # or scan(file.choose())
```

We specified a file name on the working directory. If that is not where the data file is, it can be convenient to browse for the values. The file.choose function will do just that.

Structured data

There are some convenient functions for generating structured data. For simple sets of contiguous integers the colon operator, :, may be used:

```
1:5                                            # 1 2 3 4 5

## [1] 1 2 3 4 5

1:length(whale)                                # 1 2 3 ... 10

## [1]  1  2  3  4  5  6  7  8  9 10

0:(length(whale) - 1)                          # 0 1 2 ... 9

## [1] 0 1 2 3 4 5 6 7 8 9
```

The value returned by the command a:b is a sequence from a to b with a step size of 1 and not exceeding b. We can have b being less than a, in which case the sequence counts down. The second example shows a common construct used when we want to reference all values of a data vector by its indices.[6] The third example illustrates that parentheses are needed to do arithmetic on the values of a and b, as the colon operator has higher precedence, it will be done before the subtraction unless otherwise directed through parentheses.

Sequences increasing by 1 are *arithmetic sequences*, which more generally can increase (or decrease) by a given step size, say h (which could be negative). To generate the values $a, a + h, \ldots, b = a + nh$, the seq function is convenient. Such a sequence is returned through seq(a, b, by=h), or using positional arguments just seq(a, b, h). To count by 10s to 100 we could have:

```
seq(0, 100, by=10)                             # count by 10s

## [1]   0  10  20  30  40  50  60  70  80  90 100
```

The value of b is a suggestion and if the argument for by does not allow b-a to be written as a multiple of h, b won't be included (or exceeded) in the returned vector.

---

[6]For programming, it is suggested to use the function seq_along which is basically 1:length(x), but handles more gracefully the case where the data vector is of 0-length.

To return a sequence of a given length between a and b can be done by choosing the by value appropriately, but is much easier done by specifying the desired length, through the length.out argument:

```
seq(0, 100, length.out=11)                    # counts by 10s as well

## [1]   0  10  20  30  40  50  60  70  80  90 100
```

The rep function   The rep function can be used to repeat values. The basic form is to repeat the first argument, the number of times specified in the second:

```
rep(5, times=10)                              # 10 5s

## [1] 5 5 5 5 5 5 5 5 5 5
```

That can be convenient. The values can be vectorized, for much more complicated patterns. To get the pattern 1, 2, 3 four times we have:

```
rep(1:3, times=4)

## [1] 1 2 3 1 2 3 1 2 3 1 2 3
```

Whereas to get the pattern 1, 1, 1, 2, 2, 3 we have:

```
rep(c(1,2,3), times=c(3,2,1))                 # or rep(1:3, 3:1)

## [1] 1 1 1 2 2 3
```

When the times argument is a vector of length matching the length of x, then it specifies how many times each corresponding value in x should be repeated.

Indexing

The subscript in our data set notation, $x_1, x_2, \ldots, x_n$, indicate that data has a notion of size and order. R's data vectors allow us to access and assign to parts of a data vector using indices. As in our notation, the first element is indexed by a 1.[7] There are several different ways that we can extract parts of a data vector: by numeric index, by name, by a matching-sized logical vector (c.f. Table 2.1). We will come back to the latter.

---

[7]Several computer languages use 0-based indices.

Numeric indices   The notation for indexing is the square bracket [in contrast to how functions utilize parentheses.[8] Inside the paired brackets go the specific index. The simplest case is just a single number:

```
whale                                   # all the values

## [1]   74 122 235 111 292 111 211 133 156   79

whale[1]                                # first element

## [1] 74

whale[2]                                # second element

## [1] 122

whale[11]                               # 11th element is not there

## [1] NA
```

The last command, whale[11], illustrates what happens when an attempt to access an index beyond the length of the data vector. R does not raise an error, but rather returns NA.

We are not limited to single indices, the index vector can be as long as desired:

```
whale[c(1,3,5,7,9)]

## [1]   74 235 292 211 156
```

Here we see that the indexing operation is vectorized, and all the corresponding values are returned.

There is a convenient convention of *negative indexing* which extracts *all but* the specified values. This can be quite convenient. The one caveat, we cannot use both negative and positive indices at once. This call will return all but the first element of whale:

```
whale[-1]

## [1] 122 235 111 292 111 211 133 156   79
```

Indexing by 0 returns a 0-length data vector, as does indexing by a 0-length vector. If no index is specified, as in whale[], the entire data vector is returned.

---

[8]We will see that lists also use a double square bracket, [[.

Indexing with names   For data vectors with a names attribute, we can reference values by their name, instead of position. For example, the average amount of rain in Seattle and New York is given by:

```
precip[c("Seattle Tacoma", "New York")] # which is rainier?

## Seattle Tacoma      New York
##            38.8          40.2
```

The match function is used to find where a value is in a data set. To find out the corresponding indices for these names, one could use the following construct and then index by number:

```
match(c("Seattle Tacoma", "New York"), names(precip))

## [1] 65 42
```

This illustrates the greater ease of using the names directly. There is no negative indexing with names.[9] As well, when a name does not exist, the value of NA is given:

```
precip["Seattle"]                        # needs "Seattle Tacoma" to match

## <NA>
##   NA
```

Later, we will see that we can index by a logical vector in addition to numeric indices or names.

Assignment through indexing   The assignment function [<- allows us to assign to parts of a data vector. The simplest case is when the length of the indexing vector matches the length of the assigned values. For example, to change the first value of a data set can be done through:

```
x <- c(1, 2, 3)
x[1] <- 11                               # 11 is now first value
x

## [1] 11  2  3
```

The index can be a vector, so the following works:

---

[9]The subset function which does have some features for negative indexing with names will be discussed in Chapter 4.

Command	Description
x[1]	The first element of x.
x[]	All elements of x.
x[length(x)]	The last element of x.
x[c(2,3)]	The second and third elements of x.
x[-c(2,3)]	All but the second and third elements of x.
x[0]	0-length vector of same type as x.
x[1] <- 5	Assign a value of 5 to first element of x.
x[c(2,3)] <- c(4,5)	Assign values to second and third elements of x. In assignment, recycling of the right-hand side may occur. Assignment can grow the length of a data vector.

Table 2.1: Various uses of indexing by numeric indices.

```
x[2:3] <- c(12, 13)
x

## [1] 11 12 13
```

We can extend the size of a data vector through assignment. "Holes" will be filled in by NAs:

```
x[6] <- 6
x

## [1] 11 12 13 NA NA  6
```

To reduce the size of a data vector we simply reassign the variable using a subset of the data vector, e.g. x <- x[1:2].

Recycling   When the right-hand side has fewer elements than the data vector referred to on the left-hand side, R will recycle the value on the right-hand side. This makes it easy to assign many values at once. For instance to set the last two values to 0 in x, we have:

```
x[2:3] <- 0
x

## [1] 11  0  0 NA NA  6
```

The vector 0 is recycled to c(0,0) then assigned. When the right-hand side is a single value, recycling is easy to understand. When it is not, then the right-hand side is repeated a sufficient number of times:

```
x <- 1:10
x[] <- 1:3                                        # 10 on left, 3 on right

## Warning: number of items to replace is not a multiple of replacement
length

x

## [1] 1 2 3 1 2 3 1 2 3 1
```

R gives a warning about the 3 (the length of the right-hand side) not being a multiple of 10 (the number of items referenced through x[], but does as requested).

Recycling isn't magic, just very convenient. If could be done manually with the rep function—just repeat the vector a given number of times. For recycling that is not a multiple of the desired length, we can use a construct like the following. The %% operator finds the remainder after division, and then these values are shifted by 1 to get the corresponding index:

```
n <- 10                                   # get 10 elements of x
x[1 + 0:(n-1) %% length(x)]               # use remainder for indices

## [1] 1 2 3 1 2 3 1 2 3 1
```

Data types

So far, we have used numeric and character data in the examples. Of course, there can be many other types of data possible. To organize this, R assigns a *class* attribute to most R objects and otherwise creates an implicit class for an object. The class of an object is used to determine how it should be printed.[10] The class function will return the class of an object. For most objects, this is a single character, but may be a character vector.

Numeric data types

The two main classes for numeric data are numeric and integer, though there are others, e.g. complex. Most of the time numbers are numeric. For example, to see that all of these objects are numeric:

```
c(class(1), class(pi), class(seq(1, 5, by=1)))

## [1] "numeric" "numeric" "numeric"
```

---

[10]Indeed, the class of an object decides which function definition should be used for many different functions, not just print and show.

To make an integer value, we need to work a bit: we can preallocate space for an integer data set of length n with integer(n); we can use the suffix L to force a number to be treated as an integer (e.g., 1L); we can coerce numeric values of integer type through the as.integer function.

To the casual R user, the distinction is not important. Integers are stored differently. They are precisely known, but have a limited range, roughly between $\pm 2 \cdot 10^9$. Numeric values are stored using floating point representation. This format can store much larger integer values and has a much wider range of numbers it can represent.[11]

However, floating point representation cannot store all numbers exactly. For the data sets in this text, we won't need to think about this. Though if we try to be too literal, the impact can pop up. For example, the square root of 2 is irrational and the floating point representation is an approximation. It shows in the second of these expressions:

```
sqrt(2) * sqrt(2)                          # looks right

## [1] 2

sqrt(2) * sqrt(2) - 2                       # a difference

## [1] 4.441e-16
```

In the last command, we see that there is a small difference between the two numbers—out in the 16th decimal point. This difference can be an issue when checking equality of values.

Categorical data types

R has two distinct classes for working with categorical data: factor and character. As mentioned, to distinguish the two: factors are used to classify values, character data is used to characterize values.

Character data Character data is created just by quoting values. Quotes can be matching pairs of single or double quotes, though double quotes are preferred and used to display character values. Within a quoted value a quote symbol can be used, but it must be escaped by prefixing it with a backslash (cf. ?Quotes for more details.)

---

[11]We can think of each floating point number as stored with three parts: a sign (± 1), an exponent (e, with $k = K - e$ for some K), and a precision, p. Then we can use scientific notation to represent each number through $\pm 1 \cdot p10^k$. In R—like other computer languages—scientific notation is printed with an e to indicate the exponent, e.g., 3.141593e+13 or -3.183099e-14. The format also allows for the representation of $\pm \infty$: Inf, -Inf; a value for "not a number:" NaN; and plus or minus 0.

```
c("Lincoln", "said", "\"Four score", "and seven years ago...\"")
```

```
## [1] "Lincoln"                    "said"
## [3] "\"Four score"               "and seven years ago...\""
```

Combining strings R has some useful utilities to combine strings. Two we mention are `sprintf` and `paste`. The `sprintf` function uses C-style formatting of strings in a vectorized manner. The details are in the help page. We illustrate two common uses: creating labels and padding values.

To create the labels, X1, X2, ..., X10, we could type them all in, or note they are just X with the numbers 1 through 10. This leads us to a template "X%s" where the %s is filled in with a string (the 1:10 is coerced into character data via as.character):

```
sprintf("X%s", 1:10)
```

```
## [1] "X1"  "X2"  "X3"  "X4"  "X5"  "X6"  "X7"  "X8"  "X9"  "X10"
```

There are format strings for numbers too. For example, to pad numbers so as they line up in a table when printed, we might want to pad with leading spaces. The template "%8d" indicates it will use 8 characters and expects an integer value (d). For example:

```
sprintf("%8d", c(1, 12, 123, 1234, 12345))
```

```
## [1] "       1" "      12" "     123" "    1234" "   12345"
```

The pattern "%2.6f" would be used with numeric numbers to show 6 decimal places, and pad the integer part to at least 2 spaces.

The paste function The `paste` function offers similar functionality to `sprintf`. For example, the label example would be done with:

```
paste("X", 1:10, sep="")
```

```
## [1] "X1"  "X2"  "X3"  "X4"  "X5"  "X6"  "X7"  "X8"  "X9"  "X10"
```

However, `paste` has more flexibility. This function has two named arguments sep and collapse, and can take an arbitrary number of variables.

The sep argument is used as a separator between variables:

```
paste("The", "quick", "brown", "fox", "...", sep="_")
```

```
## [1] "The_quick_brown_fox_..."
```

This default value of sep is a single space. (The paste0 function is basically paste with a fixed value of sep="".) Of course, the variables can be vectorized (not sep):

```
paste(c("Four","The"), c("Score","quick"), c("and","fox"), sep="_")

## [1] "Four_Score_and" "The_quick_fox"
```

Recycling will be used to match vector lengths. The result of the above is a vector. The collapse arg can be given to combine the elements of the resulting vector with the given delimiter. This example takes a named vector and produces a different output:

```
x <- c(one=1, two=2, three=3)
out <- paste(names(x), x, sep=":", collapse=", ")
sprintf("[ %s ]", out)

## [1] "[ one:1, two:2, three:3 ]"
```

The above is a start on writing an R object in a common string-based format (JSON). It shows how we can compose the two functions, which is sometimes most convenient to get a desired output.

R has many other utility functions for manipulating character data available to the programmer. We won't have need of these in this text, but note that the useful add-on package stringr gives them a more uniform interface and adds some conveniences.

Factors   A factor can be made from a character vector with the factor function. For example, an experiment might have twelve cases, each in one of three categories. Here we can create a variable to indicate which:

```
x <- paste("X", rep(1:3, 4), sep="")
y <- factor(x)
y

##  [1] X1 X2 X3 X1 X2 X3 X1 X2 X3 X1 X2 X3
## Levels: X1 X2 X3
```

The values in y print differently than those in x, which would be enclosed in quotes. As well, the "levels" of y are printed.

The *levels* of a factor are a list of all possible categories for the data in the factor. They need not all be represented in a particular factor, but when we create a factor through factor the default choice is simply the collection of unique values. (We can specify more through the levels argument.) The current levels of a factor are returned by the levels function.

The presence of levels makes working with factors a bit different. We can not assign a value into a factor unless it matches a level. Trying to will raise a warning and insert a value of NA:

```
y[1] <- "X4"

## Warning: invalid factor level, NA generated

y

## [1] <NA> X2  X3  X1  X2  X3  X1  X2  X3  X1  X2  X3
## Levels: X1 X2 X3
```

Adding a level   To add a level to a factor can be done with the levels<- function. The following would allow us to assign X4 to the first element of y:

```
levels(y) <- c(levels(y), "X4")        # add "X4" to existing levels
y[1] <- "X4"
y

## [1] X4 X2 X3 X1 X2 X3 X1 X2 X3 X1 X2 X3
## Levels: X1 X2 X3 X4
```

The right-hand side of the assignment can be used to change the names for the labels:

```
levels(y) <- paste("label", 1:4, sep="")
y

## [1] label4 label2 label3 label1 label2 label3 label1 label2
## [9] label3 label1 label2 label3
## Levels: label1 label2 label3 label4
```

Or can be used to collapse levels. Here we collapse a state factor to a region factor (the state.region variable could also be used here):

```
y <- factor(state.name[1:5])
y

## [1] Alabama    Alaska    Arizona    Arkansas   California
## Levels: Alabama Alaska Arizona Arkansas California

levels(y) <- c("South", "West", "West", "South", "West")
y

## [1] South West  West  South West
## Levels: South West
```

The above shows one way to drop a level. Another way is to specify the factors. For example, placing the [1:5] after calling factor produces 50 levels for our data vector of 5 states. We can specify levels by calling factor with specific levels listed:

```
y <- factor(state.name)[1:5]
y                                        # 50 levels

## [1] Alabama    Alaska  ～ Arizona    Arkansas   California
## 50 Levels: Alabama Alaska Arizona Arkansas ... Wyoming

factor(y, levels=y)                      # levels are actual values

## [1] Alabama    Alaska     Arizona    Arkansas   California
## Levels: Alabama Alaska Arizona Arkansas California
```

Generating new factors Related to rep, but used for generating factors which are repeated, is the gl function. Suppose our task is to create a factor with 5 copies of red, green, and blue. We could type this in:

```
r <- "red"; b <- "blue"; g <- "green"
factor(c(r,r,r,r,r,g,g,g,g,g,b,b,b,b,b))

##  [1] red   red   red   red   red   green green green green green
## [11] blue  blue  blue  blue  blue
## Levels: blue green red
```

Or we could have used rep(c(r, g, b), c(5,5,5)).

Either isn't too terrible, but such repetition begs for a more automated approach. The gl function does this, we specify the number of levels (3), how often they are to be repeated (5 times), and optional labels:

```
gl(3, 5, labels=c("red", "green", "blue"))

##  [1] red   red   red   red   red   green green green green green
## [11] blue  blue  blue  blue  blue
## Levels: red green blue
```

We can also specify a length apart from the default of $n \cdot k$ which will cause truncation or recycling.

Another common set of derived factor levels come from an interaction between two variables. As a statistical concept, this is addressed later. Regarding the value, the interaction between a factor f1 and f2 is a new factor (basically an ordered pair of the two) with the new levels being all possible pairs of the levels.

For example, consider this reduced set of data from the Cars93 (MASS) data frame:

```
m <- head(Cars93)
out <- m$Origin : m$AirBags
out

## [1] non-USA:None              non-USA:Driver & Passenger
## [3] non-USA:Driver only       non-USA:Driver & Passenger
## [5] non-USA:Driver only       USA:Driver only
## 6 Levels: USA:Driver & Passenger USA:Driver only ... non-USA:None

levels(out)

## [1] "USA:Driver & Passenger"     "USA:Driver only"
## [3] "USA:None"                   "non-USA:Driver & Passenger"
## [5] "non-USA:Driver only"        "non-USA:None"
```

Not all the possible pairs are represented in the new factor, out, but these constitute the levels.

Ordered factors   Factors represent categorical data of any type. For categorial data which is ordered, R has the specialization of factor: ordered factors. Passing the argument ordered=TRUE to the factor constructor will indicate that the specified levels are ordered. The distinction is most important when modeling with factors, but the order does indicate how the levels should be printed.

Date and time types

Working with dates and times is made more convenient using a special data type. While R has some built-in features to work with dates and times, the lubridate package will be described here, as it simplifies the usage. It is installed with the UsingR package, but will need to be loaded. This package introduces the notion of "instants," "durations," and "intervals" of time. We concern ourselves with some basics, learning how to make and manipulate instants of time.

The following finds the current time, reports its class, then calls an accessor function to extract the month:

```
require(lubridate)
current_time <- now()              # current time
class(current_time)                # cf. ?POSIXct

## [1] "POSIXct" "POSIXt"

as.numeric(current_time)           # seconds since Jan 1, 1970
```

```
## [1] 1.399e+09

month(current_time, label=TRUE)          # what month?

## [1] May
## 12 Levels: Jan < Feb < Mar < Apr < May < Jun < Jul < ... < Dec
```

The above shows that now works with R's built-in classes for handling times. The POSIXct class records a time as the number of seconds since the start of 1970, possibly negative. The lubridate function month converts this into a month. Other such functions are year, day, hour, minute, ... The POSIXlt class stores times in a list holding these pieces.

Besides using now, we can create times other ways. The parse_date_time function is one. Times and dates are often presented in a variety of different formats, so R needs some help parsing them from a string to a date object.

The following time is in a common format: "15-Feb-2013 07:57:34" with order day, month, year and then hours, minutes, and seconds being obvious to a human mind. The only thing missing is a time zone. To read this string in as a date, we need to give hints as to what the fields are. For that, there are codes. The six values above are coded with: "dbYHMS". Only b is unexpected— and necessary—as the "m" for month is used by minutes. For the list of these codes consult ?parse_date_time.

```
x <- "15-Feb-2013 07:57:34"
y <- parse_date_time(x, "dbYHMS")
year(y)                                  # can get pieces

## [1] 2013
```

The convenience functions ymd and ymd_hms can also do this task.

Another way to create new times is to modify existing ones. As times can be internally thought of as just a number of seconds since some origin, it is clear how we would do arithmetic with times. To turn back the clock one day can be done with:

```
now() - days(1)

## [1] "2014-05-01 13:01:15 EDT"

now() - hours(24)                        # same thing

## [1] "2014-05-01 13:01:15 EDT"
```

Logical data

R uses TRUE and FALSE to represent Boolean or logical data.[12] Logical data is produced by many R functions, for example the "is" functions:

```
is.na(1)

## [1] FALSE

is.numeric("one")

## [1] FALSE

is.logical("false")                        # "false" is character

## [1] FALSE
```

But most common, is the use of the comparison operators—<, <=, ==, !=, >=, >—to produce logical values:

```
3 < pi

## [1] TRUE

"one" == 1                                 # 1 is first coerced to "1"

## [1] FALSE

sqrt(2) * sqrt(2) == 2                      # floating point gotcha

## [1] FALSE
```

The last example shows that comparing floating point numbers with == can be problematic, as mathematically equal values will not always be equal within R due to round-off error. R provides a function all.equal to test for near equality. The latter would best be done in this verbose manner:

```
isTRUE(all.equal(sqrt(2) * sqrt(2), 2))

## [1] TRUE
```

The comparison operators are vectorized. Basically, the comparison question is asked element by element, with recycling happening as needed:

---

[12]The variables T and F are possible to use for TRUE and FALSE. As these are not guaranteed to be overwritten by some other binding, they are best to avoid.

```
whale <- c(74, 122, 235, 111, 292, 111, 211, 133, 156, 79)
whale > 100
```

```
## [1] FALSE  TRUE  TRUE  TRUE  TRUE  TRUE  TRUE  TRUE  TRUE FALSE
```

```
whale == 111
```

```
## [1] FALSE FALSE FALSE  TRUE FALSE  TRUE FALSE FALSE FALSE FALSE
```

The operators ! (for not), & (for and), and | (for or) can be used to combine values.[13] The following shows how to filter by more than one condition:

```
whale < 100 | whale > 200
```

```
## [1]  TRUE FALSE  TRUE FALSE  TRUE FALSE  TRUE FALSE FALSE  TRUE
```

```
whale > 100 & whale < 200
```

```
## [1] FALSE  TRUE FALSE  TRUE FALSE  TRUE FALSE  TRUE  TRUE FALSE
```

The functions any, all, which, **and** %in%   There are many useful functions for working with logical vectors. The any and all functions answer whether any of the values are TRUE or if *all* the values are true. For example, are any values more than 300?

```
any(whale > 300)
```

```
## [1] FALSE
```

Are all the values more than 50?

```
all(whale > 50)
```

```
## [1] TRUE
```

The which function returns the indices of the TRUE values in a logical vector. For example:

```
which(whale < 100 | whale > 200)
```

```
## [1]  1  3  5  7 10
```

---

[13]There are also the shortcut comparison operators && and ||. These are used in programming to more quickly determine logical conditions and are meant to produce a single TRUE or FALSE value.

If there are no TRUE values, a 0-length integer vector is returned.

The %in% function tests whether an element is in a collection, returning TRUE or FALSE if it is. It is an infix operator, used as follows:

```
292 %in% whale                          # is 292 in whale?

## [1] TRUE
```

This query could be asked other ways, for example using == as follows compares 292 to each value in whale and answers if they are equal. If any are, the following output will be TRUE:

```
any(292 == whale)

## [1] TRUE
```

However, %in% is defined in terms of the match function which allows the first argument to be a vector. It returns the index of the first match or NA:

```
match(c(292, 293), whale)

## [1]  5 NA
```

Coercion   Logical values can be coerced to numeric values by as.numeric, where TRUE becomes 1 and FALSE becomes 0. This can be confusing, but is commonly used. For example to add up the number of whale beachings more than 200 can be done by counting or programmatically with:

```
sum(whale > 200)

## [1] 3
```

This is a common idiom and leverages the fact when logical values are used with arithmetic operations the values are silently coerced to numeric. Being explicit through sum(as.numeric(whale>200)) is less convenient.

We can also coerce numbers to logical with as.logical. Only 0 maps to FALSE, NaN maps to NA, and others to TRUE. The characters "T", "TRUE", "True", "true" become TRUE, corresponding values become FALSE, else the value is coerced to NA.

Indexing   In our discussion of indexing a vector, we left for now the use of a logical vector for indexing. If vec is a logical vector *with the same length* as x, then x[vec] will refer to only those variables for which vec is TRUE.

This proves remarkably useful. A typical example might to be extract those values larger than the mean value:

```
whale <- c(74, 122, 235, 111, 292, 111, 211, 133, 156, 79)
whale[ whale > mean(whale) ]

## [1] 235 292 211 156
```

This next expression returns data exceeding 1 standard deviation from the mean, in either direction:

```
whale[whale < mean(whale)-sd(whale) | whale > mean(whale)+sd(whale)]

## [1]  74 235 292  79
```

Another common idiom is to use logical indexing to remove NA values:

```
hip_cost <- c(10500, 45000, 74100, NA, 83500, 86000, 38200, NA,
              44300, 12500, 55700, 43900, 71900, NA, 62000)
hip_cost[ !is.na(hip_cost) ]                    # store as variable ...

##  [1] 10500 45000 74100 83500 86000 38200 44300 12500 55700 43900
## [11] 71900 62000
```

• Example 2.1: Recoding values
Often, an obviously too large number, such as 999 for a weight, is used to code NA. For example, the babies (UsingR) package does this. What is obvious to the eye, is not to the computer, so we recode these to NA.

```
x <- babies$dwt                            # Weight of dad.
x[ x == 999 ] <- NA                        # 999 is used for NA ,
range(x, na.rm=TRUE)                       # avoid NA values

## [1] 110 260
```

                                                                                    ••

• Example 2.2: Using one variable to filter another
The kid.weights (UsingR) data set has several variables including height and age (in months) recording height and weight for several subjects. To extract the heights of four-year-olds, we use the age variable to subset:

```
age <- kid.weights$age
ht <- kid.weights$height
ht[ age >= 48 & age < 60]
```

```
##  [1] 38 24 40 36 36 41 38 24 40 41 45 37 36 36 39 40 36 43 33 39
## [21] 30
```

••

Problems

2.1 Enter the following data into a variable p using c

   2 3 5 7 11 13 17 19

Use length to check its length.

2.2 Al recorded his car's mileage at gust last eight fill-ups:

65311 65624 65908 66219 66499 66821 67145 67447

   Enter these numbers into the variable gas. Use the function diff on the data. What does it give? Interpret what both of these commands return: mean(gas) and mean(diff(gas)).

2.3 Let our small data set be

   2 5 4 10 8

1. Enter this data into a data vector x.

2. Find the square of each number.

3. Subtract 6 from each number.

4. Subtract 9 from each number and then square the answers.

Use the vectorization of functions to do so.

2.4 Create the following sequences:

1. "a", "a", "a", "a", "a"

2. 1, 3, ..., 99 (the odd numbers in $[1, 100]$)

3. 1, 1, 1, 2, 2, 2, 3, 3, 3

4. 1, 1, 1, 2, 2, 3

5. 1, 2, 3, 4, 5, 4, 3, 2, 1

using :, seq, or rep as appropriate.

2.5 Store the following data sets into a variable any way you can:

1. 2, 3, 5, 7, 11, 13, 17, 19

2. 1, 2, 3, 4, 5, 6, 7, 8, 9, 10 (positive integers)

3. 1/1, 1/2, 1/3, 1/4, 1/5, 1/6, 1/7, 1/8, 1/9, 1/10 (reciprocals)

4. 1, 8, 27, 64, 125, 216 (the cubes)

5. 1964, 1965, …, 2014 (some years)

6. 14, 18, 23, 28, 34, 42, 50, 59, 66, 72, 79, 86, 96, 103, 110 (stops on New York's No. 9 subway)

7. 0, 25, 50, 75, 100, …, 975, 1000 (0 to 1000 by 25s)

Use c only when : or seq will not work.

2.6 The average distance from the center is computed by $(|x_1 - \bar{x}| + \cdots + |x_n - \bar{x}|)/n$, where $\bar{x}$ is the mean of the data vector. Compute this for the rivers data set using the function sum to add the values and abs to find the absolute value.

2.7 Precedence rules are used to decide the order of evaluating operations when parentheses are not present. Look at the value produced by -1:3. What is done first - or :? Now look at 1:2*3. Which is done first : or *?

2.8 The precip data set records average annual rainfall for many different cities in the United States. It is stored as a data vector with names. Find the average amounts for the cities starting with a "J".

2.9 An experiment had 10 different trials. Create a character vector with 10 different names for the trials, e.g., "Trial 1", ...

2.10 Working with file names in R is easy, but requires the proper use of file separators, which vary depending on the operating system. For example, suppose you have the directory and file name of a file and want to get the entire file.

```
f <- system.file("DESCRIPTION", package="UsingR")
dname <- dirname(f)
fname <- basename(f)
```

To combine dname and fname into a full pathname use paste with the sep argument being .Platform$file.sep. What is the result?

2.11 The Manufacturer variable in the Cars93 (MASS) data set is stored as a factor. How many levels are there? How many different cases are there?

2.12 The Cylinders variable in the Cars93 (MASS) data set records the number of cylinders in the respective car. Why can this not be stored as a numeric value? Which cars have 5 cylinders?

2.13 The mtcars data set records information about cars from 1972. The values are coded using numbers. Recoding as factors can be more informative for the user. Recode the am variable, with 0 being "automatic" and 1 being "manual."

2.14 The Arbuthnot (HistData) data set contains information on the number of Male and Female births in London from 1629 to 1710. Using and and > determine if there was ever a year with more female births.

2.15 The negation operator ! is used to reverse Boolean values. For example:

```
A <- c(TRUE, FALSE, TRUE, TRUE)
!A
```

```
## [1] FALSE  TRUE FALSE FALSE
```

One of De Morgan's laws in R code is !(A & B) = !A | !B. Verify this with B <- c(FALSE, TRUE, FALSE, TRUE) and A as above.

2.16 In the precip data set, find all the cities with an average annual rainfall exceeding 50 inches.

2.17 For the precip data set, we can find the mean and the 25%-trimmed mean with mean(precip) and mean(precip, trim=0.25). Are any values in the data set more than 1.5 times the trimmed mean above the mean?

2.18 Consider the following "inequalities." Can you determine how the comparisons are being done?

```
"ABCDE" == "ABCDE"
```

```
## [1] TRUE
```

```
"ABCDE" < "ABCDEF"
```

```
## [1] TRUE
```

```
"ABCDE" < "abcde"
```

```
## [1] FALSE
```

```
"ZZZZZ" < "aaaaa"
```

```
## [1] FALSE

"11" < "8"

## [1] TRUE
```

2.19 You track your commute times for two weeks (ten days), recording the following times in minutes:

```
17 16 20 24 22 15 21 15 17 22
```

Enter these into R. Use the function max to find the longest commute time, the function mean to find the average, and the function min to find the minimum.

Oops, the 24 was a mistake. It should have been 18. How can you fix this? Do so, and then find the new average.

How many times was your commute 20 minutes or more? What percent of your commutes are less than 18 minutes long?

2.20 Suppose monthly sales (in 10,000s) of CDs in 2013 were

```
JAN FEB MAR APR MAY JUN JUL AUG SEP OCT NOV DEC
 79  74 161 127 133 210  99 143 249 249 368 302
```

Enter the data into a data vector cd. Through indexing, form two data vectors: one containing the months with 31 days, the other the remaining months. Compare the means of these two data vectors.

2.21 The following data records the average salary in major league baseball for the years 1990–1999 (in millions):

```
0.57 0.89 1.08 1.12 1.18 1.07 1.17 1.38 1.44 1.72
```

Use diff to find the differences from year to year. Are there any years where the amount dropped from the previous year?

The percentage difference is the difference divided by the previous year times 100. This can be found by dividing the output of diff by the first nine numbers (but not all ten). After doing this, determine which year has the biggest percentage increase.

2.2 Functions

Another very important object type in R is the function. We have seen that to manipulate data in R is done through the calling of various functions. For the most part, these functions are written themselves in R, though some are implemented at a lower level when speed is an issue.

Writing new functions in R is not too difficult, though there are subtleties. We discuss the subtleties in Appendix A. Here we discuss how to write basic functions. A function in R is just another object, in this case one with the class function.

Functions are created with the function constructor which requires a specification of the arguments and a specification of the function's body.

Arguments Arguments are enclosed in parentheses and specified through a name, as in function(x, y). Arguments can have default values, which are specified using an equals sign, as in function(x, trim=0, na.rm=FALSE). The default value of this signature for trim would be 0 and for na.rm is FALSE. Default values make it is easier for the user to call the function, as not all argument values must be specified.

Body The body of a function is an expression, typically enclosed in braces (which are optional if the expression contains only one command). Within the body, the argument names are assigned the values passed in when the function is called.[14] For example, a function to compute the mean might be defined through:

```
function(x) {
  sum(x)/length(x)
}
```

The body of this function takes the caller's input—a numeric variable is expected—adds the values with sum and divides by their number found by length. We could be more verbose, and in this case add to readability:

```
function(x) {
  total <- sum(x)
  n <- length(x)
  total / n
}
```

A function will return the last value computed in the body unless the body calls the return value with the specified return value. In the simple example above, there is no return specified, so the result of the last command is returned.

The above function definitions create a new function object. However, as they are not assigned to a name they are created then forgotten. These are

---

[14]One of the subtleties of functions in R is how some argument names are found. Here we have explicit values specified.

examples of *anonymous functions*. Such functions can be quite convenient as arguments to other functions, as we shall see later. More typical, though, is to bind the new function object to a name for later reuse. Here we use our_mean for the function's name:

```
our_mean <- function(x) {
  sum(x)/length(x)
}
our_mean(c(1,2,4,6,8,0))                           # try it out

## [1] 3.5
```

Problems

2.22 Write a function to compute the average distance from the mean for some data vector.

2.23 Write a function f which finds the average of the x values after squaring and subtracts the square of the average of the numbers. Verify this output will always be non-negative by computing f(1:10).

2.24 An integer is even if the remainder upon dividing it by 2 is 0. This remainder is given by R with the syntax x %% 2. Use this to write a function iseven that indicates if a number is even. How would you write isodd?

2.25 Write a function isprime that checks if a number x is prime by dividing x by all the values in $2, \ldots, x - 1$ then checking to see if there is a remainder of 0. The expression a %% b returns the remainder of a divided by b.

2.3 Numeric summaries

We turn our attention now to various summaries of numeric data. Data sets come in many sizes and finding the proper way to summarize them for easier understanding is important. We focus on three main areas:

Center By far the most common summary of a data set is its average value, referred to as the mean. This is a sense of the center of a data set, but is not the only one. One alternative is the value for which as many other values are bigger as are smaller, basically the middle value when they are sorted. This is the median. Other summaries are the most common value in the data set, the so-called mode. There are other possible senses of the middle still. We will discuss why some of these are better than others and how to compute them for a data set with R.

Spread The term "spread" is used here to measure the variability in a data set. We just said the mean is the most common summary of a data set, but how are we to interpret what the mean says about the data? If there is no variability, then the mean is the value of all the data. That says everything. If the variability is very large, then the mean informs us much less. Without this sense of variability, we can't be sure of our interpretation. Once quantified, we can use the value of the spread to talk about how far from center something is, as the spread gives us a natural sense of the scale. Learning to think in terms of the spread of the data, rather a fixed measurement like inches, or minutes is an important step in learning how to reason with data.

Shape It is important to realize that the "shape" of a data set can influence how much we can interpret from knowing both the center and spread. The shape here roughly refers to various things: are values larger than the mean equally likely as for values less? More likely? Less likely? Are values very far from the mean really unlikely or not so unlikely? Are there values where the measurements cluster, or are the possible values spread out? In classical inferential statistics, a primary role is played by a particular shape—the bell shape. For this, the answers are the two sides are equally likely, large values are rather unlikely and values tend to cluster near the mean. We will look to see if our sample data sets agree with this, or not.

Center

There are numerous ideas for what is meant by the term "center," a deliberately vague term. A few possible definitions could be the average, which we know is somewhere between the endpoints, or the middle value which always exists if there are an odd number of data points. We discuss both in the following.

The sample mean

Let's consider data on the weights of four-year-old children, with some values (in pounds) taken from the kid.weights (UsingR) data set.

```
wts <- c(38, 43, 48, 61, 47, 24, 29, 48, 59, 24, 40, 27)
```

The data is not sorted nor organized. Sorting data, though tedious by hand, can help us understand the values much better. With R sorting data is made easy with the sort function.

```
sort(wts)
```

```
## [1] 24 24 27 29 38 40 43 47 48 48 59 61
```

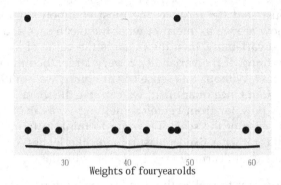

Figure 2.1: A dot plot of the data on children's weights. Such a graphic shows the data in sorted order allowing quick visual senses of both the center and the spread. Values are just drawn on the number line with repeated values being stacked.

This data is also drawn on the number line in Figure 2.1, a process that also sorts the data. From either, we can see a wide range of values.

To make more unambiguous summaries of this data set than "wide range," we need to define some specific numeric summaries. The *sample mean* is a familiar one, where we add up all the values and divide the number of values we added:

$$sample\ mean = \bar{x} = \frac{1}{n}(x_1 + x_2 + \cdots + x_n) = \frac{1}{n}\sum_i x_i.$$

The sample mean is often just called "x bar" due to its standard notation.

This computation is implemented in R in the basic mean function. For numeric arguments, this function has the argument na.rm to remove NA values before computing.

The mean for our data set gives us a sense of the average weight and is:

```
mean(wts)
```

```
## [1] 40.67
```

How to interpret the mean? There are a few useful ways. Graphically the mean is the visual balance point of the values given. Figure 2.2 shows a visual illustration for a three-point data set.

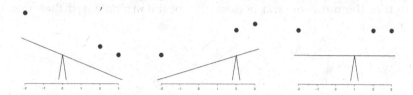

Figure 2.2: The mean is the value that balances the dot plot.

The fact that the mean balances the data also follows from the physics formula for the center of mass, to which our formula for the mean is a specialization.

A consequence of this, is that if we consider the deviations from the mean, $d_i = x_i - \bar{x}$, then these will "average out" or have mean 0:

```
devs <- wts - mean(wts)
mean(devs)                              # 0 up to rounding errors

## [1] 2.369e-15
```

Computing these deviations is also called "centering" the data set. It can be very useful to consider values in relation to the mean ("10 more than the mean", say), as opposed to just values with some measurement (62), as the latter may have less context.

This result follows mathematically from that fact that if we add a constant value (possibly negative) to each data point, then the mean of this transformed data set will be the mean of the original plus this value. In notation, if $y_i = x_i + c$ then $\bar{y} = \bar{x} + c$. When c is $-\bar{x}$ we get the above.

This observation also has other uses. When trying to compute a mean by hand, it can be very convenient to subtract off a guess for the mean value, thereby making all the values closer to 0. This makes it easier to work with them. We only need add back the subtracted amount later.

The trimmed mean The picture of the mean finding a balance between the data demonstrates also a weakness of the mean when used to represent "center." This being that one point can greatly influence the value for the mean. Visually, this involves just moving one of the data points far to the right or left. The mean will have to track this movement to keep the balance. We want a sense of center to describe the "bulk" of the data, not necessarily a single value, so modifications are sometimes made.

The trimmed mean is one such modification. A common transformation of a data set is to "Windsorize" the data, by trimming from both sides a certain percentage of the most extreme values. The trimmed mean is the mean of the data after such trimming. The `trim` argument for the `mean` function can be specified (its default is 0) to set a trimming proportion (a value in 0 to 0.5). So to trim the most top and bottom 10% of the `wts` data and then find the mean, we have:

```
mean(wts, trim = 0.10)                        # trim 10% of both ends

## [1] 40.3
```

That this trimmed mean is essentially the mean without trimming is due to the data not having a very large or small values to skew the sense of center by the mean. This need not be the case.

• Example 2.3: Income distributions are often skewed
The `exec.pay` (UsingR) data set holds values on compensation for CEOs of some American companies. For this data set, there is a large difference between the mean and the trimmed mean:

```
mean(exec.pay)

## [1] 59.89

mean(exec.pay, trim = 0.10)

## [1] 29.97
```

This difference is due to the presence of some very large compensation packages—the largest being over 41 times the average compensation. These large values cannot be balanced off by equally small ones, as the compensation cannot be negative. When one side of the data (relative to the center) carries much more "weight" than the other, then the data is referred to as "skewed."   ••

Weighted averages Discrete data—where it is expected that there will be common values—is often presented in tabulated form, and not in raw data form. As such, our formula for the mean does not immediately apply. If value y_k occurs n_k times, then we have the total sum is: $n_1 \cdot y_1 + \cdots + n_k \cdot y_k$ and the sample mean formula becomes:

$$\frac{1}{n}\sum_k n_k \cdot y_k = \sum_k \frac{n_k}{n} \cdot y_k = \sum_k w_k \cdot y_k.$$

This last expression is a *weighted average* where the weights, $w_k = n_k/n$, add to 1.

• **Example 2.4: Computing a weighted average**
The Macdonell (HistData) data set contains data on measurements of heights and finger lengths for 3,000 criminals recorded in 1906. The data is stored in three variables: height, finger, and frequency, where the latter records how often in the sample of 3,000 the combination occurred. These are frequencies, n_k, not weights, but we can compute those. We do so below to compute the average height for the individuals in the data set:

```
w <- Macdonell$frequency / sum(Macdonell$frequency) # n_k/n
y <- Macdonell$height
sum(w*y)
```

```
## [1] 5.42
```

In Chapter 4 we discuss alternatives to typing the data frame name repeatedly, as we do above. ••

The sample median

When the trimmed mean is pushed to its limits—trimming 50% of the data from each end—we are left with basically a single value, that being the "middle" value of the sorted data set. This is called the *median* value.

The sample median, M, of $x_1, x_2, \ldots, x_n$ is the middle value when there are an odd number of data points, and the average of the two middle values when there is an even number. A prescription to find the median is:

• Sort the data from smallest to largest.

• If there are $n = 2k + 1$ data points, the median is the $k + 1$st number in the sorted list. If there are $n = 2k$ data points, the median is the average of the kth and $k + 1$th values in the sorted list.

In both cases, why the median is a center is clear: there are k other values greater than or equal and k other values less than or equal the median.

The sample median is found using the median function, it too has the argument na.rm to remove NA values.

For example, the median of the values in wts is:

```
median(wts)
```

```
## [1] 41.5
```

That the mean and median for this data set are similar, says the data are more or less symmetrically distributed about the center. Where the mean is a balance point, the median is the center by count. Balancing takes into account the size of the deviations, whereas the median just balances off the signs of the deviations.

● **Example 2.5: Trimming doesn't effect the median**
Since the median is just the middle value, if the same percentage of points are taken from the top and bottom of the data then the median shouldn't change. To demonstrate, we manually trim our data and apply the median.

```
n <- length(exec.pay); trim = 0.10
lo <- 1 + floor(n * trim)               # floor drops decimal values
hi <- n + 1 - lo
median(sort(exec.pay)[lo:hi])           # compare to median(exec.pay)

## [1] 27
```

As median(exec.pay) is also 27 there is no difference.   ●●

● **Example 2.6: Mean and median can give different senses of center**
Table 2.2 contains summaries of the household net worth of U.S. households for various years. The striking thing is the difference between the two measures of center, with the median being much less than the mean.

Wealth distributions are characterized by some households having significantly more than the median value. There can't be a corresponding set of values significantly less than the median value (presumably no one can go billions of dollars in debt). This makes the balancing point move to the right compared to the median.

There are other such examples where the mean will be greater than the median—income distributions, real-estate prices, waiting times for repairs, ...—as there are reasons one side of the median is much more spread out than the other.   ●●

Measures of position

We would like to generalize the concept of the median—which basically splits the data in half—with half the data smaller and the other half larger. The pth quantile is basically the value in the data set for which $100 \cdot p$ percent of the data is less than the value and $100 \cdot (1 - p)$ is more. The median is then the 0.5 quantile.

Year	Median	Mean	Ratio
1989	79100	313600	4.0
1992	75100	282900	3.8
1995	81900	300400	3.7
1998	95600	377300	3.9
2001	106100	487000	4.6
2004	107200	517100	4.8
2007	126400	584600	4.6
2010	77300	498800	6.5

Table 2.2: Median and Mean household net worth in U.S. by year. Source Federal Reserve Board, 2010 SCF Chartbook (http://www.fas.org/sgp/crs/misc/RL33433.pdf).

The *percentiles* do the same thing, except that a scale of 0 to 100 is used, instead of 0 to 1. The term *quartiles* refers to the 0, 25, 50, 75, and 100 percentiles, and the term *quintiles* refers to the 0, 20, 40, 60, 80, and 100 percentiles.

The pth-quantile is at position $1 + p \cdot (n - 1)$ in the sorted data. When this is not an integer, a weighted average is used.[15]

The quantile function returns the quantiles. This function is called with the data vector and a value (which can be a vector) for p. We illustrate on a very simple data set, for which the answers are easily guessed.

```
x <- 0:5                              # 0,1,2,3,4,5
length(x)                             # even.

## [1] 6

mean(sort(x)[3:4])                    # median averages 3rd, 4th

## [1] 2.5

median(x)                             # clearly in middle

## [1] 2.5

quantile(x, 0.25)                     # 1 + .25(5) = 2.25.

##   25%
## 1.25

quantile(x, seq(0, 1, by=0.2))        # quintiles
```

---

[15]There are other definitions used for the pth quantile implemented in the quantile function. These alternatives are specified with the type= argument. The default is type 7. See ?quantile for the details.

```
##   0%  20%  40%  60%  80% 100%
##    0    1    2    3    4    5
```

```
quantile(x)                                        # quartiles are default
```

```
##    0%   25%   50%   75%  100%
## 0.00  1.25  2.50  3.75  5.00
```

Hinges   Closely related to the quartiles are the "hinges." These are found by applying the median to the data left of the median and to the right of the median. For example, if there are 6 data points, then the median of the first 3, the median (the average of the 3rd and 4th values), and the median of the last 3 are the hinges. These values are easier to find by hand than the quartiles, though slightly different.[16] The fivenum function returns the so-called five-number summary which includes the minimum value, the maximum value, and these three hinges. Compare the output of the following to that of quantile(x) above:

```
fivenum(x)
```

```
## [1] 0.0 1.0 2.5 4.0 5.0
```

• **Example 2.7: Wealth distribution**
Measuring income distributions by quantiles, rather than by summaries of center like the median and mean can be enlightening. We have seen that these two measures of center can be quite different due to large values to the right of the median. To see how large, the following data lists percentiles for combined income and capital gains amounts in 2011 (from income_percentiles (UsingR)):

```
income <- c("90" = 110651, "95" = 155193, "99" = 366623,
            "99.5" = 544792,  "99.9" = 1557090, "99.99" = 7969900)
income
```

```
##      90      95       99    99.5    99.9   99.99
##  110651  155193  366623  544792 1557090 7969900
```

We see from this that 90% of U.S. families earned less than $110,651 in income and capital gains, but the top 0.01% earned at least $7,969,900 (72 times that).   ••

---

[16]The boxplot function uses the hinges, as the original designer of the graphic presumably found the hinges easier to compute, yet equally informative at the level of detail of the graphic.

Other measures of center

The *mode* of a data set is defined as the value that occurs most often. This is taught in some circles as a measure of center, but not here and not in R where the mode function gives the storage mode of an object. The mode, as defined above, only applies to discrete data (continuous data rarely has ties, and then only due to round off). As well, the mode is only reasonably expected to be the center value if the data is symmetric.

That being said, computing the mode in R is an interesting challenge. The basic procedure would be to tabulate the number of times each value occurs and then search for the maximum. The table function (of which we discuss more in Section 2.4) will do the tabulation and max will find the maximum of the data set. But even that is not quite enough. Here is a recipe (we could present others) for finding the mode using some of the functions we have discussed so far.

```
table(wts)

## wts
## 24 27 29 38 40 43 47 48 59 61
##  2  1  1  1  1  1  1  2  1  1

table(wts) == max(table(wts))

## wts
##    24    27    29    38    40    43    47    48    59    61
##  TRUE FALSE FALSE FALSE FALSE FALSE FALSE  TRUE FALSE FALSE

which(table(wts) == max(table(wts)))

## 24 48
##  1  8

as.numeric(names( which(table(wts) == max(table(wts))) ))

## [1] 24 48
```

From the above, we find that both 24 and 48 share the title of most frequent weight.

Spread

The spread or variability of a data set is an important characterization. A simple measure of this could be the range of the data, which is just the distance between the smallest and largest values. In R the range is a two-value vector, for our purposes essentially the same as c(min(x), max(x)). We can compose the output with diff to give the distance, max(x) - min(x):

```
range(wts)                                    # minimum and maximum values

## [1] 24 61

diff(range(wts))                              # the distance between

## [1] 37
```

While this gives some sense of spread, it is very sensitive to just one large value. For better summaries, we look first at the most common sense of spread—that given by considering the deviations.

The variance and standard deviation

As mentioned, the deviations of a data set are just the original data centered by its sample mean: $d_i = x_i - \bar{x}$. The distance from the mean would the absolute value of the deviation, $y_i = |x_i - \bar{x}|$. The average deviation was noted to be 0, so would be a poor choice of spread. The average absolute deviation, $(1/n) \sum |x_i - \bar{x}|$ is a reasonable choice, but it proved mathematically advantageous to look not at the distance, but the squared distances. This leads to a definition for the *sample variance*, s^2:

$$sample\ variance = s^2 = \frac{1}{n-1} \sum_i (x_i - \bar{x})^2.$$

This is not quite the average of the squared distances, as we have divided by $n-1$ and not n.[17] But, clearly this carries the same interpretation. Values far from the center (the mean) will have big deviations which when squared will be even bigger. So more spread-out data sets will have larger variances. The presence of the n allows us to compare across different size samples, as otherwise large data sets would be expected to have large values of the sum due to there being more terms in the sum.

The var function will compute this summary statistics for us:

```
var(wts)

## [1] 162.6
```

Doing the computation using a vectorized approach is not so much more difficult, just overkill:

---

[17] As doing so makes this an unbiased estimate for a population variance, a point we won't pursue.

```
sum( (wts - mean(wts))^2 ) / (length(wts) - 1)
```

```
## [1] 162.6
```

Sample standard deviation If your data has units, say pounds as the `wts` data set does, then the sample variance will be in squared units. To avoid this, the sample standard deviation is defined:

$$\text{sample standard deviation} = \sqrt{s^2} = \sqrt{\frac{1}{n-1}\sum_i (x_i - \bar{x})^2}.$$

The interpretation is the same as the variance—larger values mean more spread-out data—but now the scale is appropriate.

The `sd` function computes the standard deviation. It requires a data vector with at least two values, as does `var`.

• **Example 2.8: Hip cost variability**
Recall the data set on the cost for a full hip replacement was marked by having a wide range.

```
hip_cost <- c(10500, 45000, 74100, NA, 83500, 86000, 38200, NA,
              44300, 12500, 55700, 43900, 71900, NA, 62000)
range(hip_cost, na.rm=TRUE)
```

```
## [1] 10500 86000
```

```
sd(hip_cost, na.rm=TRUE)
```

```
## [1] 24849
```

A standard deviation of nearly $25,000 dollars seems quite large, though of course we have no indication in this data of other factors that can account for such variability (location, what is being accounted for, ...). ••

The z-score The deviations, $d_i = x_i - \bar{x}$, express the data relative to its center, rather than on an absolute scale. This allows us to focus on values that are far less than the mean or far more. As we've seen, such values can have a big impact on the value of the standard deviation. But how do we talk about a big or large value relative to the others? One way is to use a scale that is driven by the data. Roughly speaking, this is like saying a value is bigger than most others, as opposed to saying a value is big. The former carries much more information.

The z-score of a data point is given by:

$$z\text{-}score = \frac{x_i - \bar{x}}{s}.$$

This gives the size of a data point in terms of its position relative to center, on a scale of standard deviation units. For example, a value of 3 means that the data point is 3 standard deviations larger than the mean value.

We can define a function to find z-scores with:

```
z_score <- function(x) (x - mean(x))/sd(x)
z_score(wts)

##  [1] -0.20912  0.18298  0.57509  1.59456  0.49667 -1.30701
##  [7] -0.91491  0.57509  1.43772 -1.30701 -0.05228 -1.07175
```

Alternatively, the scale function can be used to find the z-scores. This function will return a matrix, so we append a [,1] to extract the first column.[18]

```
scale(wts)[,1]

##  [1] -0.20912  0.18298  0.57509  1.59456  0.49667 -1.30701
##  [7] -0.91491  0.57509  1.43772 -1.30701 -0.05228 -1.07175
```

• **Example 2.9: Grading by z-scores**
Professor H. grades by z-scores. Students having a z-score greater than 1.28 will get an "A." In a small class the following student grades were recorded:

54 50 79 79 51 69 55 62 100 80

Did anyone get an "A"?

```
x <- c(54, 50, 79, 79, 51, 69, 55, 62, 100, 80)
z <- (x - mean(x))/sd(x)
x[z >= 1.28]

## [1] 100
```

---

[18]We will discuss this indexing notation for rectangular data types in Chapter 4.

Just the 100 was good enough for an "A".

Now, given these values for the mean and standard deviation, what score would be just good enough for an "A"?

```
mean(x) + 1.28 * sd(x)

## [1] 88.91
```

This formula, which reverses the z-score formula, should be read as the value which is exactly 1.28 standard deviations more than the mean. ••

Facts about z-scores At first this may seem to just make the problem more abstract, but there are real advantages. After shifting by the center and scaling by spread the data set becomes standardized, which allows data sets on different scales to be compared.

Also, there are two key facts about how unusual a large z-score can be. The first applies to "bell-shaped" data (normally distributed) which has the 68-95-99.7 rules of thumb that approximately 68% of the data will have a z-score between -1 and 1 (no more than 1 standard deviation from the mean), 95% will be within -2 and 2 and 99.7% will be within -3 to 3.

Further, for any data set (not just "bell-shaped") Chebyshev's theorem tells us that the proportion of values with an absolute z-score more than k is no more than $1/k^2$. For example, for any data set, no more than 1/9th of the data can be more than 3 standard deviations from the mean. (For "bell-shaped" data this is roughly 0.3% of the data, so much less.)

• Example 2.10: Computing a proportion

Empirically checking that Chebyshev's theorem is satisfied for a given data set requires that we compute the proportion of values that satisfy some criteria. This is a fairly common question. Let's see how it would be done in R.

For the exec.pay (UsingR) data set there are 199 values. What proportion are more than 3 standard deviations from the mean? Translating, this means the absolute value of the z-scores should be more than 3:

```
z <- (exec.pay - mean(exec.pay)) / sd(exec.pay)
out <- abs(z) > 3                      # 199 TRUE or FALSE values
sum(out) / length(z)                   # sum of logical

## [1] 0.01508
```

That is 1.5% if the z-scores are larger than 3. This is much more than would be expected for bell shaped data (we would expect 0.3%) but less than theoretically possible, which is 11.1% (1/9).

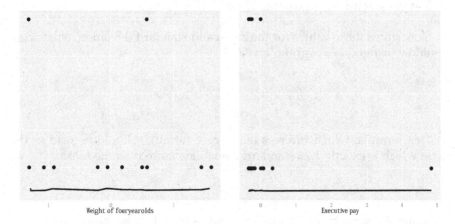

Figure 2.3: Plot of absolute z-scores for the wts data set and a subset of the exec.pay data set. There are no values larger than 2 in the wts data set, in agreement with the rule of thumb for bell-shaped data. For the executive pay data, we see a z-score nearly as large as 5, virtually impossible for bell-shaped data.

The above construct uses the sum of a logical vector as a convenient way to count the number of TRUE cases.   ••

Coefficient of variation   The standard deviation is sometimes normalized by the mean, to produce the *coefficient of variation*. This is suited for positive data, as otherwise the mean can be negative or 0, which will cause problems with the definition.

What is the advantage of this scaling? Let's look at pay. For the bulk of middle class America with roughly a $60,000 average pay, a standard deviation of $50,000 would indicate a rather large range. However, this would not be so for the executives in the exec.pay data set, as their average pay is nearly $600,000. This added context makes a difference. However, the actual standard deviation of the executives is much higher. The coefficient of variation is:

```
sd(exec.pay)/mean(exec.pay)              # coefficient of variation

## [1] 3.457
```

For this measure, values bigger than 1 indicate more relative variability than a standard "exponential" distribution, a common model for describing the lifetime of light bulbs and many other phenomenon.

The IQR

We mentioned how the range is an obvious measure of spread, but that it suffers from being effected by just one large or small value. Similarly, the standard deviation can be very sensitive to a single large or small value x_i, as the contribution from $(x_i - \bar{x})^2$ is then quite large.

Another common measure, not so effected, is the *interquartile range*, or IQR. This is the range of the middle 50% of the data. It can be seen informally through trimming the top and bottom 25% of the data, then taking the distance of the range, or more directly as the difference between Q3 and Q1 (the first and third quartiles).

The function IQR will compute this value for us. For example, the rivers data set has the lengths of 141 "major" rivers in North America. The median and IQR are given by:

```
median(rivers)                                    # center

## [1] 425

IQR(rivers)                                        # spread

## [1] 370
```

Comparison of the IQR and standard deviation is difficult, though for bell-shaped data the ratio of the IQR to the standard deviation is roughly 1.35. For the rivers data, this is not the case:

```
IQR(rivers)/sd(rivers)

## [1] 0.7492
```

This is due to some values in the data which are much larger than the rest, which heavily influences the standard deviation.

• Example 2.11: SAT rankings

Many college-bound students in the United States are familiar with the IQR, even if not by name. The ubiquitous college rankings almost always list the 25th and 75th percentiles for SAT scores. For example, Table 2.3 reproduces a table showing the first and third quartiles for accepted students at Ivy-League colleges. The IQR is Q3 − Q1. "Writing" has the smallest average IQR of 85, with "reading" the highest with 98.75. Does the data imply that nearly 25% of Yale students have perfect scores (800 on each category)?   ••

The median absolute deviation, mad   The *median absolute deviation* (mad) is another measure of variability that avoids an influence from a few very

| | Reading | | Math | | Writing | |
Institution	25%	75%	25%	75%	25%	75%
Brown	630	740	650	760	640	750
Columbia	690	780	700	790	690	780
Cornell	630	730	670	770	-	-
Dartmouth	670	780	690	790	690	790
Harvard	690	790	700	800	690	790
Princeton	700	790	710	800	700	790
U Penn	660	750	690	780	670	770
Yale	700	800	710	790	710	800

Table 2.3: SAT ranges for the "Ivy League Colleges". Data from the National Center for Educational Statistics, as found on http://collegeapps.about. com/od/sat/a/sat_side_x_side.htm.

large or small values, relative to the rest of the data. The standard deviation looked at the squared deviations, $(x_i - \bar{x})^2$; the mad considers the absolute deviations from the median, $\mid x_i - M \mid$. The median is used to gauge the center of this transformation of the data. To make the values comparable to the standard deviation for bell-shaped data, the default implementation of the median absolute deviation in R, mad, then multiplies by 1.4826.

We can compare the values for the rivers data set, where a few lengthy rivers skew the standard deviation:

```
mad(rivers)/sd(rivers)                    # 213/496
```

```
## [1] 0.4353
```

We see that the standard deviation is much larger for this variable. Whereas, for the height variable in the kid.weights (UsingR) data set, the values are comparable:

```
ht <- kid.weights$height
mad(ht)/sd(ht)                            # 11.68/10.7
```

```
## [1] 1.108
```

Shape

We've seen that different measures of center can yield similar values or very different values based on the presence of unusually large or small values. In this section we talk about the "shape" of a data set with the aim of building a vocabulary that makes it easier to understand when we should expect such differences.

The shape is also more important to understand in the context of statistical inference. When we try to understand a parent population from sample data, the tools we discuss almost always have some assumptions on the exact shape of a distribution. The most typical assumption is that the data is not only bell-shaped, but that the "bell" in theory has a precise mathematical description, this being explained later when we mention the normal distribution.

Symmetry and skew   Figure 2.4 shows two different data sets. The left one is imbalanced, the right one much less so. The left one is a sample from our exec.pay data set. That data set is characterized by some CEOs receiving very substantial payment plans, even when compared to the already substantial average. As such, there are values on the right side, that are not balanced off on the left side. Such data is said to be *skewed right*.

In contrast the data on the heights of four-year-olds is mostly *symmetric*. Some children are taller than average, some shorter than average of course, with the distribution of those bigger and smaller being roughly the same.

For symmetric data the mean and median measure center in basically the same manner. This is not so for data that is skewed right. In that case, the mean tends to be pulled to the right and is generally larger than the median. Skewed-left data tends to pull the average down.

For example, this is of importance to students as test scores over a semester are often skewed left—it is easy to have one really bad exam, but can be impossible to have an exam much better than typical, as there aren't enough points available. For a student the average test score is typically lower than the median test score. Perhaps instructors should use medians and not averages in assigning grades.

The *sample skewness* of a data set is a numeric measure of skew. There is no built-in R function, though some functions exist in add-on packages. A definition is given by the following:

$$\text{sample skewness} = \sqrt{n}\frac{\sum(x_i - \bar{x})^3}{(\sum(x_i - \bar{x})^2)^{3/2}} = \frac{1}{n}\sum z_i^3.$$

Here $z_i = (x_i - \bar{x})/s$ is the z-score for x_i. The presence of z-scores in determining the skew points out that it is a measure of shape and not center or scale.

A function to compute the sample skewness follows:

```
skew <- function(x) {
    n <- length(x)
    z <- (x - mean(x)) / sd(x)
```

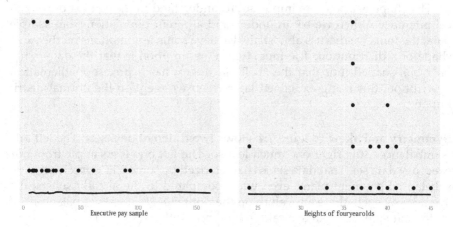

Figure 2.4: Dot plots of two data sets with different shapes. The left data set, a sample of the executive pay data set, is skewed right, the right data set, on the heights of four-year-old children, is mostly symmetric. For the symmetric data, the mean and median measure the center in a similar manner (36.7 to 38). For the skewed data this is not so (42.5 to 24).

```
    sum(z^3) / n
}
```

Applying this to the exec.pay data set shows a large positive skew and to the four-year-old heights a negative skew:

```
skew(exec.pay)

## [1] 9.579

four_year_hts <- kid.weights[kid.weights$age %in% 48:59, "height"]
skew(four_year_hts)

## [1] -1.06
```

Tails   A more subtle measure than symmetry and skew is the question of "tails." Tails characterize how much data is far from the bulk of the data. We mentioned how z-scores can quantify this by looking at size relative to the center on the scale provided by the standard deviation. The left graphic in Figure 2.5 shows z-scores of two simulated data sets on a dot plot. The lower data is long tailed, as evidenced by the larger values. Long-tailed data, even when symmetric, can cause issues with the mean and standard deviation, though trimming can prove useful.

The *sample excess kurtosis* is a measure of the tails in a data set. The following compares the deviations to the fourth power to the deviations to the second power. Longer tails will lead to larger values. The following formula subtracts 3 so that "normal" data will have kurtosis that is roughly 0.

$$\text{sample excess kurtosis} = n\frac{\sum(x_i - \bar{x})^4}{(\sum(x_i - \bar{x})^2)^2} - 3 = \frac{1}{n}\sum z_i^4 - 3.$$

The simplification shows that this depends just on the z-scores, so again is a measure of shape.

Here we compute the excess kurtosis for the height variable in the galton (UsingR) data set. This is a famous data set comparing the heights of children with that of their parent. Here we look at just the parental heights. Height measurements are usually close to "normal" so a value around 0 is expected:

```
kurtosis <- function(x) {
  n <- length(x)
  z <- (x - mean(x)) / sd(x)
  sum(z^4)/n - 3
}
kurtosis(galton$parent)              # height of parents in data set

## [1] 0.05104
```

Modes   We mentioned that the most likely value of a discrete data set is sometimes called the mode. For continuous data, it is unlikely any two values are the same, so this concept of a mode does not apply. Rather, we informally define a mode as an area where parts of the data tend to cluster.[19] The right graphic of Figure 2.5 shows at least three modes present in the galaxies (MASS) data set. Sometimes data sets are classified as *unimodal, bimodal,* or *multimodal* depending if there is one, two, or two or more modes present.

"Normal," bell-shaped data in addition to being symmetric and regular tailed, is unimodal.

Some univariate data sets may reflect the presence of underlying factors. For example, a data set on heights might best be split in two based on the gender of those measured, as on average females are shorter than males. Height data is generally symmetric and clustered around a central value, however when both females and males are mixed in, the data might show clusters around two distinct centers. As such, the presence of modes can suggest that the data should be split into groups before an analysis is done.

---

[19]In Chapter 6 where probability models are introduced for continuous data, we can define a mode more precisely as the position of a local maxima of the density function.

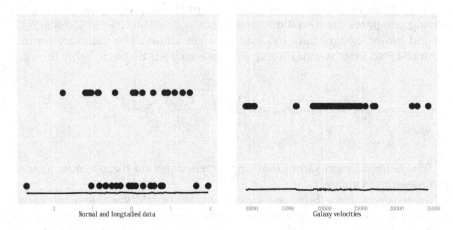

Figure 2.5: The left graphic shows stacked dot plots of z-scores of two data sets. The lower one has long tails, the top one "normal" tails. The right graphic shows the galaxies data set. The overlapping dots in the data show the presence of at least 3 clusters, corresponding to modes.

Viewing the shape of a data set

We have various graphical tools to view the shape of a distribution. While the choice of one over the other is not automatic—each generally has differing strengths and weakness—for all we should be able to glean quickly estimates of the center, spread, and shape.

Dot plots In previous graphical summaries of the data we have used dot plots to represent the data. These have been rendered as though hand drawn to emphasize they can be created without any special fuss—just make a number line containing the entire range of data and place dots for each data point. If values are repeated, we can either create stacks or jitter the data to essentially make it continuous. (Jittering adds a small random amount to each value and can be done through the jitter function.)

  If each dot is imagined to have a uniform weight, then the mean is the balancing point for these weights and the median the middle point (or average of the two middle points). Both these are easy to eyeball if the dots can be distinctly identified. As well the IQR is not too hard to eyeball, just mentally split the values into four parts and then lop off the top and bottom quarter of the points. Dot plots are also helpful in identifying the shape of a data set, in particular if it is skewed or multimodal.

  The main issues with dot plots are the trouble with repeated values and more importantly the fact that they can only be used for relatively small data sets.

The R function stripchart can make dot plots, but it makes such a visually poor graphic we won't illustrate. In Chapter 4 we will introduce other packages for making dot plots.

Stem-and-leaf plot   Related to a dot plot is the stem-and-leaf plot. Rather than represent each data value with a point on a graphic, the stem-and-leaf plot records a number (the leaf) for each point. By placing these with the proper stem the data set can be represented in an organized compact manner.

To illustrate, we have the following data for the number of points scored in a game by each member of a basketball team:

2 3 16 23 14 12 4 13 2 0 0 0 6 28 31 14 4 8 2 5

We can think of each number in terms of a 1s-digit and a 10s-digit. So 16 is a 1 and a 6. The 1 is the stem and the 6 the leaf. This allows us to reorganize the data as follows:

```
0 | 000222344568
1 | 23446
2 | 38
3 | 1
```

For example, the data value 2 becomes a 0 and a 2, and is represented by one of the 2s in the first row. We place the stems in a vertical column, and then sort the leaves. To read this then requires us to put the numbers back together. For example, the 8 in the row for a stem of 2 is the value 28 in the data set.

The presentation does many things at once: it sorts the data for us (though if doing this by hand, you may choose not to), it shows the range (in this case 0 to 31), gives us a sense of the median (just find the middle which is between 5 and 6), and shows us a rough shape (skewed right, as seen by mentally rotating the chart 90 degrees counterclockwise). Like the dot plot, a stem and leaf chart is an excellent choice for analyzing a data set by hand.

As with the dot plot the graphic suffers when there are too many data points. As well, the display does poorly when the values are too spread out or too tightly clustered. These can be addressed somewhat by adjusting the stem. Finally, each data point is represented by just two digits, so some rounding may be unavoidable.

The R function stem will produce a stem-and-leaf plot.[20] It has an argument scale to adjust the meaning of the stem. If there are too many leaves in a stem, adjusting this upwards from 1 can be helpful. For example, a stem-and-leaf plot can be made for the bumpers (UsingR) data set quite easily with:

---

[20]The stem.leaf function in the aplpack package, loaded with UsingR, enhances the basic display to provide a cumulative count from the top and bottom to assist in identifying the middle value.

```
stem(bumpers)

##
##    The decimal point is 3 digit(s) to the right of the |
##
##    0 | 68
##    1 | 333556
##    2 | 000123445
##    3 | 011233
```

From this we can see this data is mostly symmetric and not long tailed. We can also count that there are 23 values in the data set so by counting can find that the median has a stem of 2 and leaf of 1. But what is it as a number? The line "The decimal point is 3 digit(s) to the right of the |" needs to be interpreted. Mathematically it is saying the value is 2.1e3, or 2,100. (It in fact is 2,129 which gets truncated to 2,100 in the making of the summary.)

Histogram   A histogram is closely related to the stem-and-leaf graphic. If we squint and look past the numbers in the stem-and-leaf plot and just note the lengths of each row and turn this 90 degrees counterclockwise we have a rough histogram.

Histograms group individual data points, then represent them with a bar of a given area. The basic idea is to break up an interval on the number line that covers the entire set of data point into several subintervals, or "bins." For each subinterval the number of data points inside are counted and then a bar is drawn to represent these. The area of the bar is proportional to the proportion of points represented. That is, if we have 10% of the data points in some subinterval, then that bar should have 10% of the total area of the graphic.

Figure 2.6 shows two histograms of the waiting variable of the faithful (MASS) data set. The command to produce the left graphic was simply:

```
hist(faithful$waiting)
```

The histogram shows two modes. The bin size in the figure seems to be 5. With this, we can illustrate how to get the counts in each bin using some lower level R commands that can be useful in other contexts. (We could do this more directly using the invisible values output by hist.)

We first need to create the bins before we can count them. This is done with the seq command as follows:

```
bins <- seq(40, 100, by=5)
```

Then we want to count the number of values in each bin. The cut command will categorize each value in a data set by the breaks (bins) specified by a second vector:

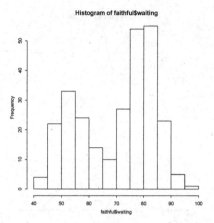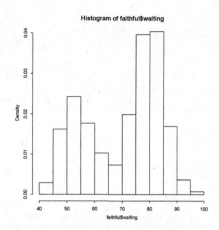

Figure 2.6: Two histograms of times between eruptions at the Old Faithful geyser in Yellowstone National Park shows two modes. The left graphic represents frequencies, the right graphic is scaled to have total area equal to 1.

```
x <- faithful$waiting
out <- cut(x, breaks=bins)
head(out)

## [1] (75,80] (50,55] (70,75] (60,65] (80,85] (50,55]
## 12 Levels: (40,45] (45,50] (50,55] (55,60] (60,65] ... (95,100]
```

We just need to tabulate these to get the counts:

```
table(out)

## out
##  (40,45]  (45,50]  (50,55]  (55,60]  (60,65]  (65,70]  (70,75]
##        4       22       33       24       14       10       27
##  (75,80]  (80,85]  (85,90]  (90,95] (95,100]
##       54       55       23        5        1
```

We see that the sub-intervals were taken to be open on the left and closed on the right. So a value of 80, say, would be in (75,80] and not (80,85]. This is also the default for hist, though the right argument can be used to change the setting. In this data set the smallest value of 43 is clearly in the bin (40,45]. If we had a smallest value of 40, then something would need to be done, either another bin or just adding it to the leftmost one.

The histogram then just represents these counts with bars. There are two standard ways. For the right graphic in the figure, the frequency is represented. The leftmost bar has height 4 in agreement with the count for (40,45].

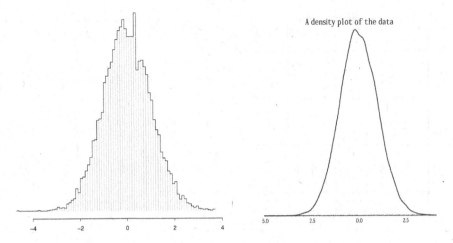

Figure 2.7: A histogram of a random sample of $n = 10{,}000$ data points and a corresponding density plot of the data. The vertical lines of the histogram are de-emphasized. From either, we can see the data is symmetric, unimodal with a mean of 0.

An alternative which is useful when using the histogram to estimate the underlying theoretical density (Chapter 6) is to scale the y-axis so the entire area is simply 1. This is the "probability" scaling, and for hist is done through the argument probability=TRUE.

The center, spread, and shape are all identifiable from a histogram. The mean value is still essentially a balancing point (as though we were to cut the histogram out of some material and tried to balance it). The median splits the area in half. The IQR is found by splitting the area into four equal-area pieces and ignoring the left and right ones. From the shape we can easily identify symmetry or skew, the modes, and in some cases, long tails.

The histogram that gets drawn depends on the choice of bin size. If the bins are too small, then the graphic becomes spiky; if too large then the graphic becomes blocky. The hist function allows the user to adjust the choice of bin selection through the breaks argument. The default is to use the Sturges algorithm to select a number of bins based on the number of data points (basically the range divided by $1 + \log(n)$). Other named alternatives are "Scott" and "FD" (cf. ?nclass.sturges or [59] for details). Alternatively, a vector of break points or number of bins can be specified, though some are taken as a suggestion only.

Density plots Imagine we have a sample and a histogram and asked a question: If we pick a data point from the sample *at random* what is the chance we pick a value in a given bin? Well, intuitively we know that this should be the number of data points in that bin divided by the total number of data

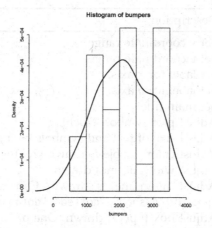

Figure 2.8: Histogram of bumpers data with a density plot layered on top.

points. It turns out that this is precisely the area of the bar representing that bin when the histogram is drawn with probability=TRUE.

So, the area of a bar above a bin represents the probability that a data point selected at random will be from this bin.

The relationship between probability and area will be described more fully in Chapter 6, as this will be how probability is defined for certain types of data. For now, we introduce a graphic that is similar to the histogram.

Imagine a scenario where we tried to graph more and more data with a histogram while taking an increasing number of bins. If the two choices are balanced out, then the resulting histogram will look something like the left graphic of Figure 2.7. In that graphic, the vertical lines of the bars are de-emphasized, as they just clutter the graphic. The relationship between area and probabilities still holds, assuming the scale is chosen so the total area is 1.

The graphic on the right is a *density plot* of the data. In some sense this is a limiting value. We have drawn the density plot in a stylized version to emphasize the graphics value as an immediate summary of a possibly large univariate data set. See Figure 2.8 for a formally rendered density plot. As with histograms, the following can be found from a density plot:

- The mean of the data is still related to the balancing point.

- The median is found by mentally dividing the figure into a left and right equal-area parts.

- The IQR is found by further dividing the areas into 4 equal-area parts and dropping the leftmost and rightmost.

- Symmetry and skew are clearly shown, as are modes which are now basically the "bumps" in the figure.

Argument	Description
xlim	Set x coordinate range.
ylim	Set y coordinate range.
xlab	Set label for x axis.
ylab	Set label for y axis.
main	Set main title.
pch	Adjust plot symbols (?pch).
cex	Adjust size of text and symbols on a graphic.
col	Adjust color of objects drawn (?colors).
lwd	Adjust width of lines drawn.
lty	Adjust how line is drawn. Can be "blank", "solid", "dashed", "dotted", "dotdash", etc.
bty	Adjust box type, if drawn. One of "o", "l", "7", "c", "u", or "]".

Table 2.4: Standard plotting arguments to modify a graphic.

The density function is used to create density plots in base graphics. This function does the computations, we use the plot function to graph them. The following shows how, though we don't display the output.

```
plot( density(bumpers) )
```

We will see the plot function used in many different guises throughout these first few chapters. For density data it produces a density plot with some defaults for the labeling and titling that can be modified through the arguments xlab, ylab, and main.

• **Example 2.12: A histogram and density plot**
A common graphic is to show both a histogram and a density plot at the same time. Constructing this graphic illustrates some subtleties of using R's base graphics. We consider again the bumpers (UsingR) data set.

The main issue we work to avoid is clipping. The initial graphic must be told to accommodate the width and height of both graphics, as by default only the first drawn graphic will determine this. As well, we need to *layer* the second graphic on the first. We do this below with the lines command.

The first task is do identify the proper size for the viewing window. Each function can return this information. For the histogram, we call hist passing in plot=FALSE to suppress its rendering:

```
b_hist <- hist(bumpers, plot=FALSE)
b_dens <- density(bumpers)
```

Functions	Description
points	Add points to a graphic.
lines	Add points connected by lines to a graphic.
abline	Add a line of the form $a + bx$, $y = h$, or $x = v$.
text	Add text to a graphic.
mtext	Add text to margins of a graphic.

Table 2.5: Functions to add layers to a graphic.

The heights of the histogram bars are stored in b_hist$density, whereas b_hist$breaks records the break points. For the density plot, the heights are available in b_dens$y, and b_dens$x holds the x values.[21]

Now when we draw the histogram, we specify the size of the viewing window directly using the xlim and ylim arguments, with each value computed from the objects just found:

```
hist(bumpers, probability=TRUE,
     xlim = range(c(b_hist$breaks,  b_dens$x)),
     ylim = range(c(b_hist$density, b_dens$y)))
lines(b_dens, lwd=2)
```

The last command lines(b_dens, lwd=2) calls the lines function to add a layer with the density estimate. The argument lwd=2 instructs R to draw the density line with twice the default line width. (Figure 2.8).

There are more arguments that can be specified to modify the drawing of a basic graphic. Table 2.4 shows several such options. Most of these are common to all the graphic-drawing functions. ●●

The construction of the density plot uses a function (called a density) and a window size (referred to as the bandwidth). Different choices produce different graphics. The bandwidth is similar to the choice of bin size in the histogram: larger values smooth out the graphic, smaller ones make it more spiky. R provides algorithms for automatic selection. The default is "nrd0". We can specify other names to the bw argument: "nrd", "ucv", "bcv", or "SJ".

Boxplots The five-number summary of a univariate data set is basically the minimum value, the maximum value, the quartiles Q_1 and Q_3, and the median. The five numbers provide a good summary of even very large data sets.

---

[21]The $ is employed as it is for data frames, even though the objects are lists.

The *boxplot* is a graphical device based on the five-number summary.[22] The basic plot looks like the left graphic in Figure 2.9:

- A box is drawn from the first quartile to the third. This represents the middle 50% of the data.

- The median is drawn indicating the center of the data and splits the box into areas, each representing 25% of the data.

- The top and bottom 25% of the data is represented by "whiskers" which stretch to the minimum and maximum values.

The right graphic in Figure 2.9 shows the conventional modification to cover *outliers*. These are defined as values more than $1.5 \times IQR$ below Q_1 or above Q_3. The whiskers are drawn to cover all points in these areas that are not outliers, leaving the outliers to be marked by points.

With this, it is easy to identify:

center The median clearly marks the center.

spread The IQR is the length of the box and a measure of spread.

shape The regions of the boxplot pair off. If one area is much longer than its corresponding pair, then the data is skewed. The right graphic on the weight variable of the kid.weights (UsingR) data set shows skew, as the upper 25% of the data represents much more of the range than the lower 25% of the data.

In addition, the reader of the graphic can clearly see outlier values. As these can have an important impact on the assumptions made for statistical inference, this information is very valuable.

What the boxplot lacks, as compared to a density plot, is the ability to identify modes.[23] However, the boxplot graphic is so useful for summarizing data, especially when comparing multiple samples, as it naturally lends itself to side-by-side comparisons.

The basic R function to draw a boxplot is boxplot. The left graphic of Figure 2.9 was created with:

```
boxplot(bumpers, horizontal=TRUE, main="Bumpers")
```

---

[22]The boxplot function uses the hinge-version of the quartiles, instead of those given by quantile, following the original design of J. W. Tukey.

[23]If this is important, consider the violin plot, as found in the vioplot package, which replaces the box with a density estimate.

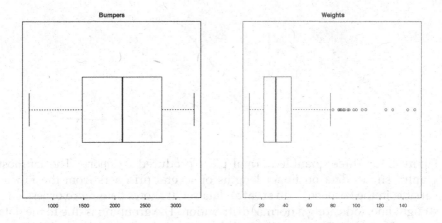

Figure 2.9: Boxplots of various data sets. The left one shows the bumpers data set, a mostly symmetric data set with no outliers. The right one, of the weight variable in the kid.weights data, shows a right skew and some outliers.

Quantile graphs The quantile plot is a graphic that allows us to compare two distributions. In the next chapter we discuss its use to compare two arbitrary distributions, for now we discuss comparing a distribution to the "normal" distribution, our previously mentioned symmetric, bell-shaped distribution.

For "normally" distributed data the 68-95-99.7 rule of thumb applies, describing how much of the data is expected to be within 1, 2, and 3 standard deviations of the mean. These imply that a z-score of -2 would be the 2.5 percentile, -1 the 16th percentile and so on. In fact, we will mention in Chapter 6 how to find any quantile of this distribution, not just these special ones.

The *quantile-normal* plot takes these theoretical quantiles and compares them to the quantiles of a sample data set. Corresponding quantiles are plotted as pairs. The cover graphic shows the process for two theoretical distributions. If the distributions have the same shape, then the points will more or less be on a straight line. If the two shapes are not similar, the points will not be collinear. Of course, samples have sampling variability, so the points would not be expected to lie exactly on a line. The key to appreciating the graphic is to understand what is more or less straight.

Figure 2.10 shows three such graphs. The first one shows data on finger lengths of criminals. This data was used by Gosset to demonstrate his work on the t-distribution, which we become familiar with in later chapters. The assumption on normality is important and this graphic shows a straight line. The others have differences.

The graphic was produced with qqnorm. In the case of the finger data set, the values are stored in tabulated format with finger recording a value and frequency the number of such measurements. This allows a more compact

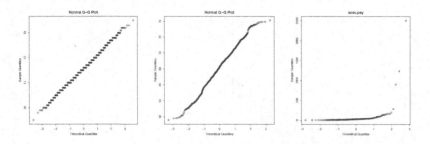

Figure 2.10: Three quantile-normal plots produced by qnorm. The leftmost graphic shows data on finger lengths of several prisoners from the finger variable in the Macdonell (HistData) data set. It shows data more or less on a straight line, indicating a normal distribution. The grouping is due to the data being discretized. The second graphic uses data on the height of children in Galton's classic study of heights. This data has slight bends on the edges, like an "S". This being due to the tails being slightly less long than the normal. The final data shows what a decidedly non-normal distribution appears like in this graphic. The executive pay data is used which is skewed right and long tailed. Such data shows a clear curve.

storage for the data, but isn't helpful for creating the desired plot. To manipulate the data into a format for qqnorm requires creating a vector with each data point recording a score.

```
x <- rep(Macdonell$finger, Macdonell$frequency)
qqnorm(x)
```

The middle graphic of Figure 2.10 is also a plot of discretized data, in this case the child heights in the Galton (HistData) data set. To produce the graphic, the data was *jittered*. That is, random noise was added to each value to break up similar values. The jitter function does this. We specified a value of 5 for factor, which indicates how much noise to add:

```
x <- jitter(HistData::Galton$child, factor=5)      # add noise
qqnorm(x)
```

This graphic shows slight flattening out at the left and right side. This is indicative of shorter tails than the normal distribution.

The rightmost plot of Figure 2.10 demonstrates both how skewed and long-tailed data look when viewed through this transformation. The skew is represented by graphics which do not have symmetry through the central point. The long tail on the right side leads to the right-hand portion of the graphic moving upwards. The left side of this graphic flattens out, indicating that the left tail is shorter than the normal.

Problems

2.26 The beatles (LearnEDA) data set records the length of songs (in seconds) by the Beatles in the time variable. Find the mean length, median length, longest length, and shortest length in minutes.

2.27 The ChestSizes (HistData) data set contains Quetelet's data on chest measurements of 5,738 Scottish Militiamen. The data is tabulated. Find the mean.

2.28 Read this stem-and-leaf plot. First find the median by hand. Then enter in the data and find the median using median.

```
The decimal point is 1 digit(s) to the right of the |

  8 | 028
  9 | 115578
 10 | 1669
 11 | 01
```

2.29 The farms (LearnEDA) data set records in the count variable the number of farms in the United States per state (in 1,000s) in 1999. Make a stem-and-leaf chart. Try to guess the mean, then verify. Which state would you guess is the outlier?

2.30 A stem and leaf chart has a stem of 2 a leaf of 3 and the message "The decimal point is 4 digit(s) to the left of the |". What is the number?

2.31 For the data sets bumpers (UsingR), firstchi (UsingR), and math (UsingR), make histograms. Try to predict the mean, median, and standard deviation from the graphic. Check your guesses with the appropriate R commands.

2.32 Fit a density estimate to the data set pi2000 (UsingR). Compare with the appropriate histogram. Why might you want to add an argument like breaks = 0:10-.5 to hist?

2.33 The data set normtemp (UsingR) contains body measurements for 130 healthy, randomly selected individuals. The variable temperature contains normal body temperature. Make a histogram. Estimate the sample mean body temperature, and then check using mean.

2.34 The data set DDT (MASS) contains independent measurements of the pesticide DDT on kale. Make a histogram and a boxplot of the data. From these, estimate the mean and standard deviation. Check your answers with the appropriate functions.

2.35 The ChestSize (HistData) data set is tabulated data showing the chest sizes of 5,738 Scottish males. Make a histogram of this data.

2.36 The paradise (UsingR) data set contains snowfall measurements (in inches) at Mt. Rainier for several years. Make a histogram of the data and comment on the shape. The data is in a time series with NA values. To get this as a numeric data set, use:

```
x <- as.numeric(paradise)          # as numbers
x <- x[!is.na(x)]                   # strip NA values
```

2.37 There are several built-in data sets on the 50 United States. For instance, state.area, showing the area of each U.S. state, and state.abb, showing a common abbreviation. First, use state.abb to give names to the state.area variable, then find the percent of states with area less than New Jersey (NJ). What percent have area less than New York (NY)? Make a histogram of all the data. Can you identify the outlier?

2.38 The lawsuits (UsingR) data set contains simulated data on the settlement amounts of 250 common fund class actions in $10,000s. Look at the differences between the mean and the median. Explain why some would say the average is too high and others would say the average is the wrong way to summarize the data.

2.39 Can you copyedit this paragraph from the August 16, 2003 *New York Times*?

> The median sales price, which increased to $575,000, almost 12 percent more than the median for the previous quarter and almost 13 percent more than the median for the period a year ago, was at its highest level since the first market overview report was issued in 1989. (The median price is midway between the highest and lowest prices.)

2.40 In real estate articles the median is often used to describe the center, as opposed to the mean. To see why, consider this example from the August 16, 2003 *New York Times* on apartment prices:

> The average and the median sales prices of cooperative apartments were at record highs, with the average up almost 9 percent to $775,052 from the first quarter this year, and the median price at $479,000, also an increase of almost 9 percent.

Explain how using the median might affect the reader's sense of the center.

2.41 The data set pi2000 (UsingR) contains the first 2,000 digits of π. What is the percentage of digits that are 3 or less? What percentage of the digits are 5 or more?

2.42 The data set rivers contains the lengths (in miles) of 141 major rivers in North America.

1. What proportion are less than 500 miles long?

2. What proportion are less than the mean length?

3. What is the 0.75 quantile?

2.43 The time variable in the nym.2002 data set contains the time to finish the 2002 New York City Marathon for a random sample of the finishers.

1. What percent ran the race in under 3 hours?

2. What is the time cutoff for the top 10%? The top 25%?

3. What time cuts off the bottom 10%?

Do you expect this data set to be symmetrically distributed?

2.44 Compare values of the mean, median, and 25%-trimmed mean on the built-in rivers data set. Is there a big difference among the three?

2.45 The built-in data set islands contains the size of the world's land masses that exceed 10,000 square miles. Make a stem-and-leaf plot, then compare the mean, median, and 25% trimmed mean. Are they similar?

2.46 The data set OBP contains the on-base percentages for the 2002 Major League Baseball season. The value labeled bondsba01 contains this value for Barry Bonds. What is his z-score?

2.47 For the rivers data set, use the scale function to find the z-scores. Verify that the z-scores have sample mean 0 and sample standard deviation 1.

2.48 Compare the three measures of spread (sd, IQR, mad) for the exec.pay (UsingR) data set. Are the values comparable?

2.49 The data set npdb (UsingR) contains malpractice-award information. The variable amount is the size of malpractice awards in dollars. Find the mean and median award amount. What percentile is the mean? Can you explain why this might be the case?

2.50 Find the coefficient of variation (standard deviation scaled by the mean) for the `rivers` data set.

2.51 The data set `babyboom` (UsingR) contains data on the births of 44 children in a one-day period at a Brisbane, Australia, hospital. The variable `running.time` records the time after midnight of each birth. The command `diff(running.time)` records the differences or inter-arrival times. Such times are often exponentially distributed. Is the coefficient of variation close to 1?

2.52 The data set `babyboom` (UsingR) contains data on the births of 44 children in a one-day period at a Brisbane, Australia, hospital. Compute the skew of the `wt` variable, which records birth weight. Is this variable reasonably symmetric or skewed?

   The variable `running.time` records the time after midnight of each birth. The command `diff(running.time)` records the differences or inter-arrival times. Is this variable skewed?

2.53 The data set `hall.fame` (UsingR) contains baseball statistics for several baseball players. Make histograms of the following variables and describe their shapes: `HR`, `BA`, and `OBP`.

2.54 Sometimes a data set is so skewed that it can help if we transform the data prior to looking at it. A common transformation for long-tailed data sets is to take the logarithm of the data. For example, the `exec.pay` (UsingR) data set is highly skewed. Look at histograms before and after taking a logarithmic transform. Which is better at showing the data and why? (You can transform with the command `log(1 + exec.pay, 10)`.) Find the median and the mean for the transformed data. How do they correspond to the median and mean of the untransformed data?

2.55 Create a quantile-normal plot for the chest-size measurements in `ChestSizes` (HistData). Do the plotted points appear to lie very close to a straight line?

2.56 The `Michelson` (HistData) data set records 100 measurements by Michelson of the speed of light (in the variable `velocity`). Make a quantile-normal graph and discuss if the graphic shows the points falling on a straight line.

2.57 The data set `cfb` (UsingR) contains a sampling of the data from a survey of consumer finances. For the variables `AGE`, `EDUC`, `NETWORTH`, and `log(SAVING + 1)`, describe their distribution using the concepts of modes, symmetry, and tails. Can you convince yourself that these distributions should have the shape they do? Why?

2.58 The `brightness` (UsingR) data set contains the brightness for 966 stars in a sector of the sky. It comes from the Hipparcos Catalogue. Make a histogram of the data. Describe the shape of the distribution.

2.59 In the Cars93 (MASS) data set, which is more skewed, the Price or MPG.highway variable?

2.60 Write a function to find the most common value of a data vector, e.g., the "mode."

2.4   Categorical data

Summarizing univariate categorial data is fairly straightforward. The basic tool is to tabulate the values and to present that information either in print or graphic form.

Tabulating data   For example, the smoke in the babies (UsingR) data set records the smoking status of a mother. The data is coded with the codes listed in the help page. We use these codes to create a factor and then tabulate:

```
x <- babies$smoke
x <- factor(x, labels=c("never", "now", "until current",
                        "once, quit", "unknown"))
table(x)

## x
##          never         now until current    once, quit
##            544         484            95           103
##        unknown
##             10
```

As is seen, the table command computes the frequency of each category and returns the values in some order, in this case the order of the levels.

A related question is: what percentage of mothers smoked in the data set? (These were mothers in the United States in the 1960s.) For that we need to divide the counts by the total sum. This can be done directly by computing the sum with sum(table(x)), or using the helper function prop.table. We create a percentage by multiplying the proportion by 100:

```
out <- table(x)
prop <- 100 * out / sum(out)            # or prop.table(out)
round(prop, digits = 2)                 # format output

## x
##          never         now until current    once, quit
```

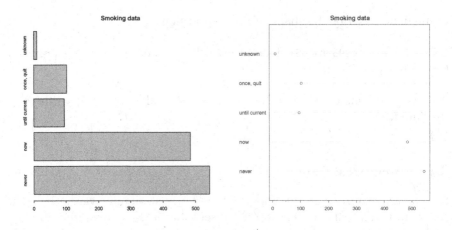

Figure 2.11: A horizontal bar chart and dot chart of the smoking data.

##	44.01	39.16	7.69	8.33
##	unknown			
##	0.81			

The above used the round function to print just two digits after the decimal point.

Bar charts   Tabulated data is often visualized through *bar charts*. Each count is represented by a bar with a proportional height. The result is similar to a histogram, though fundamentally different in that the x-axis (for vertical displays) is used to display categories, not numeric values.

The barplot function produces a simple bar chart. We pass in the tabulated data for representation. In Figure 2.11 we present both a vertical and horizontal bar chart of the smoking data. The argument horiz=TRUE will create a plot like the rightmost one in the figure. We used the following to produce the leftmost graphic:

```
barplot(table(x), horiz=TRUE, main="Smoking data")
```

Dot charts   A Cleveland dot chart is a good alternative to a bar chart. It shows the exact same information, but in a sparer manner that is easier to read. The dotchart function produces the right graphic in Figure 2.11 through:

```
dotchart(table(x))
```

Pie charts A pie chart is a common graphic to represent proportions. The proportion of each category is represented by a similar proportion of a circle through a wedge. Unlike a bar chart it is very hard to actually compare values when they are similar. To quote the help page of pie:

> Pie charts are a very bad way of displaying information. The eye is good at judging linear measures and bad at judging relative areas. A bar chart or dot chart is a preferable way of displaying this type of data.

Should a pie chart still be desired, the pie function can do so, with the command pie(table(x)).

Problems

2.61 Numeric data can be discretized through the cut function. For example, the command cut(bumpers, c(0, 1000, 2000, 3000, 4000)) will categorize the repair cost of a bumper by its rough amount. Make a table of this. Which range has the largest number of data points?

2.62 The Cylinders variable Cars93 (MASS) data set records the number of cylinders in a factor. What kind of summary does R compute for factors? Look at summary(Cars93$Cylinders) to see.

2.63 The lorem variable in UsingR contains 5 paragraphs of dummy typesetting text that has been in use for centuries. What is the most common letter used? To answer this, you can break a character value into letters by the idiom unlist(strsplit(x,"")) where x is character data.

2.64 Make a dot chart of the Cylinders variable in the Cars93 (MASS) data set.

2.65 Find an example of a table in the media summarizing a *univariate* variable. Could you construct a potential data set that would have this table?

2.66 Try to find an example in the media of a misleading barplot. Why is it misleading? Do you think it was meant to be?

2.67 Find an example in the media of a pie chart. Does it do a good job of presenting the data?

2.68 The data set central.park (UsingR) holds the coded variable WX representing bad weather (e.g., 1 for "fog", 3 for "thunder", 8 for "smoke" or "haze"). NA is used when none of the types occurred. Make a table of the data, then make a table with the extra argument exclude=FALSE. Why is the second table better?

Bivariate data

This chapter looks at bivariate data—data involving two data variables. When looking at two variables at once, we can ask questions about the relationships between them. In Figure 3.1 we have two graphs. The left graph shows a scatter plot of the data in the `faithful` data set which has recordings on the Old Faithful Geyser in Yellowstone National Park. The data records the time between eruptions and the corresponding duration of the following eruption. We've added marginal histograms of the two variables.[1] In Figure 2.6 we commented that the marginal histogram of the waiting time was bimodal. From the clustering in the scatter plot we see more—there are basically two types of eruptions: smaller ones with shorter waiting times and longer ones with longer waiting times.

The right graphic of Figure 3.1 uses the same data, but randomly shuffles the y values. This does not effect the marginal distributions, but the effect is pronounced in the scatter plot. There is now no relationship between the x and y values. The presence of a relationship can not be seen from the marginal distributions. It either is clear from the context of the problem, or will show up graphically.

With bivariate data we have a few additional questions. For paired-off data we can ask about relationships, their presence or absence, and in their presence try to characterize the relationships. The scatter plot is useful with numeric data, for categorial data we can look at contingency tables or some related graphics.

For unpaired data we can ask if various measures are similar or different: e.g., do they seem to have the same center, spread, or shape?

3.1 Independent samples

A common experimental setup involves separating a cohort into a control and a treatment group. (We return to this example with more detail in Chapter 9.) The treatment group receives some treatment, the control group does not. Afterwards, measurements for each group are taken. To control for biasing due to the separation of the cohort, random assignment is often used. This

---

[1]Following an example from the help page of `layout`.

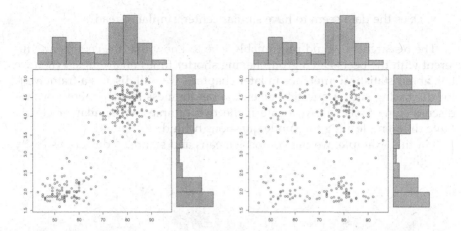

Figure 3.1: The left graphic shows a scatter plot of the waiting time between eruptions versus eruption duration for events at the Old Faithful Geyser in Yellowstone National Park. Marginal histograms are added. The right graphic is similar, but the durations were shuffled. For the two figures, the univariate marginal distributions are identical, but the left scatter plot demonstrates a strong relationship between the two variables: longer waiting times tend to be followed by longer eruptions.

leads to independence—where knowledge of one participant's performance does not hint at another's. To control for the "placebo effect", the control group often receives some treatment, but one that is not expected to have an actual effect.

For example, suppose that to investigate the question of what foods can have an impact on sports performance, a study was done with a group of subjects.[2] The treatment group was fed $3/2$ cups of beets 75 minutes prior to an activity, the control group was not. The activity was a timed run. The researcher's thinking was that the nitrates found in beets would widen the subjects' blood vessels, increase blood flow to the working muscles, hence lower the time duration of the activity. Suppose the data measured (in minutes) are given by:

```
Beets eaten:    41 40 41 42 44 35 41 36 47 45
No Beets eaten: 51 51 50 42 40 31 43 45
```

What can be said about this data? The three longest times were in the "no-beets-eaten" group, but that group also had the shortest time. In this initial exploration of the data, we might ask a few questions:

- Does the data seem to come from similar populations, that is do the two have the same shape?

---

[2]Such a study was reported in [35], though this example uses simulated data.

- Does the data seem to have similar center, similar spread?

The researchers would presumably love to know that the centers are different with the beets-eaten group having shorter times on average. When we talk about statistical inference in later chapters, we will discuss a framework for discussing such a question given the randomness involved. For now, we discuss some tools to investigate if the two samples are similar or clearly quite different, leaving the subtle questions behind.

For this example, we can compare means and standard deviations easily enough:

```
beets <- c(41, 40, 41, 42, 44, 35, 41, 36, 47, 45)
no_beets <- c(51, 51, 50, 42, 40, 31, 43, 45)
c(xbar1=mean(beets), xbar2=mean(no_beets),
  sd1=sd(beets), sd2=sd(no_beets))

##   xbar1   xbar2     sd1     sd2
## 41.200  44.125   3.706   6.813
```

We can see that the beet group was about 3 minutes faster and had less spread. Whether this difference is due to the assignment of faster runners to the beets group, or due to the eating of the beets is not answered.

As with univariate data, it is also helpful to visualize data sets. Many of the previously discussed graphics carry over.

Stem-and-leaf chart   The basic stem-and-leaf chart can be modified to display two data sets in a back-to-back manner. The built-in stem function does not do this; stem.leaf.backback from the aplpack package does:

```
stem.leaf.backback(beets, no_beets, rule.line="Sturges")

## _____
## 1 | 2: represents 12, leaf unit: 1
##          beets          no_beets
##
## _____
##                | 3* |1              1
## 2         65| 3. |
## (6)   421110| 4* |023             4
## 2         75| 4. |5              (1)
##                | 5* |011            3
##                | 5. |
##                | 6* |
## _____
## n:           10        8
## _____
```

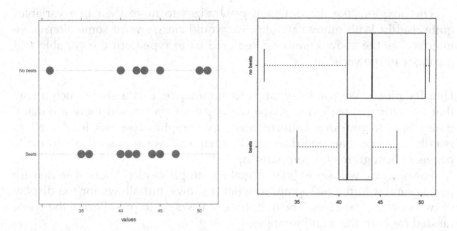

Figure 3.2: Using a dot chart (left graphic) and parallel box plots to display the simulated data on performance related to beet eating. Both graphics allow for easy comparisons of center and spread.

The chart contains much more information than the basic chart produced by stem. The left- and rightmost columns record position of the data in the data set, the stems are recorded in the middle. After focusing on the middle, it can be seen that no_beets has more spread, but the difference in centers is not so obvious.

Dot plots The left graphic in Figure 3.2 shows a dot plot of the two variables. The extension to displaying two (or more) variables is straightforward—new variables are just added as additional rows. This graphic is very useful for comparing data sets. We see similar things as we did with the back-to-back stem-and-leaf chart, but the graphic has less "chart junk" and can display larger data sets and more variables at a time.

As with the univariate case, we put off until Chapter 5 the discussion on how to create this graphic.

Boxplots The right graphic of Figure 3.2 shows the two data sets using parallel boxplots. Like a dot plot, a boxplot shows center and spread quite well, but unlike the dot plot, can be used to show larger data sets. In the figure, we can see medians differ, but both fall within the spread of the other. This concept of looking at differences in center on a scale determined by the spread is the key to statistical inference.

These boxplots were drawn with the command:

```
boxplot(no_beets, beets, names=c("no beets", "beets"),horizontal=TRUE)
```

The boxplot, like the dot plot, generalizes to more than two variables quite readily. With many variables, we would clearly want some alternative interface, as the above command required us to type both the variable and the name of the variable.

Density plots   We use histograms to summarize a data set in such a way that the center, spread, and shape are apparent even when there are many data points to consider. Unfortunately, this graphic does not lend itself so readily to more than one data set. In Chapter 5 we will see how to make panels of histograms for comparison.

For now, we will show how to make multiple density plots. The density plot is similar to the histogram in what it shows, but allows for the display of two or more data sets. The only trick is we need to pre-allocate the space needed for both the x and y ranges.

Let's consider a different data set. The michelson (MASS) data set contains measurement of the speed of light in air made by Michelson in 1879. The data set consists of five experiments. We compare the fourth and fifth below. The data is stored in a data frame with one variable Speed recording the speed and one Expt recording the experiment number. We use indexing to get our two variables:

```
speed <- michelson$Speed; expt <- michelson$Expt
fourth <- speed[expt == 4]
fifth <-  speed[expt == 5]
```

With this, we now pre-compute the densities so that we can find their respective ranges:

```
d_4 <- density(fourth)
d_5 <- density(fifth)
xrange <- range(c(d_4$x, d_5$x))
yrange <- range(c(d_4$y, d_5$y))
```

Now we plot. The first density plot is drawn using plot, the second layer added with lines. As the default labels come from the initial density plotted—and don't apply to both—we modify both the xlab and main arguments. As well, we use the lty=2 argument for the lines command, to have that line added with a different line type:

```
plot(d_4, xlim=xrange, ylim=yrange, xlab="densities", main="")
lines(d_5, lty=2)
```

The left graphic in Figure 3.3 shows the result. We see that the centers seem similar, the spread of the fourth might be slightly larger and the shape of the fourth more peaked. The important comparison here would be the center.

The figure drawn would benefit from a legend indicting which line represents which variable. The `legend` function will add a legend in base graphics. The position of the legend, the text, and the defining feature needs to be specified. The following would place the legend in the upper left with the line type marking which part of the graph corresponds to which variables:

```
legend(650, 0.008, legend=c("Fourth", "Fifth"), lty=c(1,2))
```

Quantile comparisons   The quantile-normal plot of Figure 2.10 is a graphic used to compare a distribution to the benchmark normal distribution. The quantile-quantile plot generalizes this to compare one distribution to another. The graphic computes matching quantiles and then plots them.

The right graphic in Figure 3.3 illustrates with the Michelson data just considered. The data more or less tracks a straight line, though the observed difference in the shape of the peak causes it to flatten out in the middle. The small size of the data sets involved make such irregularities hard to gauge if they are due to sampling or some systematic difference. Either way, the most important question to answer with this from a statistical inference viewpoint is whether the tails seem comparable, as they do in this case.

The graphic was produced using `qqplot`, as follows:

```
qqplot(fourth, fifth)
```

To clarify the concept, a similar graphic could have been created using the quantile function (though `approx` is actually used):

```
ps <- seq(0.05, 0.95, by=0.05)
x <- quantile(fourth, ps)
y <- quantile(fifth, ps)
plot(x, y)                                   # makes a scatter plot
```

Problems

3.1 The `ToothGrowth` data set contains measurements of tooth length (`len`) for different treatments of supplement type (`supp`). Break the data up by the two types and compare the two resulting variables. Use boxplots.

3.2 For the `michelson` (`MASS`) data set, produce a graphic with a density plot of Experiments 1 and 2.

3.3 Use a `qqplot` to answer if the `rivers` and the `islands` data sets have a similar shape (both are skewed right).

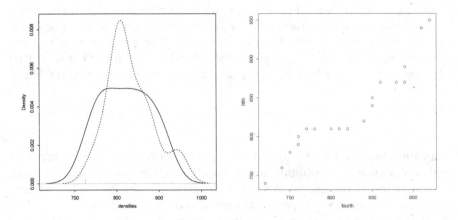

Figure 3.3: Density plots and a quantile-quantile plot of data from the fourth and fifth experiments of Michelson.

3.2 Data manipulation basics

R has two containers for data that can simplify working with bivariate (and multivariate) data. These are the list and the data frame. We will discuss these data types at greater length in Chapter 4, for now we focus just on how these objects can simplify working with multiple samples.

Lists

As previously mentioned, a data vector in R is a container to hold values, all of which are the same type. These values are stored in an order that can be referenced by index, and R provides many flexible means to index. A *list* is a generalized vector, in that it is a linearly indexed collection of elements like a vector, but generalized in that each element need not be of the same atomic type.

R provides the list constructor. Here we create a list to store the beets data.

```
beets <- c(41, 40, 41, 42, 44, 35, 41, 36, 47, 45)
no_beets <- c(51, 51, 50, 42, 40, 31, 43, 45)
b <- list(beets = beets, "no beets" = no_beets)
```

The above could have been shortened to just list(beets, no_beets), but, as with c, names can be added at construction using keyword arguments. As with data vectors, the names attributes of an object can also be referenced or assigned later through the names function.

Lists will print with their names, when present:

```
b                                        # print the list b

## $beets
##  [1] 41 40 41 42 44 35 41 36 47 45
##
## $'no beets'
## [1] 51 51 50 42 40 31 43 45
```

The dollar sign (e.g., $beets) is there to indicate the names. The choice of the dollar sign is related to the use of the dollar sign operator to access a component of a list by name. For example:

```
b$beets

##  [1] 41 40 41 42 44 35 41 36 47 45
```

This is identical to how we have been accessing variables in a data frame.

Indexing can also be done by number (or name or logical vector of matching length). The main issue is there are two results we may want when indexing: a list including the component or just the component. As such, there are two indexing notations: [to return a list and [[to return the component:

```
b[1]                                     # a list with just the first component

## $beets
##  [1] 41 40 41 42 44 35 41 36 47 45

b[[1]]                                    # just the first component

##  [1] 41 40 41 42 44 35 41 36 47 45
```

In the following chapter, we will give more details about indexing. Here the goal is to see how lists of vectors can make working with multiple samples easier.

The boxplot command has a list interface, which will use the named components of the list to label the boxplots, when present. For example, the boxplot in Figure 3.2 could have been produced with just:

```
boxplot(b)
```

• **Example 3.1: Multiple comparisons**
Several of the graphics listed generalize nicely for comparing more than two independent samples, the boxplot being one of them.

We demonstrate in a tedious manner with the michelson (MASS) data set:

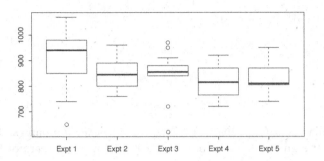

Figure 3.4: Side-by-side boxplots of the five experiments run by Michelson to measure the speed of light.

```
speed <- michelson$Speed; expt <- michelson$Expt
speeds <- list(speed[expt == 1], speed[expt == 2], speed[expt == 3],
               speed[expt == 4], speed[expt == 5])
names(speeds) <- paste("Expt", 1:5)
boxplot(speeds)
```

Figure 3.4 shows the five boxplots. We can observe that experiment one seems slightly different than the others, perhaps adjustments were made to experimental procedures between experiments.   ••

Data frames

We have mentioned that data frames are a preferred way to store many variables in rectangular manner with each variable being a different column and observations for similar cases in rows.

R provides the data.frame constructor for creating data frames, among other ways—they can be produced column-by-column with cbind and row-by-row with rbind. The usage is similar to list. For example, here we create a data frame to hold some student data:

```
student <- c("Alice", "Bob")
grade <- c("A", "B")
attendance <- c("awesome", "bad")
data.frame(student, grade, attendance)

##    student grade attendance
```

```
## 1   Alice     A     awesome
## 2     Bob     B         bad
```

Note how R performed some magic and added column names derived from the variable names. As with c and list, keyword arguments could have been specified to customize these names. Not obvious in the way the above prints, is that by default data.frame will convert character data to factors. If for some reason this is not desirable (e.g., the student name is really a characteristic and not a category), then the argument stringsAsFactors=FALSE can be specified.

Model formulas

For a data frame, the data must be rectangular and each column must contain data of the same type. To store data of differing lengths (like the data on beet eating) requires some coding. The common way to do this is similar to how the michelson data frame holds its values: it has one column recording the measurement (Speed) and another recording the experiment number (Expt).

Example 3.1 showed a cumbersome way—subsetting—to force this storage type into a list. However, as this is a common storage mechanism, many R functions have an interface that makes the format easy to work with, R's *model formula interface*. Formulas are used to express relationships between variables for many different functions, and the interpretation of a formula can vary depending on its use.

The basic form of a model formula is

```
response(s) ~ predictor(s)
```

One or more variables can be specified for both the response or the predictor. This is done with a syntax that is mostly consistent across its many different uses in R. The language is consistent with the usage, as typically we have one or more variables used to predict a value in response. Another common term for the predictor variables is *covariates*.

For now, we stick to the basic case where the response is a numeric variable and the predictor is a factor (the form x ~ f). Later in this chapter, we look at numeric versus numeric (y ~ x) and extensions to two factors.

The typical interpretation of x ~ f is the data in x is *split* into groups determined by the factor f. The calling function is then applied to each group.

For example, in the michelson data set, the Expt variable is a factor, and Speed a numeric value. The boxplot function does exactly what we would like with this: split the data into groups and summarize each group with a boxplot. Except for the labeling, Figure 3.4 could have been made more simply with:

```
boxplot(Speed ~ Expt, data=michelson)
```

The data argument In several previous examples, we have commented how it can be tedious to type in the data frame name to access one of its variables. The data argument allows this specification to happen just once. Within the model formula, variable names will be searched first within the data frame, and then the enclosing environment. In the command above, values bound to Speed and Expt will be found within the data frame michelson.

The subset argument There is an additional subset argument common to the modeling interface that can be used to filter out which cases are to be displayed. Again, variables will be looked up within the data frame specified to data first.

For example, to restrict just to the last 10 (of the 20) runs per experiment, we could create boxplots with:

```
boxplot(Speed ~ Expt, data=michelson, subset=Run %in% 11:20)
```

Rather than testing by Run > 10, we test for inclusion in the values 11:20, as Run is kept as a factor in the data frame and is not numeric.[3]

The model formula template Many of R's function have a model formula interface, each following this basic template:[4]

$$\boxed{\text{action}}\ (\ \boxed{\text{response}}\ \sim\ \boxed{\text{predictor(s)}}\ ,\ \text{data=}\ \boxed{\text{dataset}}\ )$$

Also, subset is a standard argument.

The add-on mosaic package extends the formula interface to many of the functions providing numeric summaries, such as mean and sd. The Hmisc package also extends the summary function to produce many of these values when called on a model formula.

The plot generic The plot function is a generic function, with methods that provide a good graphical summary of an object. When the object is a formula of the type x ~ f side-by-side boxplots of the x values are drawn with groups decided by the factor f, as in this case the x ~ f model formula represents the question of asking how x varies across the groups formed by f.

As such, except for labeling, Figure 3.4 could have been produced with just:

---

[3]R is relaxed in how it decides when a factor matches a value. Factors (e.g., Run) are coerced to character data and then these and the values to match against (e.g., 11:20) are promoted to a common type for comparison purposes (e.g., character in the example).

[4]This template idea is borrowed from a presentation by R. Pruim, a coauthor of the mosaic package.

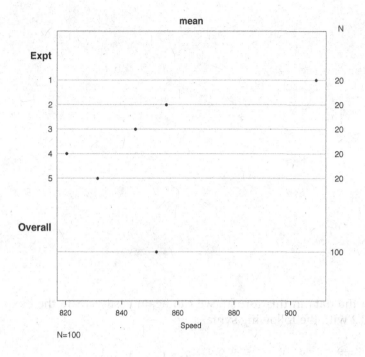

Figure 3.5: Plot of output of summary call on a basic formula. This graphic highlights the comparison of the means for each group with the overall grand mean for all the data.

```
plot(Speed ~ Expt, data=michelson)        # boxplots: Expt is a factor
```

Compare this output to the dot charts (Figure 3.5) produced by

```
out <- summary(Speed ~ Expt, data=michelson)
plot(out)
```

Stacking data   R has several facilities for reshaping of data. The functions stack and split are two of the simplest to use.

The stack function is used to concatenate, or combine, variables with a similar measurement into one vector along with a matching factor indicating where the original variable came from.

It is easily used with a named list to produce the desired output:

```
b <- list("beets" = beets, "no beets" = no_beets)
stacked <- stack(b)
```

```
str(stacked)                              # two-variable data frame

## 'data.frame': 18 obs. of  2 variables:
##  $ values: num  41 40 41 42 44 35 41 36 47 45 ...
##  $ ind   : Factor w/ 2 levels "beets","no beets": 1 1 1 1 1 1 1 1 1 1

headtail(stacked)                         # visual of data frame

##     values      ind
## 1       41    beets
## 2       40    beets
## 3       41    beets
##      ...
## 15      40 no beets
## 16      31 no beets
## 17      43 no beets
## 18      45 no beets
```

With the data in this format, we can again easily create the boxplots of
Figure 3.2 with the following syntax:

```
plot(values ~ ind, data=stacked)
```

Splitting data   While R provides numerous functions that employ the model
interface, it can be easier to implement a new function using lists than using
formulas. In Chapter 4 we mention many methods for iterating over the elements of a list that simplify the task of applying some summary to different
groups. R has an easy way to *split* data up by a grouping variable into a
named list, where the components are used to store the groups.

The split function does this task. For example, in a previous example we
did the following:

```
speed <- michelson$Speed; expt <- michelson$Expt
speeds <- list(speed[expt == 1], speed[expt == 2], speed[expt == 3],
               speed[expt == 4], speed[expt == 5])
names(speeds) <- paste("Expt", 1:5, sep=" ")
```

The split function does the same with a less redundant approach:

```
speeds <- split(michelson$Speed, michelson$Expt)
names(speeds) <- paste("Expt", 1:5, sep=" ")
```

As above, the split function essentially undoes the stack operation. However, the split function can be used more generally. The first argument, x,

can be a vector, as used above, or a data frame. In the latter case the components of the split object are individual data frames. Further, the second argument, f, can be a factor, as above, or a list of factor names. In the latter case all present interactions of the factors are used to split the data.

We will return to this topic when we describe the split-apply-combine paradigm in Chapter 4.

Problems

3.4 Three students record the time spent on homework per class. Their data is

```
Marsha 25  0 45 90  0
Bill   30 30 30 30
Holly  15  0 90  0
```

Use a list to store these values. Then create a boxplot to compare.

3.5 A group of nursing students take turns measuring some basic assessments. Their data is:

	Temp	Pulse	Systolic	Diastolic
Jackie	98.2	96	134	90
Florence	98.6	56	120	80
Mildred	98.2	76	150	95

Create a data frame of these values. Will plot and boxplot produce same graphic?

3.6 After loading Hmisc, what kind of summary is produced for a model formula of the type x ~ f? To see, investigate the output of

```
require(Hmisc)                    # loaded by UsingR
library(MASS)
summary(Speed ~ Expt, michelson)
```

3.7 The split function will coerce its second argument to a factor then split. This is useful if the grouping variable is stored as a numeric value. Verify this by splitting the mpg variable in the mtcars data set by the cyl variable.

3.8 The second argument to split can be a list of factors. The result is that all interactions (possible combinations) are used for the groups. In the ToothGrowth data set, growth (len) is measured for two types of supplements (sup) and three doses (dose). Split this len value into 6 groups.

3.9 The use of a cell phone while driving is often thought to increase the chance of an accident. The data set reaction.time (UsingR) is simulated data on the time it takes to react to an external event while driving. Subjects with control == "C" are not using a cell phone, and those with control == "T" are. Their time to respond to some external event is recorded in seconds.

Create side-by-side boxplots of the variable reaction.time for the two values of control. Compare the centers and spreads.

3.10 For the data set twins (UsingR) make a boxplot of the Foster and Biological variables. Do they appear to have the same spread? The same center?

3.11 The data set stud.recs (UsingR) contains 160 SAT scores for incoming college students stored in the variables sat.v and sat.m. Produce layered density plots of the data. Do the two data sets appear to have the same center? Then make a quantile-quantile plot. Do the data sets appear to have the same shape?

3.12 The data set normtemp (UsingR) contains normal body temperature measurements for 130 healthy individuals recorded in the variable temperature. The variable gender is 1 for a male subject and 2 for a female subject. Break the data up by gender and create side-by-side boxplots. Does it appear that males and females have similar normal body temperatures?

3.3   Paired data

The proportions of the human body have long been of interest to humankind. For example, Leonardo da Vinci's famous drawing *Vitruvian Man* (Figure 3.6) of a man with outstretched arms and legs within a square and circle is based on the ideal of human proportions described by the ancient Roman architect Vitruvius.

Another example is in *Gulliver's Travels* by Jonathan Swift (1726) where this passage is found:

> Then they measured my right Thumb, and desired no more;
> for by a mathematical Computation, that twice round the Thumb
> is once round the Wrist, and so on to the Neck and the Waist,
> and by the help of my old Shirt, which I displayed on the Ground
> before them for a Pattern, they fitted me exactly.

In da Vinci's drawing we see that the height of a person is theorized to be precisely the length of the arm span. For Swift, the assumption is that for a given person, the relationship between thumb and wrist circumference should be roughly the same (twice) for all men, similarly for wrist and neck, and so on.

Figure 3.6: Leonardo's *Vitruvian Man* demonstrating the proportions of the human body. (Original image from the Wikipedia Commons.)

Both examples are based on the "idealized" body type. Of course, bodies show a wide range of variation so to explore such a relationship would require us to look at many different subjects. We can then summarize this data about a relationship between two numeric variables either numerically or graphically.

In addition to the arguments that accompany the model formula interface, there are many extra arguments that can be passed to the plot function, as well as other plotting functions. We list some in Table 3.1.

Let's consider the fat (UsingR) data set which contains measurements of different body dimensions for a cohort of 252 males. The goal of that data collection was to see if some prediction of a body fat index could be made from variables which can be measured with a tape measure. (Body fat measurements require some special equipment otherwise.) For this example, we are just interested in relationships.

A listing of the variables in the data set shows:

```
names(fat)

##  [1] "case"      "body.fat"   "body.fat.siri"
##  [4] "density"   "age"        "weight"
##  [7] "height"    "BMI"        "ffweight"
## [10] "neck"      "chest"      "abdomen"
## [13] "hip"       "thigh"      "knee"
## [16] "ankle"     "bicep"      "forearm"
## [19] "wrist"
```

Command	Description
main	Title to put on the graphic.
xlab	Label for the x-axis. Similarly for ylab.
xlim	Specify the x-limits, as in xlim=c(0,10), for the interval $[0,10]$. Similarly for ylim.
type	Type of plot to make. Use "p" for points (the default), "l" (ell not one) for lines, and "h" for vertical lines.
bty	Type of box to draw. Use "l" for "L"-shaped, default is "o", which is "O"-shaped. Details in ?par.
pch	Plot character used. This can be a number or a single character. Numbers 0 – 18 are similar to those of S, though there are more available beyond that range.
lty	When lines are plotted, specifies the type of line to be drawn. Details in ?par.
lwd	The thickness of lines. Numbers bigger than 1 increase the default.
cex	Magnification factor. Default is 1.
col	Specifies the color to use for the points or lines.

Table 3.1: Useful arguments for various plotting commands.

The neck and wrist variables present us the opportunity to make a simple test of the assumption of Swift's Lilliputians.

First, we compare mean neck size to the mean wrist size two different ways:

```
mean(fat$neck) / mean(fat$wrist)          # ratio of means

## [1] 2.084

mean(fat$neck/fat$wrist)                   # mean of ratios

## [1] 2.084
```

For the sample data these ratios are roughly twice.[5] The question of whether the difference from exactly twice can reasonably be attributed to the inherent randomness involved in selecting a cohort, or to a systematic difference is left to our discussion of statistical inference.

The data under consideration is naturally paired off. We may write the data set as $(x_1,y_1),(x_2,y_2),\ldots,(x_{252},y_{252})$. This suggests plotting the data as point on a Cartesian plot with a scatter plot. The plot function will do so (the left graphic in Figure 3.7):

---

[5]It is a coincidence that the two ratios appear equal, in general this isn't true and in this specific case, they actually differ in fourth decimal point.

Function	Description
points	Add points to a graphic.
lines	Add line segments to a graphic.
abline	Add a line to a graphic.
text	Locate text in body of a graphic.
mtext	Locate text in margins of a graphic.
title	Add main title, x label or y label to a graphic.
legend	Locate legend in body of a graphic.

Table 3.2: Plotting functions to add to an existing graphic.

```
plot(fat$wrist, fat$neck)
```

The plots shows a relationship. Certainly, the value of exactly twice can't be expected to hold for each individual, as the points do not fall on a line with slope 2. It may be the case that on average this is so.

As mentioned, the plot function is a generic function and can be called many different ways. When called this way (plot(x,y)) the first variable is plotted on the x axis, the second on the y. When called with a model formula of the type y ~ x (numeric modeled by numeric) then the y variable is a response (dependent) variable and the x the predictor (independent) variable. The same graphic could be produced with:

```
plot(neck ~ wrist, data=fat)
```

The words "response" and "predictor" are similar to common usage. For example, the Lilliputians knew the thumb size and predicted the rest of the measurements based on that (getting a bit lucky in the process).

Again, the subset argument may be used to restrict which cases are used in creating the plot. The following command will only consider those subjects in their twenties when drawing the scatter plot:

```
plot(neck ~ wrist, data=fat, subset = 20 <= age & age < 30)
```

Whereas, the following command excludes the erroneous height measurement (29.5 is likely meant to be 69.5) in a comparison of wrist circumference to height (the right graphic of Figure 3.7):

```
plot(height ~ wrist, data=fat, subset=height > 50)
```

Correlation

The left graphic in Figure 3.7 shows the data falling around a straight line, which is to be expected intuitively as the measurement of wrist and neck cir-

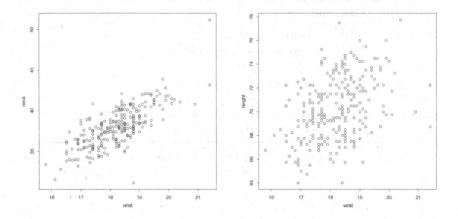

Figure 3.7: Scatter plots showing a relationship between wrist and neck circumference (left graphic) and wrist circumference and height (right graphic).

cumferences are related. The right graphic shows something similar, larger wrists tend to correspond to taller people. However, the data is more scattered, as might be expected. (Even the Lilliputian's of Swift might have guessed that if you double enough times where there is some error in measurement your errors will propagate on average, though how exactly they propagate would likely have been unknown to them in 1726.)

The *correlation* is a numeric summary for how closely related are the measurements of two numeric variables when they are in a linear relationship. Perfect correlation would mean the data fall precisely on a straight line. No correlation roughly says the values are scattered.

To investigate the definition of correlation, it is useful to shift the center of the scatter plot to the point specified by the mean of the x variable and the mean of the y variable. We can graphically add crosshairs centered on such a point and a marker with the abline and points functions. The left graphic of Figure 3.8 shows the results.

```
x <- fat$wrist[1:20]; y <- fat$neck[1:20] # some data
plot(x, y, main="Neck by wrist")
abline(v = mean(x), lty=2)                 # dashed vertical line
abline(h = mean(y), lty=2)                 # dashed horizontal line
points(mean(x), mean(y), pch=16, cex=4, col=rgb(0,0,0, .25))
```

As with the other functions in Table 3.2, a call to abline adds a layer to an existing graphic. In particular, abline function is used to draw a line across a graphic. A vertical line is specified using "v", a horizontal line with "h", and a line with intercept through "a" and slope "b". The lines function can be

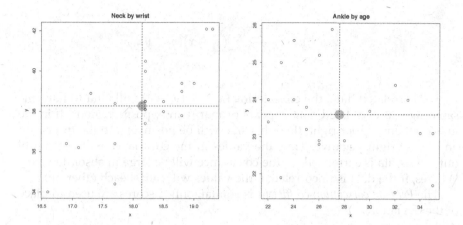

Figure 3.8: Scatter plots with their center framed by cross hairs. The data in the left graphic is correlated, as seen by the clustering of the points into two of the four regions, the data on the right is uncorrelated, as the points are scattered throughout the four regions.

used to similar effect, but abline has a more direct interface for this specific task.

The points command is used with 16 specifying a solid circle for the plot character, a size of 4 times the usual, and color specified with RGB values and a 25% alpha transparency (so that any overlapping points bleed through).

The point of this graphical display is to isolate the data into four regions determined by those with x values above and below the mean of all the x values and those with y values above and below the mean of all the y values. For the correlated data shown, most of the data appear in two opposing quadrants of the four, in this case the values are primarily both lower than average or both greater than average, with just a few mixing that pattern up.

The right graphic of Figure 3.8 plots the age of the subject against their ankle circumference. Whereas we might expect a certain widening in the middle as a person ages (waist circumference), ankles generally don't show the effects of a person gaining weight. Hence, the graphic would not be expected to show much of a relationship, and it doesn't. In this graphic, the points don't seem to cluster in just two of the four regions.

The covariance and correlation between two values quantifies the difference between points scattered in the four regions versus points scattered in two opposite regions. The *covariance* between two data sets x and y is given by:

$$\text{cov}(x,y) = \frac{1}{n-1}\sum(x_i - \bar{x})(y_i - \bar{y}).$$

For correlated data, the cross terms $(x_i - \bar{x})(y_i - \bar{y})$ will tend to have the same sign, as correlated data appears primarily in opposite regions. If in the lower left and upper right, then this sign will be positive; if in the upper left and lower right negative. Thus the values in the covariance will not cancel out if the data is correlated, so the covariance will be large in absolute value. Wheres, if the data is uncorrelated, the values will cancel each other out.

The *Pearson correlation coefficient* is similar, only z-scores are used in place of the deviations:

$$\text{cor}(x,y) = \frac{1}{n-1}\sum\left(\frac{x_i - \bar{x}}{s_x}\right)\left(\frac{y_i - \bar{y}}{s_y}\right) = \text{cov}(x,y)/(s_x s_y).$$

Dividing by $n-1$ ensures that the value is between -1 and 1, values which occur only when the data fall on a straight line with negative or positive slope. When data is perfectly uncorrelated, the correlation is 0.[6]

We can see from the formula that the correlation is symmetric, in that we could switch the order of the x and y variable. Further, the fact that the correlation is based on z-scores—which remove the effects of both center and spread by standardizing—makes the correlation intrinsic to the shape of the data.

The cov and cor functions compute the covariance and correlation. They do not have a formula interface. We can use them as follows:

```
cor(fat$wrist, fat$neck)                    # strongly correlated

## [1] 0.7448

cor(fat$wrist, fat$height)                  # mildly correlated

## [1] 0.3221

cor(fat$age, fat$ankle)                     # basically uncorrelated

## [1] -0.1051
```

---

[6]This is in a linear sense, data can be related and have 0 correlation, such as would happen by symmetrically placing points on the unit circle.

When these functions are called with a data frame (or matrix) the pairwise correlations of each pair of columns is given. This is returned in a symmetric grid with the ith row and jth column holding the correlation between the ith and jth column.

Spearman correlation coefficient The Pearson correlation coefficient measures the strength of a linear relationship. For monotonic relationships, there are other, similar measures, such as the Spearman correlation coefficient.

Consider the Animals (MASS) data set that records average body weight (body) and brain size (brain) for several species, some quite extinct. We would expect that larger bodies would be paired off with larger brains, leading to a positive correlation closer to 1 than 0. A quick check shows our assumption is way off:

```
cor(Animals$body, Animals$brain)

## [1] -0.005341
```

The issue is the presence of Brachiosaurus, a rather large, dumb dinosaur. A scatter plot shows this point as an outlier. The lone outlier has a large "x" value and small "y" value, so the product $(x_i - \bar{x})(y_i - \bar{y})$ is a large negative number, large enough to cancel the rest of the contributions which are mostly positive.

To see all of those with negative cross terms we compute:

```
body <- Animals$body; brain <- Animals$brain
cross_prods <- (body - mean(body)) * (brain - mean(brain))
Animals[cross_prods < 0, ]

##                    body   brain
## Dipliodocus       11700    50.0
## Asian elephant     2547  4603.0
## Horse               521   655.0
## Giraffe             529   680.0
## Human                62  1320.0
## Triceratops        9400    70.0
## Brachiosaurus     87000   154.5
```

(Why are humans in this list?)

We could remove the dinosaur species and recompute. They are clearly outliers and died off, perhaps, as a species because of this imbalance. Instead we transform the data. Transforming can make large values not have such a big impact on our numeric measures.

A common transform, historically dating back to at least Fechner in 1800s is to rank the data, that is placing the data in order and assigning a rank,

first, second, ... The rank function[7] will do this task. It has various methods
to assign ranks when the data is tied. Here we rank the data (with the default
ties.method="average" to break ties) and then take the correlation:

```
cor(rank(body), rank(brain))

## [1] 0.7163
```

This approach is common enough, that R's cor function has an argument
method to specify it:

```
cor(body, brain, method="spearman")

## [1] 0.7163
```

The Spearman correlation coefficient is then a measure of association.
Data which is monotonically related (increasing or decreasing relationship)
will have perfect Spearman correlation. The application of this measure is not
restricted to just linear relationships, as the Pearson correlation interpretation
is. (There is also the Kendall correlation coefficient (method="kendall") which
also uses the ranked data. We illustrate this in Section 3.4.)

• **Example 3.2: Correlated averages with replication**
The data set ToothGrowth contains measurements on the effects on tooth
length (len) based on varying dosages of vitamin C for a cohort of guinea
pigs. The idea being that more vitamin C in the diet would promote tooth
growth. We can check this numerically by observing that the correlation is
indeed positive:

```
cor(ToothGrowth$dose, ToothGrowth$len)

## [1] 0.8027
```

For each level of dosage, there are several experimental units, as the experiment was replicated. This is a common design. It is also common for
researchers to average then take the correlation of the averaged data. (Indeed, the data in the last example on animal brain weights was averaged
over species.) This could change the correlation significantly. In the following
we split the data into the three dosage groups and then compute the correlation for these dosage values and the group averages (again, in the next
chapter we introduce idioms that make this sort of computation over groups
much easier):

---

[7]The xtfrm function is also used to rank values in R.

```
l <- split(ToothGrowth$len, ToothGrowth$dose)
group_means <- c(mean(l[[1]]), mean(l[[2]]), mean(l[[3]]))
cor(c(0.5, 1, 2), group_means)
```

```
## [1] 0.9574
```

The value of 0.95 for the aggregated data is higher than the value 0.80 for the individual data. In general correlations formed from averages are typically closer to 1 or -1 than when all the data is considered individually. It is important to keep this in mind when evaluating a reported correlation. ••

• Example 3.3: Correlation is not causation

The terms *correlation* and *causation* are often confused, but it is important to keep in mind they are not the same. As early as 1950, articles were appearing linking smoking with cancer, but establishing actual causation was an issue that took much time. Industry forces worked very hard for many years trying to de-couple the two terms.

One problem with trying to establish that a correlation between two variables is due to causation is the presence of *lurking variables*, a confounding variable that correlates with both the response and predictor. We use a data set on SAT scores from the 1990s to illustrate how taking into account a lurking variable can completely change an interpretation.

The question here is the relationship between average teacher pay and student SAT scores. The data set SAT (UsingR) contains measurements by U.S. state from the 1990s.[8] We would expect that states that offer higher pay have an increased chance of attracting more effective teachers who would generally have students with higher SAT scores. So *a priori* we might expect there to be a positive correlation between the salary variable and the total variable (which records average SAT scores per state). A quick check shows this is not the case:

```
cor(SAT$salary, SAT$total)
```

```
## [1] -0.4399
```

The negative correlation can be seen from the overall shape of the scatter plot in Figure 3.9.

This unexpected result demands a closer look. The data set also includes a variable perc recording the percentage of students who took the SAT. If we make a scatter plot and use different symbols for states with a percentage less than 10, and for states with a percentage greater than 40 we see something else.

---

[8]This data is discussed in an example by Kleinman and Horton at http://sas-and-r.blogspot.com/ and attributed to [28] therein.

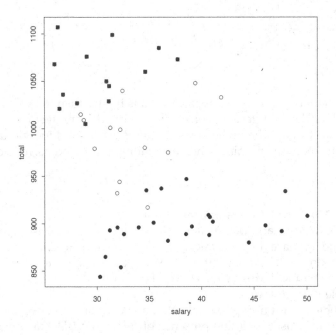

Figure 3.9: Total SAT scores by average teacher pay for each of the fifty United States. The pattern for all the points shows a negative correlation. Different plot characters are used to indicate the statewide percentage of test takers, with low percentage states marked with boxes and a high percentage of states marked with filled-in circles. When each of the three subgroups is considered individually the correlation is positive.

```
plot(  total ~ salary, SAT)
points(total ~ salary, SAT, subset = perc < 10, pch=15)  ## square
points(total ~ salary, SAT, subset = perc > 40, pch=16)  ## solid
```

Figure 3.9 shows that the correlation for each subgroup will be positive, in line with our intuition. To numerically compute these correlations we use a somewhat cumbersome approach.

```
total <- SAT$total; salary <- SAT$salary; perc <- SAT$perc
less_10 <- perc < 10
more_40 <- perc > 40
between <- !less_10 & !more_40
c(less    = cor(total[less_10], salary[less_10]),
  between = cor(total[between], salary[between]),
  more    = cor(total[more_40], salary[more_40]))
```

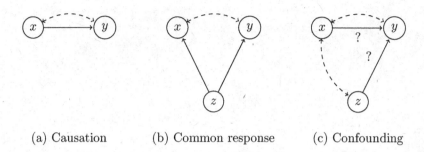

(a) Causation   (b) Common response   (c) Confounding

Figure 3.10: Moore and McCabe's figure showing various explanations for an observed association. The dashed arrows show an association, the solid arrow a cause-and-effect-link. Here x is explanatory, y a response, and z a lurking variable.

```
##     less between     more
##   0.2588  0.2225   0.3673
```

We see, counterintuitively, that all the correlations for the subgroups are positive, yet the overall correlation is negative. This is an example of a phenomenon called *Simpson's paradox* where some trend that exists for subgroups changes when the data is aggregated.   ••

Association: correlation is not causation   Figure 3.10 shows a canonical figure of introductory statistics from [42]. This graphic helps anchor the concept that there are many ways to describe an association between a response variable y and an explanatory variable x.

To elaborate, let's consider the complicated case of gun ownership and violence through a natural experiment. In 1996 the Australian government passed the National Firearms Agreement (NFA) in response to a tragic massacre in Tasmania the same year. Part of this agreement was the implementation of a gun buyback program. Per-state data read from a graphic is:[9]

```
buyback <- c(ACT=1500, NSW=2500, WA=2700, Qld=3700, Vic=4250,
             SA=4200,  NT=5000,  Tas=7500)
change <- -c(.2, 1,.9, 2.5, 1.0, 1.6, 2.5, 3.4)
m <- data.frame(buyback, change)
m
```

[9]This comes from *Did gun control work in Australia?* by Dylan Matthews posted on August 2, 2012 at http://www.washingtonpost.com. The data is attributed to [45], but originated in [50].

```
##      buyback change
## ACT    1500    -0.2
## NSW    2500    -1.0
## WA     2700    -0.9
## Qld    3700    -2.5
## Vic    4250    -1.0
## SA     4200    -1.6
## NT     5000    -2.5
## Tas    7500    -3.4
```

The buyback variable records the number of guns bought back per 100,000 residents in a state, the change variable is the change in suicide rate per 100,000 residents over the years 1990 – 95 and 1998 – 2003. This data has a strong negative correlation:

```
cor(buyback, change)
```

```
## [1] -0.881
```

Clearly, there is some association between these variables, but is there a causal relationship? A causal relationship would be represented by (a) of Figure 3.10. There the association is directly due to the change in the x variable, in this case fewer guns directly leads to fewer firearm suicides. The assumption being "By making the most lethal weapons (firearms) less available, the number of violent crimes will fall and the average lethality of those crimes will also decline" [50].

However, other factors can be at play. Perhaps the Australian public simply was in shock over the tragedy of the event and this made some value their life more than they did prior to the massacre event. If so, the buyback program would then make no difference in the suicide rate, but the numbers of guns bought back would be in response to the tragedy and the reduction of suicides could be attributed to a national response to the tragedy. That is, a third variable could have an effect on the other two. This would be an example of two associated variables being associated as they are in a *common response* to a third lurking variable.

Often a more nuanced set of factors are involved. In [5] a question is raised of whether interventions lead to displacement and method substitution. In this case, does the suicide rate by other means compensate for that lost to self-inflicted firearm wounds taking into account the current cultural trends?

In Figure 3.10 (c) represents *confounding* where the effects on the response by two or more variables cannot be distinguished from each other. In this example, the variable of method substitution may account for the decline in y, the cultural trends may account for the drop in y, or it could be the intervention x or some combination.

Baker and McPhedran write in their abstract "When compared with observed values, firearm suicide was the only parameter the NFA may have

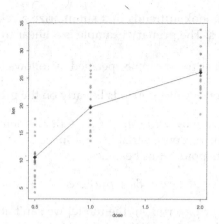

Figure 3.11: Plot of tooth length versus vitamin C dosage for measured guinea pigs in the ToothGrowth data set. The group means are connected with line segments. The slope of each line can be interpreted as the expected mean increase per unit increase in vitamin C dosage.

influenced, although societal factors could also have influenced observed changes." This leaves open the issue of confounding. It is worth pointing out that this analysis was cited by [30] in an article titled *How to find nothing* as an example of difficulties of comparing actual occurrences with *counter-factual* assumptions, in this case the assumption that death rate trends would continue indefinitely in a linear manner.

Trends

Let's pick up with the discussion on the ToothGrowth data set, where we found the correlation of the group averages. Rather than do that again, we plot line segments connecting the group averages:

```
plot(len ~ dose, data=ToothGrowth, pch=16, col=rgb(0,0,0,.25))
points(c(0.5, 1, 2), group_means, cex=1.5, pch=18)
lines(c(0.5, 1, 2), group_means)
```

Figure 3.11 shows the graphic. If the means of each group represent the expected tooth length for a given dosage, then the slope of the line segments connecting the different groups can be interpreted as the change in expected length for a unit change in vitamin C dosage.

If the underlying relationship were truly linear, then the same slope would hold across all changes in dosage and we could summarize the relationship quite succinctly in terms of a single rate of change in the expected mean response.

In this section we look at trends that summarize a relationship between two numeric variables. The primary example is a linear trend.

A model for a linear trend can be specified as follows:

The mean response value depends linearly on the predictor value.

Mathematically we may write $\mu_{y|x} = \beta_0 + \beta_1 x$, where the notation $\mu_{y|x}$ means the mean of the response variable for a specified value of the predictor x. For individual data points, this becomes:

$$y_i = \beta_0 + \beta_1 x_i + \epsilon_i,$$

where the ϵ_i are the error terms. In the model we can't actually observe the errors, as we don't know the values of the β coefficients. Rather we make assumptions about them. Were the errors always 0, then for a given value of x, all the y values would be the same so the average would depend just on x. In a statistical model, a common assumption is that *on average* the values of the ϵ_i are 0. This is interpreted to mean if there were many such errors averaged it would be really close to 0, and in the long run exactly 0. The language "mean response value" then is in this sense where we assume we could take arbitrarily many error terms. Whereas, in any data set, we just have a finite-sized sample.

The method of least squares   Let (x_i, y_i) denote our data and suppose we have a line $y = b_0 + b_1 x$ summarizing the relationship $\mu_{y|x}$. Then if x and y are in a linear relationship, we can define the value $\widehat{y}_i = b_0 + b_1 x_i$ to be the value with $(x_i, \widehat{y}_i)$ lying on the line. The *residual* or *residual error* (cf. Figure 3.12) is given by

$$\text{residual} = y_i - \widehat{y}_i.$$

If we think in some sense of the line describing the trend between x and y then the line predicts the y values in some way, so this could be seen in the following canonical format of

$$\text{residual} = \text{observed} - \text{expected}.$$

This is one of many examples in this text of centering—that is, looking at values relative to our sense of what is expected.

The *least squares regression line* is defined in terms of the residuals: it is the line which minimizes the squared residuals.[10] The choice doesn't try to fit

---

[10]R does not actually compute the values involved in this manner, as there is an underlying geometric interpretation that lends itself to faster, more numerically stable solutions.

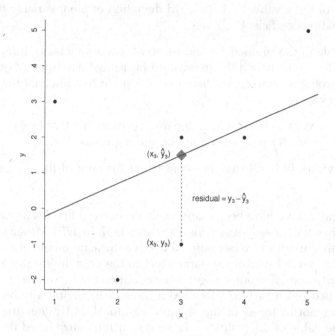

Figure 3.12: Least squares regression line superimposed on a data set. The residual is the signed vertical distance from a point to the line. For the illustrated point, this is negative.

exactly the most possible points, but rather shares the errors amongst points to create the smallest sum.

We denote the least squares regression line as $\hat{y} = \hat{\beta}_0 + \hat{\beta}_1 x$. The "hat" is widely used to indicate an estimate for an unknown parameter. Solving the minimization problem yields these values for $\hat{\beta}_0$ and $\hat{\beta}_1$:

$$\hat{\beta}_1 = \frac{\sum(x_i - \bar{x})(y_i - \bar{y})}{\sum(x_i - \bar{x})^2} = cor(x,y)\frac{s_y}{s_x},$$
$$\hat{\beta}_0 = \bar{y} - \hat{\beta}_1\bar{x}.$$

These two formulas reveal some interesting facts:

- The slope of the regression line, $\hat{\beta}_1$, is related to the correlation, though scaled to fit the scale of the problem. That is, a change of 1 standard deviation in the x direction corresponds to a change in the expected mean of the y value by 1 standard deviation of the y variable *times* the correlation coefficient.

- The role of the predictor x and response variable y is not interchangeable. Trying to switch the two would basically be a different optimization problem—horizontal distances of a point to a line, not the vertical distances.

- The intercept is derived from the more central fact that $(\bar{x}, \bar{y})$—the center of the scatter plot—is a point on the regression line.

Less obvious, but still true, is the fact that the sum of the residuals will always be 0.

Historically there have been many ways to choose a line (estimate the βs) that describes a data set. According to Stigler [55], in 1750 Mayer was confronted with a wealth of observational data on measurements of the moon. To fit a linear model with three parameters to the data, he simply averaged out arbitrarily chosen groups to get three equations in three unknowns that he solved. Boscovich in 1757 proposed a method involving a minimization problem but not in terms of the squared residuals, but rather the sum of the absolute value of the residuals. Legendre in 1805 introduced the idea of minimizing the squared residuals.

The lm function   In Chapter 11 we discuss statistical inference for the simple linear regression model. This model is specified in R's model formula notation as `response ~ predictor`. The `lm` function is used to fit *linear models*, in particular it covers this simple case of a single predictor. The basic interface is common to the model interfaces we've seen so far:

```
res <- lm(response ~ predictor, dataset, subset=...).
```

To see how `lm` is used, we look at the data set `heartrate` (`UsingR`). This data set records an age (`age`) and maximum heart rate (`maxrate`) for a hypothetical cohort of people. The rule of thumb in many gyms is that the maximum heart rate is related linearly to a person's age by the formula $220 - age$. Mathematically, this is a line with slope of -1 and intercept of 220.[11]

For this data, we can find the least squares coefficients with R:

```
res <- lm(maxrate ~ age, data=heartrate)
res
```

---

[11] According to the citation in the `heartrate` help page, the simple formula was determined "arbitrarily" by looking at ten studies and graphically estimating the values for a slope and intercept.

```
##
## Call:
## lm(formula = maxrate ~ age, data = heartrate)
##
## Coefficients:
## (Intercept)          age
##     210.048       -0.798
```

We see from the output the estimated intercept is less than 220 and the estimated slope not quite −1, though both estimates are within the ballpark of the oft-used rule of thumb.

The interpretation of the slope in this case is the change is the predicted mean maximum heart rate declines by 0.798 beats per minute for a unit (one year) change in age.

The res object only prints a small portion of the information contained in it. To tease out more, R provides various *extractor functions*. These are generic functions with implementations for many different modeling functions besides lm. Table 11.1 has a listing. We discuss just a few here, putting off the others for Chapter 11.

Visualizing the regression line   We can easily add the regression line to a scatter plot with the abline function:

```
plot(maxrate ~ age, data=heartrate)
abline(res)                                      # add line to graphic
```

Residuals, fitted values, and predictions   The residuals are computed for each point and returned by the resid function (an abbreviation for residuals). We mentioned that the residuals sum to 0, as can be empirically verified up to round-off error through:

```
sum(resid(res))
```

```
## [1] -1.776e-15
```

The residuals are related to the fitted values through residual = actual − fitted. R also provides a fitted function. We can check that this defining relationship holds:

```
actual <- heartrate$maxrate
diffs <- resid(res) - (actual - fitted(res))
max(diffs)                                       # 0 save for round-off
```

```
## [1] 1.199e-14
```

The fitted value gives the predicted mean value of the response variable for the specific value of x in the data. For other values of x where a prediction for the mean response is desired, the predict function can be used. Before illustrating, we show that predict is simply a convenience in this case, as it isn't hard to do the work directly.

To estimate the expected mean maximum heart rate for 30- and 40-year olds simultaneously we have:

```
age <- c(30, 40)
coef(res)[1] + coef(res)[2] * age        # beta_0 + beta_1 x

## [1] 186.1 178.1
```

This approach can get tedious in the case with many predictor variables. The predict method simplifies it. To use this function, a data frame with named variables matching those in the model formula is needed. This allows predict to disambiguate values when there are multiple covariates or predictor variables.

```
predict(res, data.frame(age=c(30, 40)))

##      1     2
## 186.1 178.1
```

Diagnostic plots   The plot method for linear models (e.g., plot(res)) will show diagnostic plots that make it possible to assess the assumptions needed for statistical inference. This topic will be discussed in Chapter 11.

Transformations

In talking about the correlation for the data set comparing the body weight and brain weight of various animals, we presented an example where it was best to transform the data before applying a statistical summary, in that case the correlation. With linear models this can often be the case.

Common transforms are based on powers, such as squaring or taking a square root, or logarithms, if the variables are positive.

For example, rather than rank the data on body and brain weights, we look instead at their transformation by taking logs. Figure 3.13 shows the graphic produced by these commands:

```
plot(log(brain) ~ log(body), data=Animals)
## label the outliers
idx <-  log(Animals$body) > 6 & log(Animals$brain) < 6 # select
text(log(Animals$body[idx]), log(Animals$brain[idx]), # label
     labels=rownames(Animals)[idx], pos=2)
```

Three values fall outside the otherwise strong linear relationship in the data, as labeled in the figure. Fitting the linear model with these values excluded could be done with:

```
lm(log(brain) ~ log(body), data=Animals, subset=!idx)

##
## Call:
## lm(formula = log(brain) ~ log(body), data = Animals, subset = !idx)
##
## Coefficients:
## (Intercept)      log(body)
##        2.150          0.752
```

Interpreting these values is a bit different. The basic model has become

$$\log(y_i) = b_0 + b_1 \log(x_i) + \text{error}_i$$

or in the original scale:

$$y_i = e^{\text{error}_i} \cdot e^{b_0} x_i^{b_1}.$$

This is roughly saying that the mean response in the *brain* is a multiple of the body weight squared ($e^{0.752}$) and the errors are multiplicative, not additive.

Mathematical operators and model formulas   The model formula interface overloads the usual arithmetic operators, +, -, *, and /. That is, these operators are given different interpretations when used in the formula context. These interpretations are useful when the models have more than one explanatory variable. To have the operators assume their familiar mathematical meanings, we must wrap the call within the I function, R's "as is" operator.

• **Example 3.4: Body mass index**
The body mass index (BMI) is a measure of an individual's fitness designed by Quetelet in the 1800s, but still very much part of the everyday vernacular. BMI is a simple measure computed by taking an individual's weight and dividing by their height squared using units of kilograms and meters. This comparison allows the same scale to be used for people of varying heights. It also implies that we should expect weight and height squared to exhibit a linear relationship. For that we can investigate. We use the weight and height data in the kid.weights (UsingR) data set, after restricting the data to five-year-olds.

```
idx <- 60 <= kid.weights$age & kid.weights$age < 72
```

The data is in pounds and inches, so we want to convert by dividing weight by 2.2 and height by 2.54/100. Our first attempt won't work:

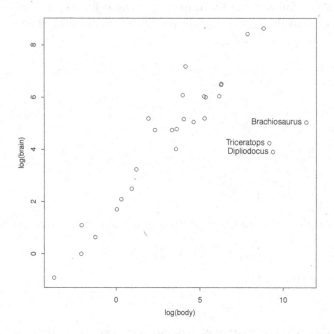

Figure 3.13: The brain weight of animals plotted as a function of the body weight after taking logarithms. There are three prominent outliers from the otherwise well-defined linear relationship labeled by their row name.

```
plot(weight/2.2 ~ (height*2.54/100)^2, data=kid.weights, subset=idx)
## Error: invalid model formula in ExtractVars
```

The left-hand side of the formula is fine, the complaint is about the right-hand side. As both * and ^ have different meanings from their mathematical usage in the model syntax, we need to wrap the result within I:

```
fm <- weight/2.2 ~ I(height*2.54/100^2)
```

With the formula assigned to a variable, we can use it to plot both the data and the regression line:

```
plot(fm, data=kid.weights, subset=idx)
res <- lm(fm, data=kid.weights, subset=idx)
abline(res)
res

##
```

```
## Call:
## lm(formula = fm, data = kid.weights, subset = idx)
##
## Coefficients:
##         (Intercept)  I(height * 2.54/100^2)
##                10.6                   927.8
```

The large slope to the regression line is a bit surprising, as the values should be similar. Whether it is an artifact of sampling variability or something more is of interest.   ••

Alternative trend lines

There are different trend lines that could be substituted for the least squares regression line.

Standard deviation line   Figure 3.14 shows data used by Pearson in his pioneering studies of correlation and regression.

The regression line is drawn in a solid line. In addition, the line with slope s_y/s_x going through the point $(\bar{x}, \bar{y})$ is drawn. This is called the SD line by Freedman et. al. [23] (a similar graphic graces the text's cover). The standard deviation line allows us to easily compare values by their z-score: points whose components have equivalent z-scores will appear on the line, whereas if a point is above this line, say, then the z-score of the y value is more than the z-score of the x value.

The slope of the regression line is the correlation coefficient ($r = 0.50$) times s_y/s_x, so the slopes will differ by r. If the regression line is used for prediction (say to predict a son's adult height by the father's height) then this fact implies for the taller-than-average fathers their sons are predicted to be taller than average, but their predicted z-score will be less by a factor of r. In this sense, there is a regression towards the mean.

Minimizing vertical distance to the line   The least squares regression line finds the values of b_0 and b_1 that minimize the sum $\sum_i (y_i - (b_0 + b_1 x_i))^2$. Similarly, we can ask, as Boscovich did, to find b_0 and b_1 that minimize $\sum_i |y_i - (b_0 + b_1 x_i)|$. This is a special case of quantile regression related to the median.

The quantreg package implements this procedure through its rq function. We illustrate by following an example from that package using their engel (quantreg) data set. This has a variable foodexp for annual food expenditure and income for annual income for a data set collected in Belgium. The special case of quantile regression we seek is found with the rq function using the parameter tau=0.5. We see below how different R modeling functions employ a similar interface:

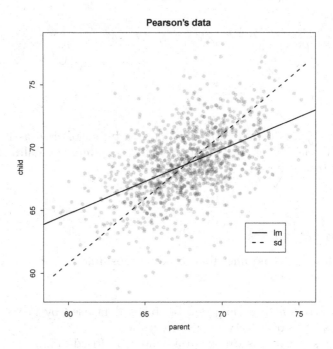

Figure 3.14: Plot of son's height by father's height for data used by Pearson. In overlay are a regression line (solid) and a standard deviation line. The regression line predicts the son's height based on the father's, the standard deviation line marks values with equal z-scores. The regression effect is the predicted z-score for the son will be closer to 0 than the father's z-score.

```
require(quantreg)                        # also loaded with UsingR
data(engel)                              # not loaded by default
f <- foodexp ~ income
res.lm <- lm(f,           data=engel)
res.rq <- rq(f, tau=0.5, data=engel)
plot(f, data=engel)
abline(res.lm, lty=1)
abline(res.rq, lty=2)
legend(4000, 1500, c("lm", "rq"), lty=1:2)
```

In the right graphic of Figure 3.15 the data has been split into 250-franc ranges and a boxplot is drawn for each subset. Layered on top is the quantile regression line, in some sense, it tracks the medians for each group.

Minimizing distance to the line   An alternate minimization would be to minimize the distance to the line (not the vertical distance or vertical distance

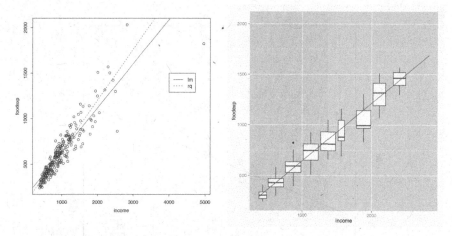

Figure 3.15: Left graphic shows two trend lines drawn fitting the relationship between food expenses and income for the data in the engel data set. The lines are similar, but the least squares regression line (solid one) is more attracted to the outlier than the median regression line (dashed one). The right graphic shows the same data split into 250-franc groups that are presented in boxplots (using ggplot2). The quantile regression line in overlay tracks the median for the groups.

squared). This problem is related to principal component analysis, a statistical technique used for data reduction. Figure 3.16 shows the first principal component along with the regression line. The first principal component is chosen to account for as much of the variability as possible. In 1901, Pearson [47] described the importance of this line for problems when both x and y are subject to randomness and neither a natural predictor. This is unlike the regression line, where the terminology of dependent and independent variables suggests the independent variable is known and the dependent depends on the value.

Robust regression lines   Like other statistics we've seen so far—such as the mean, the standard deviation, and the Pearson correlation—the regression estimates are sensitive to one or two "large" values. In this case, large is in the sense of far from the pattern. An example would be points for the dinosaurs in the Animals data set. The MASS package introduces several alternatives to lm that are robust to one or more large values. The rlm function is one such. It basically replaces the function to be optimized: $\sum(y_i - (b_0 + b_1x_i)^2)$ with one not so sensitive to large values: $\sum\rho(y_i - (b_0 + b_1x_i)^2)$. The default ρ is "Huber's M-estimator", which is like x^2 for small values and $|x|$ for large. The mathematical details are handled by the authors of MASS, who provide an interface similar to that of lm.

We compare model fits for lm and rlm as follows (Figure 3.17):

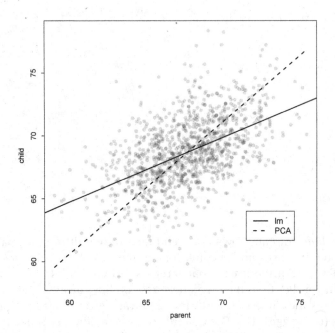

Figure 3.16: Regression line with first principal component direction drawn. The principal component direction minimizes the total perpendicular distance to the line and is in the direction explaining maximum variability.

```
f <- log(brain) ~ log(body)
res.lm  <-  lm(f, data=Animals)
res.rlm <- rlm(f, data=Animals)
plot(f, data=Animals)
abline(res.lm,  lty=1)
abline(res.rlm, lty=2)
legend(-2, 8, legend=c("lm", "rlm"), lty=1:2)
```

The figure shows the influence of the three points representing the dinosaur species, which we can get with:

```
which(resid(res.lm) < -2)

##    Dipliodocus   Triceratops Brachiosaurus
##              6            16            26
```

The regression line is pulled towards that cluster of values, the robust regression line is not so much.

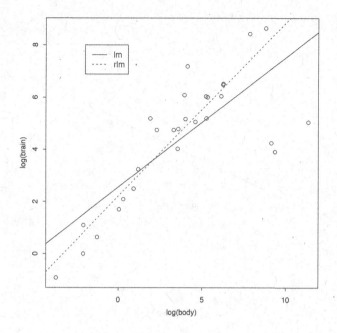

Figure 3.17: Log-log plot of brain size based on body mass for several species of animals. The simple regression line is pulled towards the three outliers, whereas the robust regression line fit by rlm is not.

Smoothers   The regression line is computed by taking into account all the data at once. A local fit is one where the predicted value depends only on nearby values. The loess function will fit a polynomial function locally, the default is a second degree polynomial. Like other modeling functions, it is employed in a manner similar to lm.

Here we follow [58] and investigate the original Galton data on heights of children and their parents. The data is in the Galton (HistData) data set.

```
f <- child ~ parent
plot(f, data=Galton, col=rgb(0,0,0, alpha=0.25))
res.lm <- lm(f, data=Galton)
abline(res.lm, lty=1, lwd=2)
##
res.loess <- loess(f, data=Galton, degree=1)  # line, not quadratic
rng <- seq(64, 73, length=20)
newdata <- data.frame(parent=rng)
predicted <- predict(res.loess, newdata)       # predicted  by loess
lines(predicted ~ rng, lty=2, lwd=2)
```

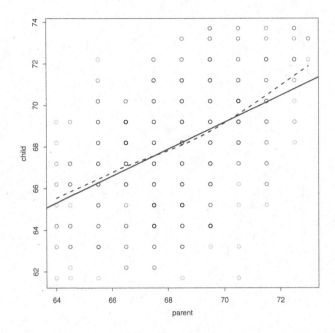

Figure 3.18: Plot of Galton data with regression line and loess line (dashed). The loess line shows a slight curve. The data is overplotted with transparency, as an alternative to jittering.

Figure 3.18 shows the fitted curve, which was computed using predict in an identical manner as could be done for an object created by lm. The authors in the cited article comment on the bend in the figure, pointing out that it may be attributed to the mixing of genders in the data.

Problems

3.13 For the homedata (UsingR) data set, make a histogram and density estimate of the multiplicative change in values (the variable y2000/y1970). Describe the shape, and explain why it is shaped thus. (Hint: There are two sides to the tracks.)

3.14 The data set normtemp (UsingR) contains body measurements for 130 healthy, randomly selected individuals. The variable temperature measures normal body temperature, and the variable hr measures resting heart rate. Make a scatter plot of the two variables and find the Pearson correlation coefficient.

3.15 The data set fat (UsingR) contains several circumference measurements for 252 men. The variable body.fat contains body fat percentage, and the variable BMI records the body mass index. Make a scatter plot of the two variables and then find the correlation coefficient.

3.16 The data set twins (UsingR) contains IQ scores for pairs of identical twins who were separated at birth. Make a scatter plot of the variables Foster and Biological. Based on the scatter plot, predict what the Pearson correlation coefficient will be and whether the Pearson and Spearman coefficients will be similar. Check your guesses.

3.17 The state.x77 data set contains various information for each of the fifty United States. We wish to explore possible relationships among the variables. First, we make the data set easier to work with by turning it into a data frame.

```
x77 <- data.frame(state.x77);
```

Now, make scatter plots of Population and Frost; Population and Murder; Population and Area; and Income and HS.Grad. Do any relationships appear linear? Are there any surprising correlations?

3.18 The data set nym.2002 (UsingR) contains information about the 2002 New York City Marathon. What do you expect the correlation between age and finishing time to be? Find it and see whether you were close.

3.19 For the data set state.center do this plot:

```
plot(y ~ x, data=state.center)
```

Can you tell from the shape of the points what the data set is?

3.20 The batting (UsingR) data set contains baseball statistics for the 2002 Major League Baseball season. What is the correlation between the number of strikeouts (SO) and the number of home runs (HR)? Make a scatter plot to see whether there is any trend. Does the data suggest that in order to hit a lot of home runs one should strike out a lot?

3.21 Try to establish the relationship that twice around the thumb is once around the wrist. Measure some volunteers' thumbs and wrists and fit a regression line. What should the slope be? While you are at it, try to find relationships between the thumb and neck size, or thumb and waist. What do you think: Did Gulliver's shirt fit well?

3.22 The data set fat (UsingR) contains ten body circumference measurements. Fit a linear model modeling the circumference of the abdomen by the circumference of the wrist. A 17-cm wrist size has what predicted abdomen size?

3.23 The data set wtloss (MASS) contains measurements of a patient's weight in kilograms during a weight-rehabilitation program. Make a scatter plot showing how the variable Weight decays as a function of Days.

1. What is the Pearson correlation coefficient of the two variables?

2. Does the data appear appropriate for a linear model? (A linear model says that for two comparable time periods the same amount of weight is expected to be lost.)

3. Fit a linear model. Store the results in res. Add the regression line to your scatter plot. Does the regression line fit the data well?

4. Make a plot of the residuals, residuals(res), against the Days variable. Comment on the shape of the points.

3.24 The data frame x77 contains data from each of the fifty United States. First coerce the state.x77 variable into a data frame with

```
x77 <- data.frame(state.x77)
```

For the following, make a scatter plot with regression line:

1. The model of illiteracy rate (Illiteracy) modeled by high school graduation rate (HS.Grad).

2. The model of life expectancy (Life.Exp) modeled by the murder rate (Murder).

3. The model of income (Income) modeled by the illiteracy rate (Illiteracy).

Write a sentence or two describing any relationship. In particular, do you find it as expected or is it surprising?

3.25 The data set batting (UsingR) contains baseball statistics for the year 2002. Fit a linear model to runs batted in (RBI) modeled by number of home runs (HR). Make a scatter plot and add a regression line. In 2002, Mike Piazza had 33 home runs and 98 runs batted in. What is his predicted number of RBIs based on his number of home runs? What is his residual?

3.26 In the American culture, it is not considered unusual or inappropriate for a man to date a younger woman. But it is viewed as inappropriate for a man to date a *much* younger woman. Just what is too young? Some say anything less than half the man's age plus seven. This is tested with a survey of ten people, each indicating what the cutoff is for various ages. The results are in the data set too.young (UsingR). Fit the regression model and compare it with the rule of thumb by also plotting the line $y = 7 + (1/2)x$. How do they compare?

3.27 The data set diamond (UsingR) contains data about the price of 48 diamond rings. The variable price records the price in Singapore dollars and the variable carat records the size of the diamond. Make a scatter plot of carat versus price. Use pch=5 to plot with diamonds. Add the regression line and predict the amount a one-third carat diamond ring would cost.

3.28 To gain an understanding of the variability present in a measurement, a researcher may repeat or replicate a measurement several times. The data set breakdown (UsingR) includes measurements in minutes of the time it takes an insulating fluid to break down as a function of an applied voltage. The relationship calls for a log-transform.

Plot the voltage against the logarithm of time. Find the coefficients for simple linear regression and discuss the amount of variance for each level of the voltage.

3.29 The motors (MASS) data set contains measurements on how long, in hours, it takes a motor to fail. For a range of temperatures, in degrees Celsius, a number of motors were run in an accelerated manner until they failed, or until time was cut off. (When time is cut off the data is said to have been *censored.*) The data shows a relationship between increased temperature and shortened life span.

The commands

```
data(motors, package="MASS")
plot(time ~ temp, pch=cens, data=motors)
```

produce a scatter plot of the variable time modeled by temp. The pch=cens argument marks points that were censored with a square; otherwise a circle is used. Make the scatter plot and answer the following:

1. How many different temperatures were used in the experiment?

2. Does the data look to be a candidate for a linear model? (You might want to consider why the data point (150,800) is marked with a square.)

3. Fit a linear model. What are the coefficients?

4. Use the linear model to make a prediction for the accelerated lifetime of a motor run at a temperature of 210°C.

		Child	
		buckled	unbuckled
Parent	buckled	56	8
	unbuckled	2	16

Table 3.3: Seatbelt usage for parent and child in California.

3.4   Bivariate categorical data

Paired bivariate data that involves categorical data is summarized in a different manner than numeric data. Still the basic question remains: is there a relationship between the variables?

Tables

Bivariate categorical data is often presented in the form of a (two-way) *contingency table*. The table is found by counting the occurrences of each possible pair of levels and placing the frequencies in a rectangular grid. Such tables allow us to focus on the relationships by comparing the rows or columns. Later, statistical tests will be discussed to analyze whether the distribution for a given variable depends on the other variable.

Two-way tables from summarized data

Our data may come in a summarized or unsummarized format. The data entry is different for each. If the data already appears in tabular format and we wish to analyze it inside R, how is the data keyed in? To illustrate: an informal survey of seat-belt usage in California examined the relationship between a parent's use of a seat belt and a child's. The data appears in Table 3.3. A quick glance at the table shows a definite relationship between the two variables: the child's belt being buckled is greatly determined by the parent's.

We can enter these numbers into R in a variety of ways, as illustrated next. Data vectors were created using the c function. One simple way to make a table is to combine data vectors together as rows (with rbind) or as columns (with cbind).

```
rbind(c(56,8), c(2,16))          # combine rows

##       [,1] [,2]
## [1,]   56    8
## [2,]    2   16

cbind(c(56,2), c(8,16))          # bind as columns
```

```
##       [,1] [,2]
## [1,]   56    8
## [2,]    2   16
```

Combining rows (or columns) of numeric vectors results in a *matrix*—a rectangular collection of numbers. We can also make a matrix directly using the matrix function. To enter in the numbers we need only specify the correct size. In this case we have two rows. The data entry would look like this:

```
seatbelts <- matrix(c(56, 2, 8, 16), nrow=2)
```

The data is filled in column by column, though the setting the argument byrow=TRUE will instruct matrix to create a matrix in a row-by-row manner.

Alternately, we may enter in the data using the edit function (or fix). This will open a spreadsheet (if available) when called on a matrix.

Thus the commands

```
x <- matrix(1)              # need to initialize x
x <- edit(x)                # will edit matrix with spreadsheet
```

will open the spreadsheet and store the answer into x when done. The 1 will be the first entry. We can edit this as needed.

It is also possible to enter the data into a text file and import it as a matrix from there.

Adding names to a matrix It isn't necessary, but a matrix with row and column names conveys more information when printed. The rownames and colnames functions will do so. As we are using these functions to modify the attributes of the matrix, the functions appear on the left side of the assignment.

```
rownames(seatbelts) <- c("buckled","unbuckled")
colnames(seatbelts) <- c("buckled","unbuckled")
seatbelts

##            buckled unbuckled
## buckled        56         8
## unbuckled       2        16
```

More generally, the dimnames function can set both row and column names at once and additionally allows the specification of variable names. A list is used to specify these. For this usage, the variable names and values are given in name=value format. The row variable is the first dimension, the column the second:

```
dimnames(seatbelts) <- list(parent=c("buckled","unbuckled"),
                            child=c("buckled","unbuckled"))
```

As a convenience, if the matrix is constructed with rbind, then names for the row vectors can be specified in name=value format. Furthermore, column names will come from the vectors if present.

```
x <- c(56,8); names(x) = c("buckled","unbuckled")
y <- c(2,16)
rbind(buckled=x, unbuckled=y)  # names rows, columns come from x

##              buckled unbuckled
## buckled           56         8
## unbuckled          2        16
```

Two-way tables from unsummarized data

With unsummarized data, two-way tables can be made with the table function, as in the univariate case. If the two data vectors are x and y, then the command table(x,y) will create the table.

• **Example 3.5: Is past performance an indicator of future performance?**
A common belief is that an "A" student in one class will be an "A" student in the next. Is this so? The data set grades contains the grades students received in a math class and their grades in a previous math class.

```
headtail(grades)

##       prev grade
## 1       B+    B+
## 2       A-    A-
## 3       B+    A-
##       ...
## 119  F     D
## 120  A     A-
## 121  A     A
## 122  B     B

table(grades$prev, grades$grade)          # also: table(grades)

##
##           A    A-   B+   B    B-   C+   C    D    F
##    A.    15    3    1    4    0    0    3    2    0
##    A-     3    1    1    0    0    0    0    0    0
##    B+     0    2    2    1    2    0    0    1    1
```

```
## B    0   1   1   4   3   1   3   0   2
## B-   0   1   0   2   0   0   1   0   0
## C+   1   1   0   0   0   0   1   0   0
## C    1   0   0   1   1   3   5   9   7
## D    0   0   0   1   0   0   4   3   1
## F    1   0   0   1   1   1   3   4  11
```

A quick glance at the structure of the table indicates that the current grade relates quite a bit to the previous grade (most entries are on or above the diagonal). Of those students whose previous grade was an A, fifteen got an A in the next class; only three of the students whose previous grade was a B or worse received an A in the next class.   ••

Marginal distributions of two-way tables

A two-way table involves two variables. The distribution of each variable separately is called the *marginal distribution*. The marginal distributions can be found from the table by summing down the rows or columns. The sum function won't just work, as it will add all the values. Rather, we need to apply the sum function to just the rows or just the columns. In Chapter 4 we discuss how the apply function that can do this. For now, the margin.table function conveniently performs the task. Just remember that the margin value 1 is for rows and 2 is for columns.

For the seat-belt data stored in seatbelts we have:

```
seatbelts

##             child
## parent     buckled unbuckled
##    buckled      56         8
##  unbuckled       2        16

margin.table(seatbelts, margin=1)      # row sum is for parents

## parent
##   buckled unbuckled
##        64        18

margin.table(seatbelts, margin=2)      # column sum for kids

## child
##   buckled unbuckled
##        58        24
```

The two marginal distributions are similar: the majority in each case wore seat belts.

Alternatively, the function `addmargins` will return the marginal distributions by extending the table. For example:

```
addmargins(seatbelts)

##              child
## parent      buckled unbuckled Sum
##    buckled       56         8  64
##    unbuckled      2        16  18
##    Sum           58        24  82
```

Looking at the marginal distributions of the grade data also shows two similar distributions:

```
tbl <- with(grades, table(prev, grade))
margin.table(tbl, 1)

## prev
##  A   A-  B+   B   B-  C+   C   D   F
## 28   5   9  15   4   3  27   9  22

margin.table(tbl, 2)

## grade
##  A   A-  B+   B   B-  C+   C   D   F
## 21   9   5  14   7   5  20  19  22
```

The grade distributions, surprisingly, are somewhat "well-shaped."

Conditional distributions of two-way tables

We may be interested in comparing the various rows of a two-way table. For example, is there a difference in the grade a student gets if her previous grade is a B or a C? Or does the fact that a parent wears a seat belt affect the chance a child does? These questions are answered by comparing the rows or columns in a two-way table. It is usually much easier to compare proportions or percentages and not the absolute counts.

For example, to answer the question of whether a parent wearing a seat belt changes the chance a child does, we might want to consider Table 3.4.

From this table, the proportions clearly show that 88% of children wear seat belts when their parents do, but only 11% do when their parents don't. In this example, the rows add to 1 but the columns need not, as the rows were divided by the row sums.

For a given row or column, calculating these proportions is done with a command such as `x/sum(x)`. But this needs to be applied to each row or col-

		Child	
		buckled	unbuckled
Parent	buckled	0.88	0.12
	unbuckled	0.11	0.89

Table 3.4: Proportions of children with seat belt on.

umn. The convenient prop.table function does so. Again, we specify whether we want the conditional rows or columns with a 1 or a 2.

For example, to find out how a previous grade affects a current one, we want to look at the proportions of the rows.[12]

```
prop.table(table(grades$prev, grades$grade), margin=1) * 100
```

```
##
##          A   A-  B+   B  B-  C+   C   D   F
##    A    54   11   4  14   0   0  11   7   0
##    A-   60   20  20   0   0   0   0   0   0
##    B+    0   22  22  11  22   0   0  11  11
##    B     0    7   7  27  20   7  20   0  13
##    B-    0   25   0  50   0   0  25   0   0
##    C+   33   33   0   0   0   0  33   0   0
##    C     4    0   0   4   4  11  19  33  26
##    D     0    0   0  11   0   0  44  33  11
##    F     5    0   0   5   5   5  14  18  50
```

Comparing the rows, we see that the grade distributions are quite different. It is apparent that the previous grade has a big influence on the current grade.

The xtabs function

The xtabs function provides an alternative to table, where the structure of the table is specified with a model formula.

We look at the Fingerprints (HistData) data set containing data collected in 1915 by Waite. For many different hands, he counted whether a fingerprint had a whorl, a small loop, or neither. For each hand, there can be no more than 5 whorls and loops in total. This is why many combinations return NA:

```
headtail(Fingerprints)
```

```
##      Whorls Loops count
```

---

[12]We set the digits options to 1 to change the output.

```
## 1           0      0      78
## 2           1      0     106
## 3           2      0     130
##
##      ...
## 33          2      5      NA
## 34          3      5      NA
## 35          4      5      NA
## 36          5      5      NA
```

To display this data set in a contingency table would require some data manipulation. One way would be to use rep to create vectors for Whorls and loops for each case:

```
idx <- !is.na(Fingerprints$count)         # issue with NA in rep
whorls <- rep(Fingerprints$Whorls[idx], Fingerprints$count[idx])
loops <- rep(Fingerprints$Loops[idx], Fingerprints$count[idx])
table(whorls, loops)

##          loops
## whorls   0    1    2    3    4    5
##     0    78  144  204  211  179  45
##     1   106  153  126   80   32   0
##     2   130   92   55   15    0   0
##     3   125   38    7    0    0   0
##     4   104   26    0    0    0   0
##     5    50    0    0    0    0   0
```

But something so fundamental should have a simpler interface. As we saw in Chapter 3 the table function can do basic tabulations. Here we discuss the xtabs function, which provides a formula interface for tabulating data.

```
xtabs(count ~ Whorls + Loops, Fingerprints)

##          Loops
## Whorls   0    1    2    3    4    5
##     0    78  144  204  211  179  45
##     1   106  153  126   80   32   0
##     2   130   92   55   15    0   0
##     3   125   38    7    0    0   0
##     4   104   26    0    0    0   0
##     5    50    0    0    0    0   0
```

This particular data set was used in a discussion about whether two variables depend on each other outside of a restriction, such as in this case Whorls + Loops <= 5.

The left-hand side of the formula is needed when the data is already tabulated. If this is not the case, it can be left blank. For example, to look

at the distribution of type of car by origin in the Cars93 (MASS) data set we
would specify the formula as:

```
xtabs( ~ Origin + Type, Cars93)                    # no LHS, so tallies

##          Type
## Origin    Compact Large Midsize Small Sporty Van
##   USA         7     11     10     7      8    5
##   non-USA     9      0     12    14      6    4
```

The xtabs function maps a data frame into a contingency table. The
as.data.frame function reverses this mapping.

The right-hand side specifies the cross-classifying variables to be used
separated by +. The symbol . is a shorthand notation for all variables in the
data set not specified on the left-hand side.

ftable The number of cross-classifying variables determines the dimensions of the table. If one, the data is displayed in a row; if two, the data
is displayed in a grid with derived names. For three or more, the data will be
presented in slices. The order of the variables in the formula determines the
slicing. The first two are used to display slices in a two-way table, with the
slices determined by the remainders.

These slices can be hard to read. The ftable function can "flatten" contingency tables. The flattened table creates rows and columns for each combination of the row.vars and col.vars (which usually have reasonable defaults,
so aren't specified). In the call below, the order of the col.vars specification indicates how the data is nested. Here the third variable (Passengers) is
nested with the first (Origin):

```
tbl <- xtabs( ~ Origin + Type + Passengers, Cars93)
ftable(tbl, row.vars=2, col.vars=c(1,3))

##          Origin     USA                  non-USA
##          Passengers 2  4  5  6  7  8      2  4  5  6  7  8
## Type
## Compact             0  0  5  2  0  0      0  1  8  0  0  0
## Large               0  0  0 11  0  0      0  0  0  0  0  0
## Midsize             0  0  6  4  0  0      0  2  9  1  0  0
## Small               0  2  5  0  0  0      0  6  8  0  0  0
## Sporty              1  7  0  0  0  0      1  5  0  0  0  0
## Van                 0  0  0  0  4  1      0  0  0  0  4  0
```

The output of xtabs shows counts. If a comparison of proportions is desired, the prop.table function can be used. Below we compare rows (by specifying margin=1) for a three-way contingency table that has been flattened
through the default choice or row variables.

```
price.range <- cut(Cars93$Price, c(0,  20, 70))
tbl <- xtabs( ~ price.range + Origin + Cylinders, data=Cars93)
out <- ftable(tbl, row.vars=3)
100 * prop.table(out, margin=1)
```

```
##               price.range  (0,20]           (20,70]
##               Origin            USA non-USA      USA non-USA
## Cylinders
## 3                           0.000 100.000    0.000   0.000
## 4                          44.898  44.898    0.000  10.204
## 5                           0.000  50.000    0.000  50.000
## 6                          35.484   6.452   29.032  29.032
## 8                          14.286   0.000   71.429  14.286
## rotary                      0.000   0.000    0.000 100.000
```

The above created a new categorical variable price.range with the cut command. This allows us to look at the differences between the number of cylinders US and non-US made cars had based on price range. In this old data, the US cars had engines with more cylinders.

Graphical summaries of two-way contingency tables

As with univariate data, barplots can be used effectively to show the data in a two-way table. To do this, one variable is chosen to form the categories for the barplot. Then, either the bars for each level of the category are segmented, to indicate the proportions of the other variable, or separate bars are plotted side by side (cf. Figure 3.19).

The barplot function will plot segmented barplots when its first argument is a two-way table (specify beside=TRUE to get side-by-side bars). Levels of the columns will form the categories, and the sub-bars or segments will be proportioned by the values in each column.

To illustrate, the two graphics in Figure 3.19 are created with the following code, where the seatbelts variable holds the data as a table object:

```
barplot(seatbelts, xlab="Parent", main="Child seat-belt usage")
barplot(seatbelts, xlab="Parent", main="Child seat-belt usage",
        beside=TRUE)
```

A legend can be added to the barplot with the argument legend.text=TRUE, or by specifying a vector of names for legend.text.

For the seat-belt data, if we wanted the parents' distribution (the rows) to be the primary distribution, then we need to flip the table around. This is done with the transpose function, t, as in barplot(t(seatbelts)).

Sometimes a relationship is better presented as proportions than counts (this will only modify the scale of the y axis). To do this, we apply prop.table prior to the barplot.

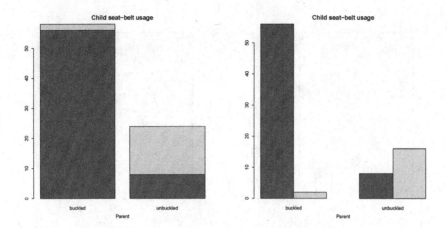

Figure 3.19: Segmented and side-by-side barplots showing distribution of child's seat-belt usage depending on whether parent is buckled or unbuckled.

Mosaic plots

A *mosaic plot* is an extension of the ideas of a barplot that makes it suitable for viewing relationships between two or more categorical variables.

For a single variable, the mosaic plot is nothing more than a segmented bar plot, where the area associated with each segment is proportional to the count for that category. The left graphic of Figure 3.20 shows a mosaic plot for a single variable, in this case the gender (Sex) of the passengers on the Titanic. The larger number of males is readily apparent.

Of more interest, is when two variables are represented in a mosaic plot, as in the right graphic of Figure 3.20, where the additional variable of survival is added. The areas for each gender are in turn segmented by survival. We see that the adage "women and children first" seems to have held, as clearly more than half the women survived, yet less than half the men.

The mosaic plot can also represent additional variables. The graphic in Figure 3.21 adds the passenger class information. In first class, nearly all the females survived, as indicated by the sliver of area for females in first class that did not survive (the upper left part of the graphic). This was not quite the case for the female passengers in third class. The width of those third-class females that did not survive is narrower than those who did, but the area is larger. Area is the proper comparison, as this is the table for women:

```
titanic <- as.data.frame(Titanic)
xtabs(Freq ~ Survived + Class, data=titanic, subset=Sex=="Female")

##           Class
## Survived 1st 2nd 3rd Crew
```

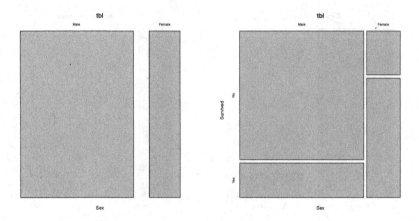

Figure 3.20: Mosaic plots of one and two variables. The first variable is used to segment a unit square, the second segments each rectangle so that the area for each cell is proportional to the frequency of the corresponding cell in the contingency table.

```
##      No    4  13 106    3
##      Yes 141  93  90   20
```

The commands above took the data set Titanic, which is a contingency table and coverts it to a data frame so that we can apply xtabs, without having to work directly with the table object. This is useful, as the function to generate mosaic plots, mosaicplot, displays tabular output. The commands to make the figures in Figure 3.20 were as follows:

```
tbl <- xtabs(Freq ~ Sex, titanic)
mosaicplot(tbl)
tbl <- xtabs(Freq ~ Sex + Survived, titanic)
mosaicplot(tbl)
```

Adding a third variable is done in a similar manner.

```
tbl <- xtabs(Freq ~ Sex + Survived + Class, titanic)
mosaicplot(tbl)
```

Figure 3.21 shows the resulting graphic. Here we note how the variables are assigned: The first variable (Sex) again splits the x axis, the second variable (Survived) again splits the y axis. The third (Class) then splits each of the four blocks. We can compare the proportion of each block at the second level by how the class variable splits it up. For example, looking at the female variable (far right) and comparing how the first-class segment for both

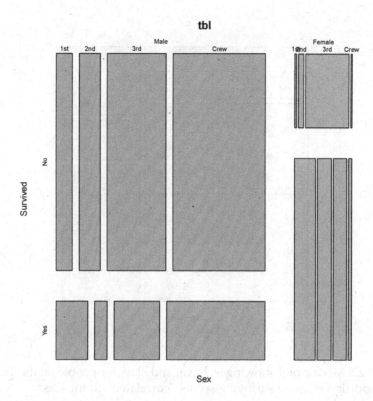

Figure 3.21: Mosaic plot of Titanic data showing gender and survival distribution split by passenger class. This view shows differences in survival by class for female passengers.

not survived (top) and survived, we see that a much larger proportion for first-class passengers in the survived, and a much larger proportion of the third-class passengers in not survived.

Different orderings give different insights. Putting Class first (not shown) makes it easier to focus on the classes. This makes it easy to see how the survival rate changed based on class and gender, especially for the female passengers.

```
mosaicplot( xtabs(Freq ~ Class + Sex + Survived, data=titanic))
```

Measures of association for categorical data

Consider the mosaic plot of passenger class/crew and survival generated by (Figure 3.22):

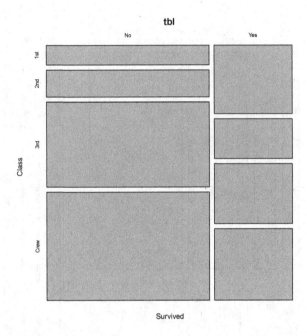

Figure 3.22: Mosaic plot showing survival and class for people on the Titanic. The proportion of those surviving seems "correlated" to the class.

The segmentation of the Survived variable by Class is quite different. This indicates that the variables are "correlated"—the value of one depends on another.

Correlation is in quotes, as the Pearson correlation is a summary of two numeric variables, and this data is categorical. Measures of numeric data may not translate to categorical data, but many ideas do. We mention a few measures contained in the books [1] and [2].

The variables Class and Survival are not numeric, but they are naturally ordered, with class being in the order Crew, 3rd, 2nd, then 1st; and survival being in order no and yes.

We can make ordered factors out of the data as follows:

```
Survived <- rep(titanic$Survived, titanic$Freq)
Survived <- ordered(Survived)
Class <- rep(titanic$Class, titanic$Freq)
Class <- ordered(Class)
head(Class)

## [1] 3rd 3rd 3rd 3rd 3rd 3rd
## Levels: 1st < 2nd < 3rd < Crew
```

The above expands the tabulated data into a record for each passenger, then orders the factor using the default ordering.[13] The last line shows the first few elements of Class and the levels, which are separated by <, as this is an ordered factor.

Ordered factors can be coerced into numeric data through as.numeric with the order intact. This allows the taking of the correlation:

```
cor(as.numeric(Survived), as.numeric(Class), method="kendall")
```

```
## [1] -0.2245
```

The negative correlation is due to the ordering of the class, with 1st being a 1 and crew a 4, so indeed this is consistent with the observation from the graphic: the survival rate is associated with the class.

Kendall τ The method="kendall" argument was used with cor, as the *Kendall τ* statistic is computed. This statistic is a measure of association between data that can be ranked, such as ordered factors. The basic statistic is:[14]

$$\tau = \frac{\text{Number of concordant pairs} - \text{Number of discordant pairs}}{n(n-1)/2}.$$

A pair of observations (x_i, y_i) and (x_j, y_j) is concordant if the ranks are in agreement (both x_i and y_i are higher ranked than their counterpart of both lower ranked). Figure 3.23 shows a small 5 point data set with one point highlighted. For that point, their are 2 concordant points (the two in the lower left) and 2 discordant point (the upper left and lower right). If the same picture were repeated for all 5 points, we would have the following count:

$$\frac{(0+1+1+2+0)-(4+3+3+2+4)}{5 \cdot 4} = -0.6.$$

(Here we double count each pair of points, so the 1/2 in the formula is not used.)

As with the Pearson correlation, the value is between -1 and 1 with 0 being a perfect non-association.

---

[13]The default order depends on the locale of the machine, so may not match that here.

[14]This formula is referred to as τ_A. There are also modifications referred to by τ_B and τ_C which account for ties in the ranked data.

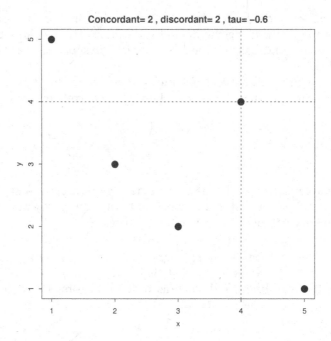

Figure 3.23: Scatter Plot of ranked data with focus on one point. For this value, there are 2 points in the third quadrant (concordant) and one each in the second and fourth quadrant (discordant), a difference of 0. Moving the crosshairs over all n points and adding the differences gives the numerator in the Kendall correlation.

The chi-squared statistic   The *chi-squared* statistic is a common summary of a table, in fact it is computed by the summary function when called on a table object:

```
f <- Freq ~ Survived + Class
tbl <- xtabs(f, data=titanic, subset=Sex=="Female")
summary(tbl)

## Call: xtabs(formula = f, data = titanic, subset = Sex == "Female")
## Number of cases in table: 470
## Number of factors: 2
## Test for independence of all factors:
##   Chisq = 131, df = 3, p-value = 4e-28
```

This yields a value of Chisq of 131 (a rounding of 130.7). What is this number? It is related to the residuals for the table. As usual the residuals are the observed frequencies f_o (the actual table values) minus the expected frequencies f_e. The expected values are found from the marginal values. In the table above, the expected value for the upper-left cell is the proportion of times a value is in the upper row times the proportion of times a value is in the left-most column times the total number of observations. This can be found with:

```
margin.table(tbl,1)

## Survived
##  No Yes
## 126 344

ptop <- 126/sum(margin.table(tbl,1))
margin.table(tbl, 2)

## Class
##  1st  2nd  3rd Crew
##  145  106  196   23

pleft <- 145/sum(margin.table(tbl,2))
fe <- ptop * pleft * sum(tbl)
fo <- tbl[1,1]
c(fo=fo, fe=fe, residual=fo - fe)          # print

##          fo        fe residual
##        4.00     38.87   -34.87
```

That is a large residual, as far fewer people in first class perished than would have been expected given the totals. The computation of the expected would be tedious to do this way. Using the chisq.test function (which we return to when we consider statistical inference on contingency tables), the above can be done at once with:

```
chisq.test(tbl)$expected

##          Class
## Survived   1st   2nd    3rd   Crew
##       No 38.87 28.42  52.54  6.166
##      Yes 106.13 77.58 143.46 16.834
```

As with other residuals, their total sum is 0. The chi-squared statitistic, is performed by the following:

$$chi\text{-}squared\ statistic = \sum \frac{(f_o - f_e)^2}{f_e}.$$

The "o" refers to observed and the "e" to expected, so the term in the numerator is a residual squared. Returning to our data, this formula can be computed with the following:

```
fo <- tbl
fe <- chisq.test(tbl)$expected
(fo - fe)^2 / fe

##          Class
## Survived     1st      2nd      3rd    Crew
##       No  31.2839  8.3642  54.3817  1.6256
##      Yes  11.4587  3.0636  19.9189  0.5954

sum((fo - fe)^2 / fe)

## [1] 130.7
```

The expected values are found by the marginal distributions, though we use an expedient above to get them. The marginals do not consider the interplay, or association, between the two variables. If the residuals are small, then there is little association. Whereas if the sum above is large, then the variables are associated. Exactly what is meant by "small" and "large" will have to wait until Chapter 10. (Though in this case the value is very large.)

The deviance Related to the chi-squared statistic, in that it is computed from similar inputs is the deviance, G^2, found by $G^2 = 2\sum f_o \ln(f_o/f_e)$, which for this data is:

```
2 * sum(fo * log(fo/fe))                      # no fe can be 0

## [1] 142.7
```

Similar to the chi-squared interpretation, large values are indicative of an association.

Association between 2 by 2 tables, odds ratio For 2 by 2 tables, a different measure of association can be formed, the *odds ratio*. Consider, the table

```
xtabs(Freq ~ Sex + Survived, titanic)

##          Survived
## Sex          No  Yes
##    Male    1364  367
##    Female   126  344
```

The probability of a male surviving was $367/(1364 + 367) = 0.21$. The odds of a male surviving is the ratio of the probabilities of survival divided by the probability of not surviving: $(367/(1364 + 367))/(1364/(1364 + 367)) = 367/1364 = 0.27$. The odds of a female surviving were higher: $344/126 = 2.73$. To compare odds, a ratio proves convenient. The odds ratio of a female surviving compared to a male would be: $(344/126)/(367/1364) = 10.15$. This is interpreted as a randomly chosen female passenger or crew member was more than 10 times as likely to have survived than a randomly chosen male on board the ship.

That this is large, naturally leads us to think that survival and gender were related, or associated. Had this value been near 1 we might think differently.

Problems

3.30 Find an example of a two-way contingency table in the media. Identify the two variables and summarize the data that is presented.

3.31 The data set coins (UsingR) contains the number of coins in a change bin and the years they were minted. Do the following:

1. How much money is in the change bin?

2. Make a barplot of the years. Is there a trend?

3. Use cut to construct a barplot by decade.

4. Make a contingency table of the year and the value. Does it look like the two variables are associated? Why?

3.32 The data set UScereal (MASS) contains information about cereals on a shelf of a United States grocery store. Make a table showing the relationship between manufacturer, mfr, and shelf placement, shelf. Are there any obvious differences between manufacturers?

3.33 The help page for mosaicplot demonstrates the data set HeadEyeColor, which records sex, Hair color, and Eye color for 592 statistics students. The data set comes as a flattened table, so simply passing the object to mosaicplot will create the plot. (Or, as demonstrated, passing shade=TRUE, as in mosaicplot(HairEyeColor, shade = TRUE), will produce a colored version.)

Make the plot, Why does the help page note "there are more blue-eyed, blonde females than expected?"

Multivariate data

A typical data analysis process consists of several steps: data cleaning, data transformations, exploratory visualizations, numerical summaries, initial modeling, and perhaps follow-up modeling. Learning how to efficiently work with R's data frames makes these tasks much more direct and manageable. In this chapter, we look briefly at some different structures for storing data, before a longer discussion on data frames. We then look at several means to implement the "split-apply-combine" process to summarize a data set. We finish with a discussion about some basics of importing and exporting data. In the next chapter we discuss some of R's functionality for graphically presenting multivariate data.

4.1   Data structures in R

In the first three chapters we have discussed various data structures for R: a data vector, a list, a matrix, and a data frame. In this section, we elaborate more on data frames first by describing more on matrices and then showing that data frames benefit from having an interface that allows for their usage in a manner similar to both matrices and lists.

Vectors   As mentioned in Chapter 1, a data vector is a container for a collection of values of the same type, such as numeric, integer, or character. The items in a data vector are indexed in a 1-based manner, and access to these values is done with [, which can be used to get and set values. Further, data vectors can have attributes, in particular names. Names can offer a more convenient access.

To symbolically represent a data vector we will use [a1 a2 ... an].

Lists   In Section 3.2 we mentioned that lists are generalized vectors. They are like vectors in that their elements are ordered and indexable, but unlike vectors each element need not be of the same type. For lists we have both [and [[defined. The former returning a list, the latter referencing the component. We've seen the use of $ as a convenience for [[when working with lists that have a names attribute.

We will refer to lists symbolically with notation like [[a1 b2 ... cn]] to indicate they are ordered, but need not have the same type.

Matrices A rectangular collection of values of the same type can be stored in a matrix. The built-in data set state.x77 holds 8 numeric facts for each of the 50 United States from the 1970s. As all the values are numeric, the data can be stored in a matrix. The first 3 rows are seen through:

```
head(state.x77, n=3)

##            Population Income Illiteracy Life Exp Murder HS Grad
## Alabama          3615   3624        2.1    69.05   15.1    41.3
## Alaska            365   6315        1.5    69.31   11.3    66.7
## Arizona          2212   4530        1.8    70.55    7.8    58.1
##            Frost    Area
## Alabama       20   50708
## Alaska       152  566432
## Arizona       15  113417
```

As can be seen from above, matrices can have row and column name attributes. Access to a cell or block of cells can again be achieved with [. The notation is extended to [i, j] where i references the row(s) and j references the column(s) of interest:

```
state.x77[1,2]                 # first row, second column

## [1] 3624

state.x77[1, 3:4]              # first row, third and fourth columns

## Illiteracy   Life Exp
##       2.10      69.05
```

As with vectors, an index can be a single numeric index, a range of numbers, a vector of numbers, empty (for all numbers), a vector of row or column names, or a logical vector of the appropriate size. The latter makes it easy to subset data in a matrix. For example, the following restricts the data set to the first 6 columns and to states with small populations:

```
state.x77[state.x77[,"Population"] < 500, 1:6]

##            Population Income Illiteracy Life Exp Murder HS Grad
## Alaska            365   6315        1.5    69.31   11.3    66.7
## Vermont           472   3907        0.6    71.64    5.5    57.1
## Wyoming           376   4566        0.6    70.29    6.9    62.9
```

The row and column names of a matrix are accessed with the functions rownames and colnames. There are also assignment versions. For example, here we see the column names of the state.x77 data set:

```
colnames(state.x77)

## [1] "Population" "Income"     "Illiteracy" "Life Exp"
## [5] "Murder"     "HS Grad"    "Frost"      "Area"
```

There are various ways to create matrices. We discussed some in Section 3.4, where a matrix was used to hold a contingency table. Some of this is a repeat of that discussion.

The rectangular structure of a matrix and the uniformity of type lends itself to thinking about them in terms of either their rows or their columns. The rbind function can be used to *bind* rows together to make a matrix. We just need to ensure that the rows we specify have the same lengths. For example:

```
rbind(1:5, c(1, 1, 2, 3, 5))            # 2 by 5 matrix without names

##      [,1] [,2] [,3] [,4] [,5]
## [1,]    1    2    3    4    5
## [2,]    1    1    2    3    5
```

The cbind function does the binding along columns; again we need the vectors that we bind together to have a common length:

```
m <- cbind(1:3, c(1.1, 1.2, 1.3), c(1, 1, 2)) # a 3 by 3 matrix
colnames(m) <- c("x", "y", "z")               # or cbind(x=..., ...)
m

##      x   y z
## [1,] 1 1.1 1
## [2,] 2 1.2 1
## [3,] 3 1.3 2
```

Internally, R stores matrices as vectors along with dimension information. This storage mechanism can make it convenient to use a single (linear) index. The following illustrates how the values are stored column by column:

```
m[1:6]                                 # first 6 entries of m

## [1] 1.0 2.0 3.0 1.1 1.2 1.3
```

We can also just give a vector a dimension attribute to create a matrix, which is what the matrix function will do. Here we reshape the numbers 1:10 into a matrix with two rows and five columns, first by column, then by row:

```
matrix(1:10, nrow=2)

##      [,1] [,2] [,3] [,4] [,5]
## [1,]    1    3    5    7    9
## [2,]    2    4    6    8   10

matrix(1:10, nrow=2, byrow=TRUE)

##      [,1] [,2] [,3] [,4] [,5]
## [1,]    1    2    3    4    5
## [2,]    6    7    8    9   10
```

Vectors can be given more than 2 dimensions, resulting in an *array*. Arrays have the array constructor.

Object size   The size of a matrix (or array) is returned by the dim function. (There is also an assignment function that allows for the reshaping of a vector, matrix, or array.) The length function will treat the matrix as a vector and return the number of elements. For matrices, the special functions nrow and ncol give the number of rows and columns.

Data frames   We have worked with data frames so far using their list-like notation $ to access individual variables. Internally, data frames are stored as a list of variables, all of the same length (possibly 0) with row and column name attributes. The row names must be unique, though duplicate column names are possible with some work, we suggest using unique names.

The fact that the rows are all of the same length, allows the values to be arranged in a rectangular manner; each column is a separate variable. This is similar to a matrix, though the type of each variable need only be consistent along any given column and not necessarily along the rows.

This allows the matrix notation to extract and refer to elements in a data frame. Here we extract out the first three rows and the first five columns of the data set Cars93 (UsingR):

```
DF <- Cars93[1:3, 1:5]
DF

##   Manufacturer  Model    Type Min.Price Price
## 1        Acura Integra   Small      12.9  15.9
## 2        Acura  Legend Midsize      29.2  33.9
## 3         Audi      90 Compact      25.9  29.1
```

As with matrices, we can reference rows or columns by name. For example:

```
DF[ , "Price"]

## [1] 15.9 33.9 29.1
```

This notation asks for all the rows and the columns which match by name Price. The result is similar to what DF$Price returns—a data vector.

The fact that indexing a data frame can return a data vector is a design decision. It is the behavior that in most interactive cases is desired. If a 1-column data frame is desired, the argument drop=FALSE can be specified:

```
DF[ , "Price", drop=FALSE]

##    Price
## 1  15.9
## 2  33.9
## 3  29.1
```

This is the same as if we had used the list notation for accessing a list with the Price variable:

```
DF["Price"]

##    Price
## 1  15.9
## 2  33.9
## 3  29.1
```

Of all of these, the notation DF$Price is most used in the sequel, though we will see different alternatives to avoid typing the data frame name more than once per command.

Problems

4.1 The data set UScereal (MASS) contains data on cereals sold in the United States in 1993. For this data set, answer the following questions:

1. How many rows does the data frame have? Columns?

2. How many different manufacturers are included?

3. How many vitamin categories are included?

4. How many cereals had a sugar level above 10?

5. What is the mean calorie value for cereals with more than 5 grams of fat? Less than or equal to 5?

6. What is the mean calorie value for cereals on the middle shelf (2)?

4.2 R uses lists for many purposes behind the scenes. For example, the output of lm(mpg ~ wt, data=mtcars) returns a list. Create this object, then answer the following:

1. How many components does this list have?

2. What are the names of the components?

3. What kind of data is held in the residuals variable?

4.3 There are many ways to get at a variable within a data frame. Which of these return the same object: mtcars$cyl, mtcars["car"], mtcars[["car""]], mtcars[,"car"], and mtcars[,"car", drop=FALSE]?

4.4 List access by name can sometimes utilize partial matching, where if there is no ambiguity if only the first part of a name is specified, then R will allow access. Which of these will match the cyl variable of the mtcars data set: mtcars$c, mtcars$cy, mtcars["cy"], mtcars["cy", exact=TRUE], mtcars[["cy"]], and mtcars[["cy", exact=FALSE]]?

4.5 There are many ways to set the variable names in R: names<-, dimnames<-, colnames<-, and setNames. Which of these will work with a data frame? You can answer by experimenting.

4.2 Working with data frames

Data frames sit between matrices and lists so we have two interfaces in which to work with them. As if that is not enough, there are also some special purpose functions that give even more convenience. Here we organize the discussion around several common tasks:

Data frame construction Data frames are lists of variables. To construct one from the variables themselves can be done with the data.frame constructor, called as:

```
data.frame(nm1=vec1, nm2=vec2, ...)
```

The above key=value style assigns column names during the construction; the row names are automatically generated. This function requires variables of the same length or variables whose length divides the longest length (in which case the variable's values are recycled).

If the data is in a matrix format and a data frame is desired, then the coercion function as.data.frame[1] can be used:

```
d <- as.data.frame(state.x77)
class(d)

## [1] "data.frame"
```

(We give an example of why this would be useful later in this chapter.)

Data frames provide convenience for many things, but at times a matrix is more desirable (matrices make some computations much faster and make some mathematical computations easier). To convert a data frame into a matrix (with coercion of variables to numeric mode), the data.matrix function is used:

```
d1 <- data.matrix(d)
class(d1)

## [1] "matrix"
```

Data lookup We've recommended accessing variables by name within a data frame using the $ notation, as opposed to the matrix notation, as it is more direct. This is often the best approach, but it can be tedious if a task requires the typing of the data frame name several times.

For example, to find out which values in a variable are within one standard deviation of the mean, an expression like the following is needed:

```
d <- mtcars[1:5,]                          # first 5 rows
mean(d$mpg) - sd(d$mpg) <= d$mpg & d$mpg <= mean(d$mpg) + sd(d$mpg)

## [1]  TRUE  TRUE FALSE  TRUE FALSE
```

Having to type the data frame name 6 times is a bit too much, even if it was shortened from mtcars. Some typing could be saved by using intermediate variables, but there is a conceptual clarity in expressing this task all at once which can have value. The with function can be used to create an enclosing environment from the specified data frame, which is searched when there are unbound variables within an expression. With this function, the above becomes more manageable:

```
with(d, mean(mpg) - sd(mpg) <= mpg & mpg <= mean(mpg) + sd(mpg))

## [1]  TRUE  TRUE FALSE  TRUE FALSE
```

---

[1]We have also used this same-named function to coerce a flattened table object back into a data frame.

This usage is similar to how R's modeling interface allows for the specification of a data set for variable lookup.

The data specified to the with function can be more general than a data frame, for example a list could be used.

Modifying cells, columns, names   The matrix notation is useful for accessing and modifying blocks of cells in a data frame. For example

```
d <- Cars93[1:3, 1:4]                    # first 3 rows, 4 columns
d[1,1] <- d[3,4] <- NA                   # set two values to NA
d
##    Manufacturer  Model    Type Min.Price
## 1          <NA> Integra   Small    12.9
## 2         Acura  Legend Midsize    29.2
## 3          Audi      90 Compact      NA
```

The assignment of NA to a factor prints differently than assigning NA to a numeric variable. We can assign more than one value at once:

```
d[1:2, 4] <- round(d[1:2, 4])            # round numeric values
```

We don't always need to replace values with the same-sized values, as recycling will be done.

Assigning across a row must be done with some care, as the columns need not be of the same type. A list can be used. Here we try to update the older Audi 90 with a newer A3:

```
d[3,c(2,4)] <- list("A3", 30)            # warning
## Warning: invalid factor level, NA generated
```

We were careful to use a list, but weren't so careful with the fact that the second column is a factor, and "A3" is not one of the levels. Let's try again after adding a few new levels.

```
levels(d$Model) <- c(levels(d$Model), c("A3", "A4", "A6"))
d[3,c(2,4)] <- list("A3", 30)
```

New rows can be added in a few ways. Here we add another Audi using matrix notation:

```
d[4, ] <- list("Audi", "A4", "Midsize", 35)
```

The 4 is easily seen to be the correct row number for adding a new row. (It is nrow(d) + 1.) If we didn't want to do such bookkeeping, we could append a new row with rbind:

```
d <- rbind(d, list("Audi", "A6", "Large", 45))
```

Adding columns can also be done using matrix notation:

```
d[, 5] <- d$Min.Price * 1.3                 # price in Euros
```

Unlike assigning new rows, we cannot specify a column index that will leave holes (empty columns) in the data frame, as it would be unclear how to fill them (there is no default column type). Specifying the column index with `ncol(d) + 1` or using cbind will avoid this issue.

However, it is usually more convenient to use the $ notation for assignment, as this can also set the column name at the same time:

```
d$Min.Price.Euro <- d$Min.Price * 1.3
```

The case (row) names of data frames (and matrices) can be accessed and set with the `rownames` function. Variable (column) names can be accessed or set through the matrix style function `colnames` or the list style `names`.

For example, to standardize the variable names to lowercase can be achieved with this construct:

```
names(d) <- tolower(names(d))
```

To change an individual name can be done with indexing, taking advantage of R's evaluation order. To update just the third column we can do:

```
names(d)[3] <- "car type"
```

It is permissible to use names with spaces, though there are some restrictions on valid names. However, using spaces makes variable reference with the $ a bit more work, at the variable name must be quoted.

The subset function Extracting blocks from a data frame can be done by specifying numeric values for the desired row and column indices. However, at times it can be more convenient to generate a specification in a different manner. A simple example, is to use the row and/or column names of the desired variables. Other examples are to generate logical expressions to indicate which row or column.

The subset function, has a few conveniences for these tasks that make it worth learning:

- For the subsetting of rows, expressions are evaluated within the data frame, as with `with`. Subsetting is done with logical expressions, not numeric indices.

- For the selection of columns, nonstandard evaluations allows us to avoid quoting variable names, to use names for ranges, and to use negative indexing with variable names to drop specific columns.

The following examples are derived from the function's help page:

```
aq <- airquality[1:5, ]                    # shorten
aq

##   Ozone Solar.R Wind Temp Month Day
## 1    41     190  7.4   67     5   1
## 2    36     118  8.0   72     5   2
## 3    12     149 12.6   74     5   3
## 4    18     313 11.5   62     5   4
## 5    NA      NA 14.3   56     5   5

subset(aq, select = Ozone:Wind)            # range of names

##   Ozone Solar.R Wind
## 1    41     190  7.4
## 2    36     118  8.0
## 3    12     149 12.6
## 4    18     313 11.5
## 5    NA      NA 14.3

subset(aq, select = -c(Month, Day))        # same result

##   Ozone Solar.R Wind Temp
## 1    41     190  7.4   67
## 2    36     118  8.0   72
## 3    12     149 12.6   74
## 4    18     313 11.5   62
## 5    NA      NA 14.3   56

subset(aq, subset = !is.na(Ozone), select=Ozone:Wind) # drop a row

##   Ozone Solar.R Wind
## 1    41     190  7.4
## 2    36     118  8.0
## 3    12     149 12.6
## 4    18     313 11.5
```

is.na, complete.cases   When subsetting a data frame, it can be convenient to drop cases where there are NA values. To drop those values for a single variable can be achieved with the is.na function. To drop the cases where any of the variables have a missing value is best done through the complete.cases.

This function returns TRUE for cases where all the variables have non-NA values, and other FALSE. For example,

```
DF <- data.frame(a=c(NA, 1, 2), b=c("one", NA, "three"))
subset(DF, !is.na(a))                          # drop first, keep second

##   a     b
## 2 1  <NA>
## 3 2 three

subset(DF, complete.cases(DF))                 # drop first, second

##   a     b
## 3 2 three
```

Transforming values   In general, most R functions are *pure functions*, in that they don't modify the arguments passed to them or depend on external state. R's design places an emphasis on integrity of the data, and unexpected modifications are avoided. However, in-place modification (or mutating) of passed-in objects can be convenient, so there are some functions that return a modified copy. The basic assignment functions, e.g., [<- are an example.

For example, suppose in our example with the Cars93 data set, the euro exchange rate went from 1.3 to 1.5, then we can update our variable with:

```
d$min.price.euro <- d$min.price * 1.5
```

Alternatively, we can use within to do this:

```
d <- within(d, {min.price.euro = min.price * 1.5})
```

Like with, within evaluates an expression (the part enclosed in braces above) in an environment formed from the data that allows for variable lookup within the data set. In addition, it examines the modifications to this environment and tries to place them into the data frame. In this case updating the variable min.price.euro. The result is the modified data frame, which needs to be assigned for the changes are to be kept.

The transform function is similar, but is more explicit with the variable transforms listed in tag=value form:

```
d <- transform(d, min.price.euro = min.price * 1.5)
```

Reshaping   Data can be presented in "wide" format. For example, we might have a data set stored as:

```
speed

##    Speed.1 Speed.2 Speed.3 Speed.4 Speed.5
## 1      850     960     880     890     890
## 2      740     940     880     810     840
## 3      900     960     880     810     780
## 4     1070     940     860     820     810
## 5      930     880     720     800     760
## 6      850     800     720     770     810
```

This is some of the data from the morley data set, which in that case is stored with three variables: Speed, Expt, and Run. Here the run is a row name, and the experiment number given in the name of the columns. Wide data, where the different measurements for some value are recorded as separate variables, is commonly used for data display. However, "long" formatted data where the different values are recorded with factors is more useful when used with R's functions. Converting from one to the other is known as reshaping the data, and R provides the reshape function for this task. (The add-on package reshape2 is an extension.)

To convert from "wide" data to "long" data using reshape requires a specification of which variables are to be combined into a single variable. In this case, all the variables are. We specify them below by name along with the desired direction:

```
m <- reshape(speed, varying=names(speed)[1:5], direction="long")
head(m)                                          # first 6 rows only

##       time Speed id
## 1.1      1   850  1
## 2.1      1   740  2
## 3.1      1   900  3
## 4.1      1  1070  4
## 5.1      1   930  5
## 6.1      1   850  6
```

The names of our data set are time which corresponds to the original Expt variable and whose values come from the varying values, Speed, and id which came from the row numbers. This could also have come from a value in the data frame. For example:

```
speed$Run <- LETTERS[1:6]
m <- reshape(speed, varying=names(speed)[1:5], direction="long")
head(m)

##       Run time Speed id
## 1.1     A    1   850  1
```

```
## 2.1   B   1    740  2
## 3.1   C   1    900  3
## 4.1   D   1   1070  4
## 5.1   E   1    930  5
## 6.1   F   1    850  6
```

Going from long data to wide data requires a specification of the variable(s) that will be broken over several columns (Speed), the variable that varies in time (the experiment, that is coded as time), and the variable(s) that identify multiple records for the same group (Run). Here is how the call to reshape puts these together:

```
reshape(m, v.names="Speed", timevar="time", idvar="Run",
        direction="wide")

##      Run id Speed.1 Speed.2 Speed.3 Speed.4 Speed.5
## 1.1   A   1     850     960     880     890     890
## 2.1   B   2     740     940     880     810     840
## 3.1   C   3     900     960     880     810     780
## 4.1   D   4    1070     940     860     820     810
## 5.1   E   5     930     880     720     800     760
## 6.1   F   6     850     800     720     770     810
```

For this particular task, a combination of stack and split could achieve something similar, but for more complicated data sets reshape can be quite useful.

Merging A data frame stores data in a tabular format, similar to how tables in data bases store data. In data base usage a join is used to combine two tables into one. For data frames, the merge function provides a similar functionality.

With some searching of the Internet, the following tables were generated. The first lists the budget and gross receipts in the United States of the five top-grossing movies in 2013. The second lists the gross receipts in millions from sales outside the U.S. The data comes from different web sites. The two tables are kept separate. However, we may wish to merge them to do comparisons.

```
domestic <- "
The Avengers,             623357910
The Dark Knight Rises,    448139099
The Hunger Games,         408010692
Skyfall,                  304360277
The Hobbit,               303003568
"
```

```
foreign <- "
The Avengers,            1511.8
Skyfall,                 1108.6
The Dark Knight Rises,   1084.4
The Hobbit,              1017.0
Ice Age,                  877.2
"
```

First, we need to read in the data into a data frame format. We discuss `read.csv` at the end of the chapter. For now, we note it reads comma separated data into a data frame. The `textConnection` function takes the data in string format and produces a "connection" that `read.csv` can read from. Though it is a technical command, this is useful when copying data from web sites.

```
df.domestic <- read.csv(textConnection(domestic), header=FALSE)
names(df.domestic) <- c("Name", "Domestic")
df.foreign <- read.csv(textConnection(foreign), header=FALSE)
names(df.foreign) <- c("name", "foreign")
```

An inner merge is the intersection of the values. In this small example, we can see there are 4 titles in common, and the output of the following command confirms this:

```
merge(df.domestic, df.foreign, by.x="Name", by.y="name", all=FALSE)

##                      Name  Domestic foreign
## 1                 Skyfall 304360277    1109
## 2            The Avengers 623357910    1512
## 3 The Dark Knight Rises 448139099    1084
## 4              The Hobbit 303003568    1017
```

An outer join is a union over all the values matching by:

```
merge(df.domestic, df.foreign, by.x="Name", by.y="name", all=TRUE)

##                      Name  Domestic foreign
## 1                 Skyfall 304360277  1108.6
## 2            The Avengers 623357910  1511.8
## 3 The Dark Knight Rises 448139099  1084.4
## 4              The Hobbit 303003568  1017.0
## 5        The Hunger Games 408010692      NA
## 6                 Ice Age        NA   877.2
```

The last two movies are not in both data sets, and have NA values where there are missing values. Specifying `all.x=TRUE` (or `all.y=TRUE`) is used to add to the intersection only those values in the left (or right) data frame that are not in both. For example, this call adds a value for The Hunger Games, as it appears in df.domestic but not Ice Age, as it does not:

```
merge(df.domestic, df.foreign, by.x="Name", by.y="name", all.x=TRUE)

##                      Name  Domestic foreign
## 1                 Skyfall 304360277    1109
## 2           The Avengers 623357910    1512
## 3 The Dark Knight Rises 448139099    1084
## 4             The Hobbit 303003568    1017
## 5       The Hunger Games 408010692      NA
```

• **Example 4.1: Data cleaning**
The babies (UsingR) data set comes from a study where missing values were coded with large numeric values, and factors coded with numbers. The task here is to clean the data up with NA values for missing values, and human readable levels for the factors. We begin doing so one variable at a time.

The id variable is a unique id. This is not numeric data, nor is it useful for grouping, so we convert it to character. Here is one way to do so:

```
babies$id <- as.character(babies$id)     # change type of variable
```

The gestation variable measures the gestation period in days, as recorded by the hospital. A typical gestation period is 9 months, more precisely 38 – 42 weeks. This data set has several values more than 45 weeks:

```
with(babies, gestation[gestation > 45 * 7])

##  [1] 999 351 999 999 999 336 999 318 320 318 329 999 328 319 323
## [16] 999 316 999 999 330 318 999 999 330 999 324 323 316 318 353
## [31] 999 338 319 321
```

The value 999 clearly is for missing data. We might be tempted to consider values more than 45 weeks to contain erroneous data, but we don't make that assumption without knowing more about the data.

We can assign an NA value to the 999 values in the data set. We do so for several variables measuring numeric values below which have a code of 999 or 99 for missing values:

```
babies <- within(babies, {
  gestation[gestation == 999] <- NA
  wt[wt == 999] <- NA
  wt1[wt1 == 999] <- NA
  dwt[dwt == 999] <- NA
```

```
  ht[ht == 99] <- NA
  dht[dht == 99] <- NA
})
```

There are several variables that should be recorded as factors, for example the smokes variable with 0=never, 1=smokes now, 2=until current pregnancy, 3=once did, not now, 9=unknown. Here we convert this data into a factor with descriptive levels:

```
babies$smoke <- factor(babies$smoke)
levels(babies$smoke) <- list("never"=0, "smokes now"=1,
                       "Until current pregnancy"=2,
                       "once did, not now"=3)
```

The values coded by 9 are converted to NA, as 9 is left out of the specification of the levels.

Similarly, we do the same for number which records the number of cigarettes smoked per day. The value of 98 is unknown, the value of 99 is unasked. We make 99 have NA:

```
babies$number <- factor(babies$number)
levels(babies$number) <- list("never"=0, "1-4"=1, "5-9"=2,
                        "10-14"=3, "15-19"=4, "20-29"=5,
                        "30-39"=6, "40-60"=7, "60+"=8,
                        "smoke, don't know"=9, "unknown"=98)
```

The date variable is the birth date where 1096=January 1, 1961. To convert into a day, we do so with the lubridate package and date arithmetic.

First, to create a variable for the start date:

```
require(lubridate)
(x <- ymd("1961-01-01"))

## [1] "1961-01-01 UTC"
```

To add 10 days to each value of x, say, we can do the following:

```
x + 10 * days(1)
## [1] "1961-01-11 UTC"
```

So to update our variable, we have

```
babies$date <- x + (babies$date - 1096) * days(1)
```

We could do more, but instead derive BMI variables for both the mom and the dad. This task is conveniently done with transform:

```
bmi <- function(wt, ht) (wt/2.2) / (ht*2.54/100)^2
babies <- transform(babies, bmi = bmi(wt1, ht),
                    dbmi = bmi(dwt, dht))
```

At this point, we have pretty much cleaned up this data set. We can ask some questions relating variables at this point. Here is a silly one where we look at all cases where the difference in BMI is large:

```
subset(babies, abs(dbmi - bmi) > 14,
       select=c(date, gestation, wt, race))

##              date gestation  wt race
## 33     1962-02-07       268 130    0
## 200    1962-08-01       291 137    3
## 532    1961-10-16       271 108    5
## 748    1962-08-13       288 174    0
## 1163   1962-05-14       291 160    0
```

••

Problems

4.6 Simplify the following commands, where the fat data set is in the UsingR package:

```
mean(fat$neck) / mean(fat$wrist)

## [1] 2.084

mean(fat$neck/fat$wrist)

## [1] 2.084
```

4.7 Use the subset function to return a data frame made from the Cars93 (MASS) data frame consisting only of non-USA cars in origin, with 4 cylinders and a maximum price of $15,000 or less.

4.3 Applying a function over a collection

We turn our attention now to constructs in R which are used to apply a function to groups formed from some splitting process.

We have seen that many R functions are vectorized. For example, the mathematical functions are. This allow us to easily subtract 3, say, from each value in some data set:

```
x <- 1:5
x - 3

## [1] -2 -1  0  1  2
```

We can visualize vectorization through this symbolic pattern

```
[x1 x2 ... xn] -> [f(x1) f(x2) ... f(xn)]
```

where the function f takes a single value and returns a single value, which in our example is `function(x) x - 3`.

Vectorization can often be much faster than alternatives, as the computation is done in C code, and not at the R level. Even when not true, vectorization and its counterparts have the benefit—once learned—that tasks are made conceptually cleaner.

The simple example of x - 3 would often be written in other languages using a for loop. In R this would take the following form:

```
x <- 1:5
res <- integer(length(x))            # allocate temporary storage
for(i in 1:length(x)) {              # iterate over indices
  res[i] <- x[i] - 3                 # compute f, do bookkeeping
}
res

## [1] -2 -1  0  1  2
```

The above code iterates over a set of numbers which are the indices for x and uses that index to look up the corresponding x value so that 3 can be subtracted from it.

This imperative style of iterating over a set, or collection, of elements using a for loop is sometimes desirable, but most R idioms work to avoid it. It would be hard to argue that the above expression is conceptually easier to parse than the declarative style of x-3. (The difference between declaring what is to be done, rather than spelling out how to do it case by case.)

If a function is not vectorized, then the Vectorize function will make it so. The Vectorize function takes a function as its input and returns a modification of the function for its output.

For example, the `median` function cannot be applied across the columns of a data frame directly, but the function can be vectorized to do so. Here we apply the new function `Vectorize(median)` to the `homedata` data set which lists assessed home values for two different eras:

```
vmedian <- Vectorize(median)                # returns a function
vmedian(homedata)

##   y1970  y2000
##  68900 251700
```

The `Vectorize` function is an example of a *higher order function*, one that takes or returns a function, in this case both. This is possible in R as functions are first-class objects. We shall see next, that there are alternate ways to vectorize a function call that are of more general usage than `Vectorize`.

Map

Vectorization is an example of a map. In the abstract, a map takes a function, `f`, and applies it to each element of a collection, returning a new collection. Symbolically, it might be viewed with

```
[a1 a2 a3 ... an] -> [[ f(a1) f(a2) f(a3) ... f(an) ]]
```

or using a list as the collection:

```
[[a1 a2 a3 ... an]] -> [[ f(a1) f(a2) f(a3) ... f(an) ]]
```

The function `f` could possibly return one of many types of objects, so we use our list notation above for a general purpose container.

In R, the `Map` function will map a function over a collection. In our examples, a collection is either a vector, or generalized vector and the elements are those indexed by `[i]` or `[[i]]`. For a list or vector this is natural. For matrices the elements come from the vector notation, and for a data frame from the list notation. This means each element for the matrix, but each column for the data frame (as happened in the `Vectorize(median)` example).

In this example, we bypass R's usual vectorization of the `sqrt` function.

```
collection <- c(4, 9, 16)
Map(sqrt, collection)

## [[1]]
## [1] 2
##
## [[2]]
## [1] 3
##
## [[3]]
## [1] 4
```

The return value of Map is a list, as the output of f can be rather general. For the sqrt function we would get a numeric vector:

```
sqrt(collection)

## [1] 2 3 4
```

Simplified output is usually preferable, and R has many variants of the Map construct that will also try to reshape the output, or combine into a simpler structure, than a list.

The sapply function The sapply function is one such example, calling simplify2array to do the work of combining the output in a pleasing manner.[2] The calling order of sapply is reversed from that of Map with the collection specified first:

```
sapply(collection, sqrt)

## [1] 2 3 4
```

Next, we split a data set by a factor to get a list, then apply the mean function to each:

```
lst <- with(ToothGrowth, split(len, supp))
sapply(lst, mean)

##    OJ    VC
## 20.66 16.96
```

This is an example of the split-apply-combine paradigm (c.f. [62]) that is very common in data analysis.[3]

The tapply function This particular construct is common enough that it has its own function, tapply:

```
with(ToothGrowth,
     tapply(len, supp, mean)          # (X, INDEX, FUN)
     )

##    OJ    VC
## 20.66 16.96
```

---

[2]The sapply function is a special case of lapply, or *list* apply, which applies a function to a collection and returns a list. The main difference is by default sapply will try to simplify the output into an array form.

[3]The popular add-on package plyr is designed to bring more order to the collection of functions used for splitting, applying, and combining data. In particular, there are different functions used to specify the type of input and the type of output.

For `tapply`, as with `split`, the grouping variable is a factor or list of factors. In the latter case, all combinations are computed before splitting:

```
with(ToothGrowth,
     tapply(len, list(supp, dose), mean)          # (X, INDEX, FUN)
     )

##        0.5     1     2
## OJ 13.23 22.70 26.06
## VC  7.98 16.77 26.14
```

If we were to try this using `split` and `sapply` directly, we would discover that `tapply` does a nicer job of reshaping the data, in this case returning a matrix of values which displays more compactly than the named vector returned by combining `split` and `sapply`.

The aggregate function   The `aggregate` function is an alternative function that provides a formula interface. This affords a more convenient syntax than that shown previously employing `with`. The output from `aggregate` is different than that of `tapply`, though similar. The formula below puts the numeric variable on the left-hand side of the formula and the grouping variable on the right:

```
aggregate(len ~ supp, data=ToothGrowth, mean)

##    supp    len
## 1    OJ 20.66
## 2    VC 16.96
```

More than one grouping variable can be used, they may be combined with +, as in `len ~ supp + dose`.

We can combine `aggregate` with our table making commands to create a function that can extend the interface for the mean to formulas:

```
mean_formula <- function(formula, data) {
  out <- aggregate(formula, data=data, mean)
  xtabs(formula, out)
}
```

To see it work, we can then do:

```
mean_formula(len ~ supp, data=ToothGrowth)

## supp
##    OJ    VC
## 20.66 16.96
```

In Appendix A we revisit this example, to show what to do so that we can call an augmentation of the above as just:

```
mean(len ~ supp, data=ToothGrowth)

## supp
##    OJ    VC
## 20.66 16.96
```

The mean function is a "reduction", as it takes a vector and reduces the dimensionality to a single number (though in R it is still a vector). This need not be the case for the functions used. The summary function provides a nice summary of different types of variables. Here we apply it to the split values of ToothGrowth:

```
lst <- with(ToothGrowth, split(len, supp))
sapply(lst, summary)

##            OJ   VC
## Min.      8.2  4.2
## 1st Qu.  15.5 11.2
## Median   22.7 16.5
## Mean     20.7 17.0
## 3rd Qu.  25.7 23.1
## Max.     30.9 33.9
```

We see that sapply can simplify this output into a nice tabular format.

The sapply function is often used with data frames to apply a function to each variable. For example, to find the mean of each variable in a data frame, this pattern can be used:

```
sapply(mtcars, mean)

##      mpg      cyl     disp       hp     drat       wt     qsec
##  20.0906   6.1875 230.7219 146.6875   3.5966   3.2172  17.8487
##       vs       am     gear     carb
##   0.4375   0.4062   3.6875   2.8125
```

(Earlier, we used the specialized Vectorize to do a similar task.)

If the data frame has NA values, the argument na.rm for mean can be specified. This is done by adding an additional argument to sapply after the function specification:

```
m <- mtcars[1:4, 1:3]
m[1,1] <- m[2,2] <- NA
sapply(m, mean)
```

```
##    mpg   cyl  disp
##     NA    NA 171.5

sapply(m, mean, na.rm=TRUE)

##      mpg     cyl    disp
##   21.733   5.333 171.500
```

Symbolically, the expression `sapply(collection, f, args...)` can be viewed as performing:

```
[x1 x2 ... xn] -> [[f(x1, args...) f(x2, args...) ... f(xn, args...)]]
```

with simplification to an array, as possible. The additional arguments are not vectorized.

The apply function   For matrices (and more generally arrays), vectorized functions are applied to each element, treating the matrix as a vector with some shape attributes. For example,

```
m <- rbind(c(1,2), c(3,4))
sqrt(m)

##        [,1]  [,2]
## [1,] 1.000 1.414
## [2,] 1.732 2.000
```

However, it may be desirable to apply a function to each column, as with data frames. Or, since rows of a matrix are also all of the same type, it may be of interest to apply a function to each row.

For summations, these tasks are common enough that there are specialized functions `colSums` and `rowSums` that do this quickly.

To illustrate, we use the following pattern which creates a matrix with five columns and three rows:

```
(m <- replicate(5, rnorm(3)))

##         [,1]    [,2]    [,3]     [,4]      [,5]
## [1,]  0.4177 -1.040 0.87860  0.4323 1.589638
## [2,]  0.9818  1.782 0.03581  2.0908 1.954652
## [3,] -0.3927 -2.311 1.01283 -1.1999 0.004938
```

This is a common pattern to generate random data for simulations (cf. Chapter 7). The `replicate` function is a convenience for this map

```
[1 2 ... n] -> [f() f() ... f()]
```

where the implied function is really just a block of commands that does not depend on the index. That is, it repeats an expression, which often has randomness involved. In this example, the expression generates 3 randomly chosen numbers from a named distribution.

For a matrix, like m, to add along the rows or columns we can use rowSums or colSums:

```
rowSums(m)                                    # add along rows

## [1]  2.278  6.845 -2.886

colSums(m)                                    # add down columns

## [1]  1.007 -1.569  1.927  1.323  3.549
```

As a matrix, m, is treated as a vector when viewed as a collection, the sapply function does not directly work for this task:

```
sapply(m, sum)

##  [1]  0.417651  0.981753 -0.392695 -1.039669  1.782229 -2.311069
##  [7]  0.878605  0.035807  1.012829  0.432265  2.090819 -1.199926
## [13]  1.589638  1.954652  0.004938
```

Rather we want to apply the function to a specified dimension of the matrix. The apply function will do so, with the margin(s) to apply the function specified in the second position (there are no named arguments, so order is crucial).

```
apply(m, 1, mean)                             # rowMeans alternative

## [1]  0.4557  1.3691 -0.5772

apply(m, 2, mean)                             # colMeans alternative

## [1]  0.3356 -0.5228  0.6424  0.4411  1.1831
```

The first is similar to all of this:

```
c(sum(m[1,]), sum(m[2,]), sum(m[3,]))

## [1]  2.278  6.845 -2.886
```

The function applied need not be a reduction. For example, to find summaries along each column, we have

```
apply(m, 2, summary)
```

```
##          [,1]   [,2]    [,3]   [,4]    [,5]
## Min.    -0.3930 -2.310 0.0358 -1.200 0.00494
## 1st Qu.  0.0125 -1.680 0.4570 -0.384 0.79700
## Median   0.4180 -1.040 0.8790  0.432 1.59000
## Mean     0.3360 -0.523 0.6420  0.441 1.18000
## 3rd Qu.  0.7000  0.371 0.9460  1.260 1.77000
## Max.     0.9820  1.780 1.0100  2.090 1.95000
```

As with sapply, we can pass in non-vectorized, named arguments to each function call by adding them to the apply call after the function specification

The special-purpose sweep function can be used in combination with apply to center or scale values. For example, this code will compute the means for each column, then subtract from each column the corresponding value:

```
xbars <- apply(m, 2, mean)
centers <- sweep(m, 2, xbars, FUN="-")  # "-" is default
centers
```

```
##          [,1]    [,2]    [,3]      [,4]    [,5]
## [1,]  0.08208 -0.5168  0.2362 -0.008788  0.4066
## [2,]  0.64618  2.3051 -0.6066  1.649766  0.7716
## [3,] -0.72826 -1.7882  0.3704 -1.640979 -1.1781
```

Going further, we can find the z-scores by dividing by the standard deviations. Here we need to specify the value FUN="/":

```
sds <- apply(m, 2, sd)
z_scores <- sweep(centers, 2, sds, FUN="/")
z_scores
```

```
##          [,1]    [,2]    [,3]      [,4]    [,5]
## [1,]  0.1188 -0.2467  0.4460 -0.005341  0.3922
## [2,]  0.9353  1.1003 -1.1454  1.002660  0.7444
## [3,] -1.0541 -0.8536  0.6994 -0.997319 -1.1367
```

This is a common enough pattern, that the function scale will perform it.

mapply We have done illustrations with functions that take a single variable and produce some output. There are functions that take two or more values and we may wish to map these.

For example, the min function finds the minimum of its arguments. It is written to combine its arguments then find the minimum:

```
min(3, 4)                                    # 3
```

```
## [1] 3
```

```
min(c(1,4), c(2,3))                          # not c(1,3) as maybe desired
```

```
## [1] 1
```

How can we get the output where the corresponding pairs are compared to see which is smaller?

Symbolically, we want to compute the following with f being our `min` function:

```
([a1 a2 ... an],[x1 x2 ... xn]) -> [ f(a1, x1) f(a2, x2) ... f(an, xn)]
```

The `Map` function can be used to do this:

```
Map(min, c(1,4), c(2,3))
```

```
## [[1]]
## [1] 1
##
## [[2]]
## [1] 3
```

The output of `Map` will return a list, but in this case a vector is more appropriate. The `mapply` function is similar to `Map` (in fact `Map` uses `mapply` in its definition) but, like `sapply`, it also attempts simplification of its output (unless it requested not to through `SIMPLIFY=FALSE`).

```
mapply(min, c(1,4), c(2,3))
```

```
## [1] 1 3
```

This particular usage is common enough that there is a special function, `pmin` to do it faster and in a safer manner.

Unlike `sapply`, with `mapply` the function goes first. This allows for an arbitrary number of collections to follow. (For example, if we had `f(x,y,z)` then we could have `mapply(f, xcollection, ycollection, zcollection)`. The collections will be recycled to a common length.

The example above using `sweep` can be done with `mapply`. In that example, the column mean is subtracted from each column. To do this with `mapply` we use the function `function(col, center) col - center` and arrange to pass in the entire column and the column mean. Here is one way:

```
our_sweep <- function(col, center) col - center
mapply(our_sweep, as.data.frame(m), apply(m, 2, mean))

##              V1       V2      V3        V4       V5
## [1,]   0.08208 -0.5168  0.2362 -0.008788  0.4066
## [2,]   0.64618  2.3051 -0.6066  1.649766  0.7716
## [3,] -0.72826 -1.7882  0.3704 -1.640979 -1.1781
```

We convert the matrix to a data frame, so that the columns are the elements of the collection, not each individual cell. While special purpose functions such as sweep may have their purpose, it is suggested that one learns
the more general idioms of R when starting out, and picking up the more
specific ones should the need arise. For applying a function to a collection,
the following are usually sufficient: sapply, mapply, and apply.

The do.call function   We saw in Chapter 3 the Spearman correlation coefficient can be computed for variables x and y from cor(rank(x), rank(y)).
This fits the pattern of finding

f(g(x1), g(x2), ..., g(xn))

Doing this elegantly with many variables is a bit tricky. The following
map is by now easy:

[x1 x2 x3 ... xn] -> [[g(x1) g(x2) g(x3) ... g(xn)]]

What we desire is a means to call a function, f say, with arguments specified by a list, not individually. The do.call function does this, replacing a
function call like

f(key1=x1, key2=x2, key2=x3, ..., keyn = xn)

with the following:

do.call(f, list(key1=x1, key2=x2, ..., keyn=xn))

Symbolically, this becomes

[[g(x1) g(x2) g(x3) ... g(xn)]] -> f(g(x1), g(x2), ..., g(xn))

With this, the construct cor(rank(body), rank(brain)) becomes:

```
body <- Animals$body; brain <- Animals$brain
do.call(cor, Map(rank, list(body, brain)))

## [1] 0.7163
```

This usage is no shorter than doing it directly, but the pattern with Map can be used with functions that take more arguments than two.

As an aside, the last call was a bit cumbersome, as when using do.call the names attribute of the list is used to name the arguments passed into the function. While this is often convenient, in this example it is not, as the names of Animals don't match the argument names of cor. However, when a component has no name, it matches by position. To strip the names from an object, they can be assigned as NULL. For example, the above could have been done with:

```
do.call(cor, Map(rank, setNames(Animals, NULL)))

## [1] 0.7163
```

Filter

Another useful concept is filtering or subsetting a collection. We have seen that R allows us to subset a vector by a logical vector. The Filter function does something similar, though with different syntax. The basic pattern is Filter(predicate, collection) which applies the predicate function to each member of the collection and returns only those elements in the collection for which the predicate is TRUE. (A predicate function is one that returns a logical variable.)

This can be used to filter out a data frame. For example, here we filter a data frame by which columns hold factors:

```
m <- Cars93[1:2, 1:15]                      # 15 columns
Filter(is.factor, m)                        # 6 are factors

##    Manufacturer   Model     Type                   AirBags DriveTrain
## 1        Acura Integra    Small                       None      Front
## 2        Acura  Legend  Midsize  Driver & Passenger            Front
##    Cylinders
## 1          4
## 2          6
```

While this provides identical results as m[sapply(m, is.factor)], the imperative style of Filter is conceptually more direct.

Reduce

The Reduce pattern formalizes many functions that are reductions, like sum or max.

Because of associativity, we can think of sum completing its task in many different ways. Here are two ways to add the numbers 1 through 4:

$$1 + 2 + 3 + 4 = ((1 + 2) + 3) + 4 = 1 + (2 + (3 + 4)).$$

Let's consider the first. It can be generalized to the following pattern, assuming $f(a,b) = a + b$:

$$f(f(f(1,2),3),4).$$

Repeatedly applying the output of binary function to a new element in a collection is called a *fold* or an *accumulation*. In this case, a left fold. Summing values is just one example.

Folds in R are implemented in the Reduce function. The default is a left fold.

To illustrate, here is a slower way than sum to add the numbers 1 through 4:

```
Reduce("+", 1:4)
```

```
## [1] 10
```

Finding a maximum of a set of numbers can be presented as a fold. We simply compare two values, keeping the largest:

```
Reduce(function(x,y) ifelse(x > y, x, y), c(6, 4, 7, 9, 10, 3))
```

```
## [1] 10
```

For an example that isn't implemented in base R (though still contrived from a statistics viewpoint), we extend a function that finds the least common multiple of two numbers, to one that finds the same for an arbitrary vector of numbers. The smallest (least) common multiple of two integers is the product of the two divided by the largest common factor. The scm function below will find the value.

```
## gcd by Euclidean algorithm, a, b integers
gcd <- function(a, b) {while (b != 0) {t = b; b = a %% b; a = t}; a}
## scm (lcm is R function name in graphics package)
scm <- function(a, b) (a * b) / gcd(a, b)
```

Here we see the function at work:

```
scm(3, 5)                              # no common primes
```

```
## [1] 15
```

```
scm(6, 9)                              # common prime
```

```
## [1] 18
```

There is no function to compute scm(2,4,7) or scm(c(2,4,7)), but this could be found from scm(scm(2,4), 7). This suggests using Reduce to extend the range of applicability:

```
Reduce(scm, 1:20)        # smallest number divisible by 1, 2, ..., 20

## [1] 232792560
```

Problems

4.8 Run the following command for the wellbeing (UsingR) data set. Which variables are negatively correlated with well being?

```
sapply(wellbeing[,-(1:2)], function(y) {
    cor(wellbeing[,2], y, use="complete.obs")
    })
```

4.9 After splitting by a factor, the different groups may have differing lengths. For example, the beatles (LearnEDA) data set contains the time for each song on various albums by the Beatles. Split the data by the album variable then count the number of songs per album.

4.10 Find the standard deviation of each variable in the mtcars data set.

4.11 Find the standard deviation for each numeric variable in Cars93 (MASS).

4.12 A stackoverflow.com question was:

> I need to subset data frame based on column type—for example from data frame with 100 columns. I need to keep only those columns with type factor or integer.

How can this be done?

4.13 The batting (UsingR) data set has baseball statistics from 2002. Can you compute the team batting average (total number of hits divided by total number of at bats)?

4.14 The batting (UsingR) data set has baseball statistics from 2002. Can you identify which players played for more than one team?

4.15 A stackoverflow.com question was:

> When I have a vector of vectors in R, how do I select a vector that contains one element from each outer vector?

For example, for this data, the output would be "1", "3", "5".

```
d <- c("1,2","3,4","5,6")
strsplit(d,",")

## [[1]]
## [1] "1" "2"
##
## [[2]]
## [1] "3" "4"
##
## [[3]]
## [1] "5" "6"
```

4.16 How could one remove all columns that contain one or more NA values from a data frame?

4.17 Write a function that summarizes the variables in a data frame with their name and their class (returned by class). (There are many ways to do so.)

4.18 A stackoverflow.com user asked for a shorter alternative to the following. (The use of which(fruits==x) to get the index being considered ugly.) Provide an alternative.

```
fruits <- c("Bananas", "Oranges", "Avocados", "Celeries?")
sapply(fruits, function(x)
   paste(x, "are fruit number", which(fruits==x)))

##                        Bananas                        Oranges
##    "Bananas are fruit number 1"   "Oranges are fruit number 2"
##                       Avocados                      Celeries?
##   "Avocados are fruit number 3" "Celeries? are fruit number 4"
```

4.19 Reduce and associativity Left and right folds are indifferent if the operation is associative. Operations like - and ^ are not though. By comparing the outputs of $2 - 3 - 4$ and 2^{3^4}, with, for example, Reduce("-", c(2,3,4)), Reduce("-", c(2,3,4), right=TRUE) determine the order in which - and ^ are performed.

4.4 Using external data

We discuss next how to use data sets within an R session that are not built-in data sets or data sets that have been typed in. Of course, there is a wide range of possibilities for storing data. In this section we touch on some of the common ones. The *R Data Import/Export* manual accompanying R has more details than we present.

Spreadsheet data

Perhaps the most common source of data are spreadsheets, such as Microsoft's EXCEL. R has some add-on packages for interacting directly with EXCEL: the xlsx package can read and write to EXCEL 2007 (xlsx) files; the gdata package provides the function read.xls for reading older xls files; the RExcel framework (http://rcom.univie.ac.at/) can integrate R with several versions of EXCEL for Microsoft Windows; and the XLConnect package offers similar features as xlsx. We send the interested reader to the documentation of those packages.

For a simple data exchange from a spreadsheet into an R session, we can use a text-file exchange. The basic idea being the spreadsheet program writes a table out to a file which is then read into R. Common formats for such data exchange are csv (comma-separated values), tsv (tab-separated values), or fwf (fixed-width format). Generally, most spreadsheets can import and export such files.

RSTUDIO has a dialog to read in these formats (Figure 4.1). Additionally, base R provides various read.* functions for processing such data. For the types just listed, there are read.csv, read.table, and read.fwf. The read.csv function is just a front end to the read.table function with certain parameters set. The "Import Dataset" feature of RSTUDIO, found on the Workspace tab, is a front end that shows the results of adjusting the parameters.

For read.table, the key arguments, among others, are header to indicate if there is any header data in the file; sep to indicate the data separator (comma, space, tab, ...); stringsAsFactors to determine if character data is read in as a factor (the default); and comment.char to specify any lines that should be treated as comments.

When using read.table directly, the file to read is typically specified by name. This can be typed in, or if more convenient, browsed for by specifying file.choose(). The name can even be a url for a web-based resource. In a previous example, we used textConnection(domestic) to create a readable text connection from a character string.

Though often used, there can be issues with exchanging data between R and spreadsheets this way: the exchange through text can lose attributes (such as dates) on the variables; the two separate data sets must be manually synchronized; and the exchange is limited to a single sheet of a spreadsheet notebook.

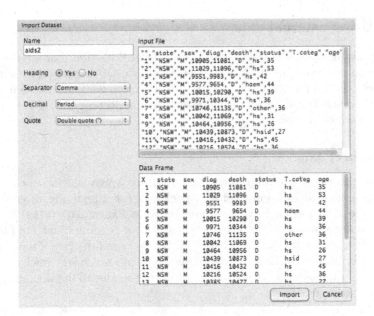

Figure 4.1: RStudio's import data set dialog. The left side controls are used to adjust the arguments to read.table.

To send tabular data from R to a spreadsheet, R provides the functions write.table or write.csv. For example, write.csv(object, file=file_name). The *R Data Import/Export* manual lists several issues to be aware of including the loss of precision, due to formatting of floating point numbers; the treatment of NA and NaN values; encoding issues; and issues with internal quotes.

Other programs   The foreign package is a recommended package, bundled with most all R installations. It allows data generated from other statistical programs to be read into R. There are readers for formats from S (version 3), SAS, Stata, Epi, MINITAB, SPSS, and Systat.

Web-based data sets

Data sets of all varieties are to be found on the Internet. Many sites provide data in some regular format, though these formats can vary widely. Other sites have data embedded in tables in a web page. We see in this section, various ways to access different types of data sets online. Some of the functionality is provided through base R, but much other is based on add-on packages, primarily contributed by Duncan Temple Lang, that extend R to work with different internet standards.

First, some examples of nicely presented data. The CDC (Centers for Disease Control and Prevention) frequently conducts the NHANES survey (http://www.cdc.gov/nchs/nhanes.htm) and provides the data in an

SAS format that can be read in with read.xport from the foreign package. The U.S. Energy Information Adminstration has gasoline price data (http://www.eia.gov/petroleum/gasdiesel/) that can be downloaded in Excel format.

To illustrate reading from an online Excel file, the following commands will read the second sheet from the spreadsheet that lists historical gas prices from the ongoing survey performed by the USEIA.[4] Through some preparatory work, it was noted that the desired data is the second sheet and has two extra lines of information at the top of the sheet. This requires the additional arguments sheet=2 and skip=2.

```
require("gdata")                              # must be installed
f <- "http://www.eia.gov/petroleum/gasdiesel/xls/pswrgvwall.xls"
gas_prices <- read.xls(f, sheet=2, skip=2)
gas_prices <- setNames(gas_prices[,1:2], c("Date", "Weekly_US"))
```

This will unfortunately leave the first column with dates that are not quite right. (The first value is "Aug 201990 1990".) As the dates are padded to use two characters, the positions of the dates are consistent. This following command uses substr to extract the date and as.Date to convert them to Date format. The resulting plot (Figure 4.2) makes use of the fact that plot will utilize this class information to make the axis labels prettier.

```
gas_prices$Date <- as.Date(substr(gas_prices$Date, 1, 10),
                           format="%b %d%Y")
plot(Weekly_US ~ Date, gas_prices, type="l")
```

Google Sheets   *Google Sheets* is an online spreadsheet app that allows for the creation, editing, and sharing of spreadsheets. This is one of several available collaborative online tools. The sharing of Google Sheets can be done as a csv file, which allows for easy inclusion into R. The only catch is the urls use secure http which is not (currently) supported by download.file, which does the work of getting the file from the internet. The add-on package RCurl, a very useful wrapper for the curl library (http://curl.haxx.se), provides the function getURL which will, by default, return the page's text for page's exposed to the internet by many different protocols, including https.

As the value returned is a string containing the text, the result can be passed to read.csv by wrapping it within textConnection. Here is an example, with a silly spreadsheet.

---

[4]Of course, web sites come and go faster than books, so the web site resources referenced herein may not exist at the time the commands are tried. In that case, the commands should fail with an informative error message.

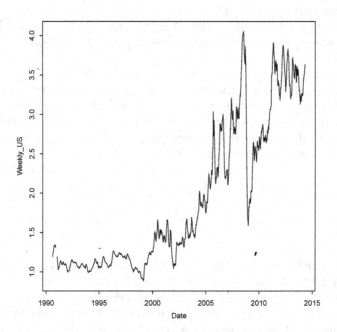

Figure 4.2: Plot of average gas prices in the United States over time.

```
key <- "0AoaQTPQhRgkqdEthU0ZZeThtcWtvcWpZUThiX2JUMGc"
f <- paste("https://docs.google.com/spreadsheet/pub?key=",
           key,
           "&single=true&gid=0&output=csv", sep="")
require(RCurl)
read.csv(textConnection(getURL(f)), header=TRUE)

##    number english spanish
## 1       1     one     uno
## 2       2     two     dos
## 3       3   three    tres
## 4       4    four  cuatro
## 5       5    five   cinco
```

The quandl package Quandl.com is a web site that indexes time-series data from numerous sources. It has millions of data sets, and even better, an open interface for downloading (and uploading) data. The Quandl package provides an interface to R users. To download a file is as easy as browsing the site to the data of interest, and noticing the code Quandl assigns.

In this example there are three codes used. The codes refer to data on the demographics of the Chinese population by broad age group. The following downloads three data sets, then calls merge twice (using Reduce) to combine the three data sets into one.

```
require(Quandl)
ch_0014 <- Quandl("WORLDBANK/CHN_SP_POP_0014_TO_ZS")
ch_1564 <- Quandl("WORLDBANK/CHN_SP_POP_1564_TO_ZS")
ch_65up <- Quandl("WORLDBANK/CHN_SP_POP_65UP_TO_ZS")
ch_all <- Reduce(function(x,y) merge(x, y, by="Date"),
                 list(ch_0014, ch_1564, ch_65up))
names(ch_all) <- c("Date", "[0,14]", "[15,64]", "[65,)")
```

After the data has been combined, there are columns with proportions for each age group. A good visualization for any given year is a segmented bar, and placing these next to each other can clearly show demographic trends. The barplot function will make segmented bar plots. To do them all at once, we need to flip the numbers in ch_all from 3 columns to 3 rows. This is achieved by the t function, but we first remove the column of dates which are subsequently formatted as column names for the resulting data set.

```
heights <- t(ch_all[,-1])
colnames(heights) <- format(ch_all[,"Date"], format="%Y")
barplot(heights, main="Proportion of [0-14], [15-64], [65,)")
```

Working with JSON data   A common format to transfer data over the internet is JSON or JavaScript Object Notation. Many sites have data that can be retrieved in this format, including Quandl.com, which is used in our example. The JSON format can be quickly parsed into R with the RJSONIO package, through its fromJSON function.

The following illustrates the usage. The JSON returned by the Quandl site is not simply tabular data, but has metadata attached. The value returned from fromJSON is a list with named elements corresponding to the tagged element of the original JSON file, for example column_names. In this case, the data is in the data component:

```
require(RJSONIO)
f <- "http://www.quandl.com/api/v1/datasets/PRAGUESE/PX.json"
out <- fromJSON(f)
out$column_names                         # names

## [1] "Date"      "Index"      "% Change"

out$data[1]                              # one from 1000s of values
```

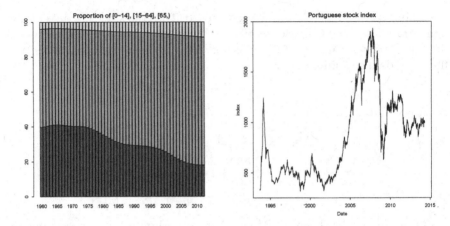

Figure 4.3: Plots of data downloaded from Quandl.com. The left graphic shows the change in time of population demographics in China. The darker shaded segment represents the proportion of the population 14 and younger. The right graphic shows a Portuguese stock index over time, including two precipitous drops in the early 1990s and 2008.

```
## [[1]]
## [[1]][[1]]
## [1] "2014-05-02"
##
## [[1]][[2]]
## [1] 1011
##
## [[1]][[3]]
## [1] 0.05
```

As can be seen, the data is not in tabular format, rather each row is represented by a list. The following convenience function (pluck) takes one component at a time, allowing us to treat the data variable by variable, which is useful for coercing the first column into a date. This makes the plot (right graphic in Figure 4.3) have nicer labels for the x axis.

```
pluck <- function(l, key) l[[key]]      # pluck from a list
px <- data.frame(Date = as.Date(sapply(out$data, pluck, key=1)),
                 index = sapply(out$data, pluck, key=2),
                 perc_change = sapply(out$data, pluck, key=3))
plot(index ~ Date, data=px, type="l", main="Portuguese stock index")
```

Parsing HTML files Not all data is so nicely presented. Many times, the data of interest is a table within an HTML page. For such data there are a variety of web-scraping techniques available to R users. The XML package provides an interface to the libxml2 (http://www.xmlsoft.org/) parser. This breaks the structure of a web page into pieces that can be traversed. Though working directly with those pieces can be daunting, the readHTMLTable function from that package does the hard work for us.

In our example, there is more than one table per page. In the second case, the desired table is the second one on the page. Here we look at two lists from wikipedia.org, one with the 40 highest grossing films in China and one for the 100 highest grossing films in Canada and the United States. We can grab both with the following:[5]

```
require(XML)
## fit in 80 characters
url_base = "http://en.wikipedia.org/wiki/"
ch <- "List_of_highest-grossing_films_in_China"
us_can <-
   "List_of_highest-grossing_films_in_Canada_and_the_United_States"
china_all  <- readHTMLTable(paste(url_base, ch, sep=""))[[1]]
us_can_all <- readHTMLTable(paste(url_base, us_can, sep=""))[[2]]
```

A natural question is what is the overlap between the two lists? There could be as many as 40 in common. A merge, where we take the titles that are common, is given by the following command.

```
in_common <- merge(china_all, us_can_all, by="Title")
## tidy up
elide <- function(x, n=20)
  ifelse(nchar(x) < n, x, sprintf("%s...", substr(x,0,n)))
rownames(in_common) <- sapply(as.character(in_common[,1]), elide)
##
in_common[, c(2,6,7)]
```

```
##                          Rank.x Rank.y Initial gross(unadjusted)
## Avatar                        1      1              $749,766,139
## Fast & Furious 6             40     92              $238,673,370
## Gravity                      36     65              $274,052,110
## Harry Potter and the...      39     20              $381,011,219
## Inception                    32     52              $292,576,195
## Iron Man 3                    8     14              $409,013,994
## Pirates of the Carib...      31     90              $241,071,802
```

---

[5]This code depends on an ordering of tables that may not be consistent as time progresses. If there are errors, consult the web pages and see which table is the one desired and adjust the indexing accordingly.

```
## The Avengers              21      3            $623,357,910
## The Hobbit: The Deso...   34     73            $258,366,855
## Titanic                    6      2            $600,788,188
## Transformers: Dark o...    4     27            $352,390,543
## Transformers: Reveng...   37     18            $402,111,870
```

Multivariate graphics

This chapter covers various standard graphics for presenting multivariate data. Up until now we have been illustrating R's base graphics for producing graphical representations. Over the years there have been additions to the base graphics interface that enhance the abilities to represent multivariate data. We discuss briefly the lattice package, a recommended package that comes bundled with R; and ggplot2, an add-on package of relatively recent origins.[1]

5.1 Base graphics

We begin with an example from Murrell [44] creating a graphic for the iris data set.[2] This famous data set of Andersen records measurements of sepal length and width and petal length and width for 50 flowers from each of 3 species. The following creates a scatter plot comparing length of width of the sepal (left graphic of Figure 5.1):

```
with(iris,
     plot(Sepal.Length, Sepal.Width,
          pch=as.numeric(Species), cex=1.2))
legend(6.1, 4.4, c("setosa", "versicolor", "virginica"),
       cex=1.5, pch=1:3)
```

In addition to plotting (x,y) pairs, the code above uses the argument pch=as.numeric(Species) to force different plot characters for each of the 3 species. In the legend call, the argument pch=1:3 is employed. For the first, the species names are mapped into values from 1:3, as this is how the factors are stored internally. For the legend call, only the values are required, not the entire mapping.

Technical differences aside, the reason for using different plot characters is the desire to represent a third variable in the graphic, in this case the categorical variable recording the species factor. There are other ways to represent

---

[1]We do not discuss various packages that are useful for interactive multivariate graphics. Among these are rggobi, iplots (and its progeny), and cranvas.

[2]Chapter 3 of Murrell's text does a very thorough job of explaining R's base graphics.

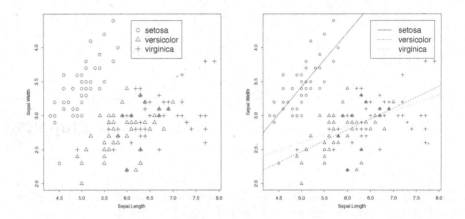

Figure 5.1: A scatterplot of sepal width by length for samples of size 50 from each of 3 species of irises. The left graphic uses different plot characters to represent the species, the right graphic employs different line types to differentiate the per-species regression lines.

categorical variables: different line types (lty) and different colors (col) are both widely used.

For example, here we add to the basic graphic a regression line computed for each of the 3 different species using different line types to indicate the species (right graphic of Figure 5.1):

```
fm <- Sepal.Width ~ Sepal.Length
plot(fm, iris, pch=as.numeric(Species))
out <- mapply(function(i, x) abline(lm(fm, data=x), lty=i),
              i=1:3, x = split(iris, iris$Species))
legend(6.5, 4.4, levels(iris$Species), cex=1.5, lty=1:3)
```

This graphic makes it apparent that the relationship is different for the "setosa" species than the others, as the slope seems quite different.

Numeric variables can be turned into categorical variables for grouping purposes. Here we use the Cars93 data set to look at the relationship between highway mileage and weight for classes of cars aggregated by their price (Figure 5.2).

```
Cars93 <- transform(Cars93, price=cut(Price, c(0, 15, 30, 75),
                 labels=c("cheap", "affordable", "expensive")))
plot(MPG.highway ~ Weight, Cars93, pch=as.numeric(price))
legend(3500, 50, levels(Cars93$price), pch=1:3)
```

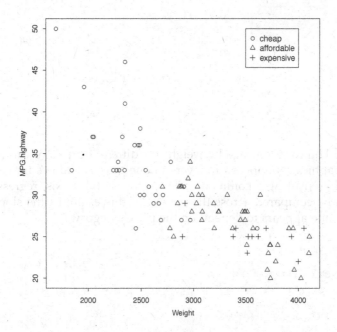

Figure 5.2: Scatterplot of highway mileage by weight for different categories of cars. The category is derived from a numeric variable using cut.

In Figure 5.2 it is difficult to identify the trend for the different groups. Figure 5.3 remedies this. Instead of three overlapping layers, we use three graphics.

The following code splits the data by the derived price variable and then creates graphics for each. To display them together, we use the mfrow setting for the par function. The specification of xlim and ylim is based on values for the entire data set, not just those in the subset. This facilitates a comparison amongst the graphs. To put this altogether, we use mapply, as it makes it convenient to pass in both the split-off piece of the data frame and the factor level, which is used in the title.

```
l <- split(Cars93, Cars93$price)
## 3 graphics in one
par(mfrow=c(1,length(l)))
## common to each graphic
fm <- MPG.highway ~ Weight
xlim <- range(Cars93$Weight)
ylim <- range(Cars93$MPG.highway)
```

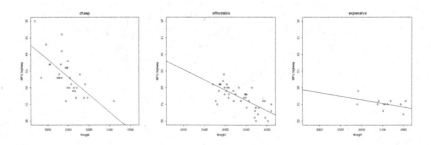

Figure 5.3: Highway mileage by weight for different prices of cars. The use of three graphics, as opposed to layers in one graphic, allows for easy comparisons. By ensuring a common scale for the x and y axis, regression-line slopes can be compared across the panels. In this example, the slope of the regression line appears to depend on the price category.

```
mapply(function(x, nm) {
  plot(fm, data=x, main=nm, xlim=xlim, ylim=ylim)
  abline(lm(fm, data=x))
}, l, names(l))
```

While the above shows that repeating graphics for different levels of a classifying factor is not terribly difficult to do, the lattice and ggplot2 graphics frameworks have facilities to make this task much easier.

Bubble charts We mention one more common way to introduce a third numeric variable into a graphic.

In Example 3.3 we looked at SAT total scores by average state-wide salary for each of fifty states, as recorded in the SAT data set. It was noted that the negative correlation did not agree with intuition due to the presence of a lurking variable. This variable can easily be seen through a *bubble chart*, where instead of different plot characters, the size of the character depends on the value. This naturally represents a third numeric variable into the graphic. The sizing should relate the area of the plotting symbol with the value of the variable, so in the following we take a square root as the cex value stretches the radius of the circle symbol. For variety, we fill the circles by using pch=16. To show overlapping circles, an alpha level is specified to the color we set for the points.

```
plot(total ~ salary, data=SAT, cex=sqrt(perc/10),
     pch=16,                                       # filled circles
     col=rgb(red=0, green=0, blue=0, alpha=0.250)) # use alpha
```

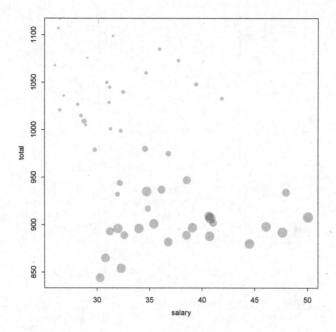

Figure 5.4: Bubble chart of total SAT scores by average teacher salary per state. The area of each bubble is proportional to the percentage of students taking the test in that state. As can be seen, similar-sized bubbles tend to cluster in an increasing pattern.

From the graphic (Figure 5.4) we can see that similarly sized bubbles form into 3 clusters, each of which can be seen to have positive correlation, despite the overall negative correlation for data as a whole.

Pairs plots A pairs plot (or *scatterplot* matrix) displays scatterplots for each pair from a list of numeric variables. The pairs function produces this graphic. In Figure 5.5 the four numeric variables of the iris data set are plotted in this manner. The diagram puts the variable name on the diagonal, as there is no sense in plotting a variable against itself. To read this diagram, the ith row and jth column has a scatter plot consisting of the ith and jth variables. (The jth row and ith column has the same plot with the order of the variables switched.) For example, in the graphic the third row, first column is a graph of Sepal.Length on the x axis and Petal.Length on the y axis.

From this graphic, we can observe correlations between many variables at once. In the figure we also colored the points by their species. The plots also demonstrate the clustering by species that is in this data set. To produce the graphic, we dropped the Species factor from the data to be plotted with the

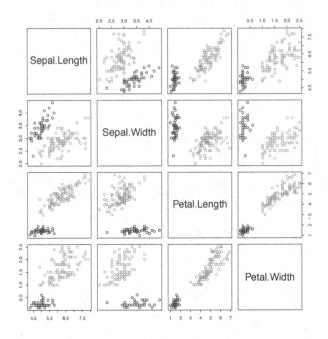

Figure 5.5: Pairs plot of the four numeric variables in the iris data set. The clustering by species is evident with the additional coloring.

aid of Filter (or alternatively iris[, -5]) and used those values instead for coloring the points:

```
species <- iris$Species
values <- Filter(is.numeric, iris)
pairs(values, col=species)
```

Parallel coordinate plots   A parallel coordinate plot is a way to compare many variables, perhaps on totally different scales, for many different cases. Each variable is laid out along the x axis, as is typical of a barplot for categorical data. For each variable, the cases are ranked and then scaled to fit on the entire range of the y axis. For each case, the different points are connected together by a line. In our example, there are 50 cases, so 50 possibly overlapping lines are drawn. This graphic can be overwhelming, but how it is used is to then highlight certain cases so that their relative ranks can be compared.

For example, Figure 5.6 uses the state.x77 data matrix which lists 8 measurements for each of the 50 United States. The state of New York is highlighted, which shows a relatively large population and a relatively small area,

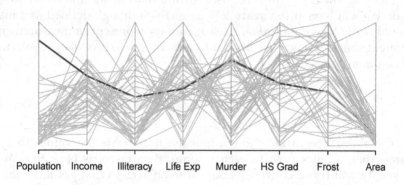

Figure 5.6: A parallel coordinates graph for the state.x77 data set. The state of New York is set off with a darker line. This graphic allows us to track the relative ranks of the case over several different variables.

among others. The graphic was produced with the parcoord function of the MASS package. A basic graphic is easy to produce, the extra arguments involved are there to set off the specified case. To do so, we employed logical indexing, and then utilized its implicit conversion to 0 or 1 when used in a numeric context (adding 1 to get the right indexing).

```
x <- state.x77
case <- "New York"
ind <- rownames(x) == case
parcoord(state.x77,
         col=gray(c(0.8, 0.2))[ind + 1],   # light/dark gray
         lwd=(1:2)[ind + 1], las = 2)
```

Heatmap matrix   A common graphical trick is to use variations in color to represent a numeric variable. For the choice of a color range, R has many functions. Below we use gray.colors to create a grayscale range. The RColorBrewer package defines many other color ranges.

The heatmap function[3] uses color variation in a heatmap, representing each value in a matrix with a color and arranges the colors in a grid labeled by the row and column names. The default graphic will also layer a dendrogram (a representation of clustering), which we avoid by a choice of argument below.

---

[3] A nice tutorial of this topic can be found at http://flowingdata.com/2010/01/21/how-to-make-a-heatmap-a-quick-and-easy-solution/.

The basic `heatmap` function uses all the data in the matrix to assign a color. We will look at the `state.x77` data which though defined as a matrix, should really be a data frame, as each column represents a measurement on different scales. As such, we first rank each column to produce measurements on common scales:

```
x <- sapply(as.data.frame(state.x77), rank)
rownames(x) <- rownames(state.x77)
```

The above does a lot, though quickly: it converts the values to a data frame, then applies `rank` to each column, then converts back to a matrix, as `sapply` will try to convert its output to an array. What got lost—the row names—are added back in manually, as they are utilized by the `heatmap` function.

The following call produces the heatmap, the specification `Rowv=NA` and `Colv=NA` turns off the addition of dendrograms:

```
heatmap(x, Rowv=NA, Colv=NA,
        scale="column",               # scale columns
        margins=c(8, 6),              # leave room for labels
        col=rev(gray.colors(50)))     # darker -> larger
```

Figure 5.7 shows the graphic. The data presented is similar to the parallel coordinates plot above, but this graphic makes it easier to track individual cases without highlighting, whereas the parallel coordinates plot makes known individual cases stand out. Here we can observe that New York is on the large side for population and income (darker colors), and the low side for most other measures.

Problems

5.1 For the `UScereal` (MASS) data set, create a scatter plot of `calories` modeled by `sugars` using the `shelf` variable to create different plot characters. Add a legend to indicate the shelf number. Is there any pattern?

5.2 For the `iris` data set create a parallel coordinate plot for the 4 categorical variables, with each species having a different color. Does one species seem to have different properties?

5.3 For the data set `UScereal` (MASS) make a pairs plot of the numeric variables. Which correlation looks larger: fat and calories or fat and sugars?

5.4 For the data set `batting` (UsingR) make a bubble plot of home runs hit (HR) modeled by hits (H) where the scale factor for each point is given by `sqrt(50)/10`. Is there any story to be told by the size of the points?

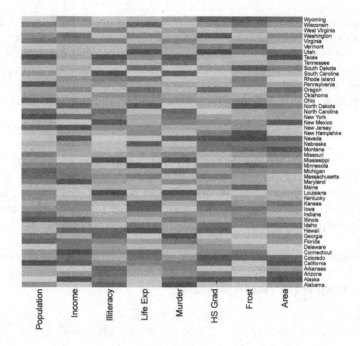

Figure 5.7: Heatmap of state.x77 data set. The ranks of each column are represented through a color with darker meaning bigger.

5.2 Lattice graphics

The lattice package is a recommended package that comes standard with most R distributions. The package author [52] has a book describing its possibilities in detail. The package builds on the underlying grid framework [44]. The design of lattice is inspired by Cleveland's work [10] on the trellis package for S-PLUS. The lattice package can be complicated, we only mention a few of its features. The playwith add-on package provides a user interface for playing with the many features.

The package describes itself with (cf. ?Lattice):

> It is a powerful and elegant high-level data visualization system with an emphasis on multivariate data.

The elegance comes from its use of R's formula interface to specify the splitting of data. The basic formula is extended to include conditioning variables and takes the form

    y ~ x | g1 + g2 + ...

The conditioning variables (g1, g2, ...) are used to split the data. Graphs for different sets of values are presented in different panels, with common scales for comparison. The conditioning variables are often factors (categorical variables) but may also be numeric variables, in which case *shingles* are formed which subset the data using slightly overlapping intervals. (Similar to what can be done with cut, but different, as the same case can appear in more than one shingle.)

To replicate Figure 5.3, we first try the most straightforward usage:

```
xyplot(MPG.highway ~ Weight | Price, data=Cars93)
```

In place of plot, we use lattice's xyplot. Lattice has a different, though similar, naming scheme for various standard graphics, including dotplot, histogram, densityplot, bwplot, qqmath, and barchart. Similar to plot, the xyplot function has arguments pch, cex, lty, lwd for controlling the basic plotting parameters. Themes can be used to create a consistent look to different plots.

The above command creates a poor graphic, as the automatic creation of shingles for the Price variable creates too many shingle levels. Instead, we use the derived price variable as before:

```
Cars93 = transform(Cars93, price=cut(Price, c(0, 15, 30, 75),
                labels=c("cheap", "affordable", "expensive")))
xyplot(MPG.highway ~ Weight | price, data=Cars93)
```

This produces a nicer graphic. An alternative to using cut is to use the shingle function or the equal.count function. The latter, might be used to derive a variable such as

```
prices <- equal.count(Cars93$Price, number=3, overlap=0)
```

where prices would have three groups of equal size.

Before finishing, we wish to modify two things: the layout to use 3 columns (not 2 rows and 2 columns) and add regression lines for each. The argument layout=c(3,1) does the first task (columns, rows is the order).

The second task can be done using the default panel function. A panel function is what is called to draw each panel. A specification such as the following plots the points (panel.xyplot) and the regression line (panel.lmline):

```
panel=function(x, y) {
  panel.xyplot(x, y)
  panel.lmline(x, y)
}
```

As this is a common task, there is a more convenient way. The type argument takes arguments describing what to plot, similar to the type argument of plot. In addition to "p" for points, there is "r" for a regression line, and "smooth" to add a loess smoother. The following then produces the graphic shown in Figure 5.8:

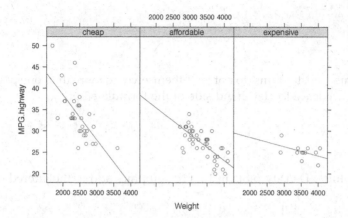

Figure 5.8: Three scatterplots of highway model by vehicle weight. The data is split into groups by a derived `price` variable. The `lattice` package uses panels to display the data by group. The use of consistent *x*- and *y*-limits affords easy comparison between the groups.

```
xyplot(MPG.highway ~ Weight | price, data=Cars93,
       layout=c(3,1),
       type=c("p", "r")
)
```

Examples of the other common plots (figures not shown) using the babies data set follow.

Dot charts Dot charts are produced by `dotplot`. If the response variable is a factor then a panel will compare responses within that factor. For example, here we look at the weight variable for babies whose dads had a certain education level:

```
dotplot(factor(smoke) ~ wt, data=babies, subset=wt < 999 & ded==3)
```

Boxplots Boxplots, like dot charts lend themselves to comparing values within a panel. Again, a factor for the response variable will place multiple boxplots within a panel, one for each level of the factor:

```
bwplot(factor(smoke) ~ wt, data=babies, subset=wt < 999)
```

Adding a conditioning variable allows three variables to be considered in one graphic. This graphic shows no obvious influence from dad's education level on the relationship between mother's smoking and birth weight.

```
dotplot(factor(smoke) ~ wt | factor(ded),
        data=babies, subset=wt < 999)
```

Histograms Histograms do not lend themselves to more than one per panel. For these, we leave the left-hand side of the formula empty:

```
histogram(~ wt | factor(smoke), data=babies, subset=wt < 999)
```

Density plots Density plots in lattice graphics add in the jittered data:

```
densityplot( ~ wt | factor(smoke), data=babies, subset=wt < 999)
```

Problems

5.5 For the michelson (MASS) data set create a dot plot of Speed by Expt. Which experiment shows more spread?

5.6 For the batting (UsingR) data set, make parallel boxplots of the batting average (H/AB) for each team. Which team had the greatest median average?

5.3   The ggplot2 **package**

The ggplot2 add-on package provides another system for plotting within R that has proven very popular. The package is discussed at length by its designer [61] and illustrated with numerous examples in [9]. Additionally, the web site http://ggplot2.org/ hosts the documentation in an easy to access format. Like lattice, ggplot2 utilizes the underlying grid graphics engine of R. However, its interface is derived from Wilkinson's grammar of graphics. This grammar is comprised of independent components which are combined together to produce a graphic. Though the permutations are endless, we will see in this section that creating the basic charts we discuss is fairly straightforward.

There are many benefits to making graphics with ggplot2, but perhaps the biggest is the visual appeal of the graphics produced. It is clear that considerable attention was given to all aspects of the rendering. As well, the use of a declarative style gives a level of abstraction to graphic production, that once learned simplifies the conceptual thinking.

Though there is a function qplot that wraps many common patterns within a single function, we focus on building graphics in layers.

Aesthetics We begin by defining a ggplot object for a data set:

```
p <- ggplot(Cars93)
```

At this point just an object is created, nothing is drawn to the screen, indeed there isn't anything to draw. To construct a graphic, we discuss two of the component blocks: aesthetics and geometries.

Aesthetics map variables in a data set into properties that can be perceived on a graph. For example, size, shape, and color are aesthetics. Additionally, values for x and y are aesthetics. Aesthetics are declared through the aes function. We will make a scatterplot of highway mileage by weight, with colors derived from the Origin variable. For this task we specify the following:

```
p <- p + aes(x=Weight, y = MPG.highway, color=Origin)
```

A few comments are in order. The + symbol is overloaded to *add* to a ggplot object. In the above, we add to p, then assign this new object to the same variable name. Within the aes function, the variable names are not quoted. This function creates unevaluated expressions that are evaluated when rendered. Transformations of the variables are permitted, e.g.,

```
aes(x=sqrt(Weight/1000), y=MPG.highway, color=Origin)

## List of 3
##  $ x      : language sqrt(Weight/1000)
##  $ y      : symbol MPG.highway
##  $ colour: symbol Origin
```

Aesthetics can also be set per layer when defining the layer.

Aesthetics are tightly bound to the data set, though they can be adjusted. For that reason, we use an alternative syntax in the following that allows the definition of the core aesthetic properties to be assigned within the ggplot call.

Geoms

The term *geoms* is short for geometric objects and refers to the functions that do the actual rendering of the data. In short these declare what should be drawn in the figure. The familiar tasks of placing points and lines on a graphic are requested by the function geom_line and geom_point.

To illustrate, we make a scatter plot of highway mileage versus weight:

```
p <- ggplot(Cars93, aes(x=sqrt(Weight/1000), y=MPG.highway,
                        color=Origin)) + geom_point(cex=3)
p                                          # show the plot
```

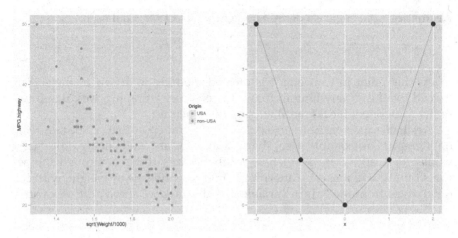

Figure 5.9: Two basic graphics produced by ggplot2. The left one uses geom_point to draw a scatterplot, the right one combines two geoms to layer points and lines.

The commands build up the plot, but the plot is produced only when the ggplot object is shown (the left graphic of Figure 5.9). This can happen at the command line by typing p, as above, or explicitly through print(p), which may be necessary inside of a function call.

A plot of $f(x) = x^2$ demonstrates the use of geom_line:

```
f <- function(x) x^2
x <- seq(-2, 2, length=100)
p <- ggplot(data.frame(x=x, y=f(x)), aes(x=x, y=y))
p + geom_line()                          # not shown
```

R's convenience function curve makes this task much easier (e.g., curve(f, -2, 2)), but the above approach is much more flexible. For geom_line it is important to note that the order the lines are connected depends on the order of the data in the pairs (x[1], y[1]), (x[2], y[2]), ..., as the rendering does nothing more than connect the dots. (As does the lines function, used with base graphics.)

Two geoms may be used at once. The following makes a simplified graph of $f(x) = x^2$ emphasizing the points used (the right graphic of Figure 5.9):

```
f <- function(x) x^2
x <- -2:2; y <- f(x)
p <- ggplot(data.frame(x=x, y=y), aes(x=x, y=y)) +
    geom_line(color="black", alpha=0.25) +      # set color attribute
    geom_point(pch=16, cex=5)                    # set plot character and size
p
```

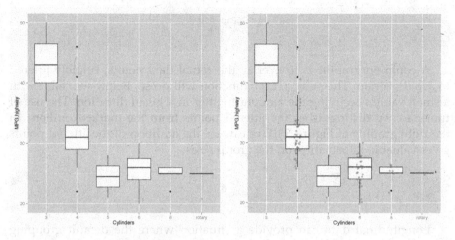

Figure 5.10: Boxplots rendered by ggplot2. The overlay of jittered points in the right graphic gives useful information as to group size that is otherwise hidden in the left graphic.

We also illustrated how aesthetic properties can be set per geom (as opposed to being mapped from the data with aes). For geom_line the color is specified by name with a transparency level specified by an alpha value between 0 and 1. Alternatively, this could have been specified through rgb(0, 0, 0, alpha=0.25), as has been done with base graphics. Though not so important in this example, setting a transparency level offers some help when over-plotting occurs due to too much data in one location.

For geom_point, we illustrated how the plot character can be set, similar to base graphics. The more commonly seen style is to use the names shape and size for these aesthetics.

Grouping

The geom_point geom maps each data point to a graphical representation. This isn't the case for graphics that reduce the data displayed, such as a histogram or boxplot. For these rendered objects correspond to *groups* of the data.

For example, to make a parallel boxplot of highway mileage for different cylinder types, the data must be grouped by the cylinder. We saw in Chapter 3 that we can explicitly do this by splitting, but we should expect a more convenient approach and ggplot2 delivers. By default, the grouping is done by the interaction of all discrete variables in the plot. The group aesthetic can specified for more complicated grouping needs. For a boxplot, the defaults mean we can specify the x aesthetic with a factor and the y aesthetic with a numeric variable. With this, we then use geom_boxplot to render each group:

```
p <- ggplot(Cars93, aes(x=Cylinders, y=MPG.highway))
p + geom_boxplot()
```

A common graphic is to overlay the actual data points, suitably jittered to avoid overlap. The geom_jitter function will do so. Below we call it with a small value specifying the amount to jitter in a given direction. The use of transparency distinguishes the jittered points from the marked outliers. In the right graphic of Figure 5.10 we can see the addition of the jittered points gives valuable information on the group sizes.

```
p + geom_boxplot() +
  geom_jitter(position=position_jitter(w = 0.1), alpha=.25)
```

Longitudinal data can provide a situation where the default grouping might not be sufficient. The morley data set recorded 100 different measurements of the speed of light. There were 20 runs for each of the 5 experiments. We can track the recordings over each run by plotting a line graph with Run on the x axis and Speed on the y axis. We can then layer this graph by different groups of the Expt variable. This task can be done by other means, but becomes very direct with ggplot2: Here we reduce the number of experiments first for clarity:

```
m <- subset(morley, Expt %in% 1:2)      # just first two experiments
p <- ggplot(m, aes(x=Run, y=Speed, group=Expt, color=Expt))
p + geom_line()                         # connect x and y with lines
```

From the graphic (Figure 5.11) we can see more variability in the measurement experiment 1 and a downward trend in the measurements of experiment 2.

Statistical transformations

Statistics summarize the data, usually reducing the dimension. For example, the mean summarizes a data set with n values with a single number. Whereas a histogram summarizes a data set with k numbers, the k being the counts within k different bins that span the range of the data.

stat_bin In ggplot2's grammer, statistical transformations (stats) summarize the data before it is rendered. The histogram is a good example. For histograms, ggplot2 provides the stat_bin function for tallying the number of values in each bin. The number of bins can be specified in many ways, the most direct is to specify a value for binwidth, which defaults to the length of the range divided by 30. A simple histogram can then be made by specifying an x aesthetic as follows:

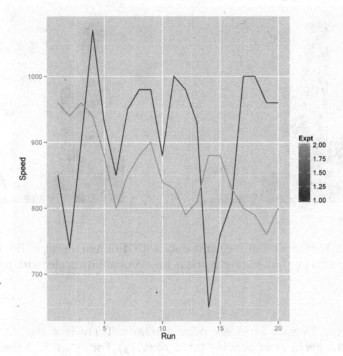

Figure 5.11: Speed by run for the first two experiments in the morley data set. To construct this graphic, the grouping was set by the Expt variable.

```
p <- ggplot(Cars93, aes(x = MPG.highway))
p + stat_bin(binwidth=5)
```

This produces the left graphic of Figure 5.12. The *y* label is count representing the number of values in each bin. To replace this scale with a proportion is possible. The y aesthetic needs to be mapped to the appropriate value. This value is not in the data, as it is derived from the statistic. Indeed, when stat_bin computes its work, it derives variables ..count.. and ..density.. which may be specified as aesthetics, as in:

```
p <- ggplot(Cars93, aes(x = MPG.highway, y=..density..)) +
  stat_bin(binwidth=5)
p
```

The explicit use of stat_bin is not necessary. The easier-to-remember geom_histogram function combines stat_bin and geom_bar (for drawing bars, such as bar charts) to draw histograms.

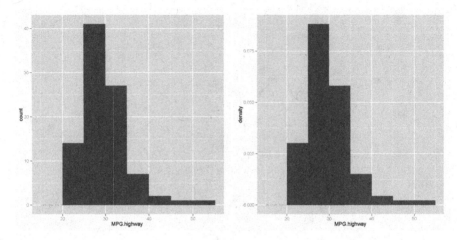

Figure 5.12: Histograms of the same data with different y scales. The left one displays counts per bin, the right one is on a probability scale, with total area adding to 1.

stat_density   Density plots are done similarly. There is a statistic that is called by the geom geom_density (stat_density). For example, a density estimate can be added to a histogram with this pattern:

```
p <- ggplot(Cars93, aes(x=MPG.highway, y=..density..)) # scale y
p + geom_histogram(alpha=0.5) + geom_density()
```


stat_smooth   A trend line is a statistical summary of a bivariate relationship. As discussed in Chapter 3, there are many different trend lines that can be added to scatterplot, including a least-squares regression line. The stat_smooth function can compute this, along with many other trend lines.

The basic usage just requires the addition of the geom_smooth, which uses stat_smooth as its default statistic:

```
p <- ggplot(Cars93, aes(x=Weight, y=MPG.highway)) + geom_point()
p + geom_smooth()
```

The leftmost graphic in Figure 5.13 shows the resulting graphic. The trend line is drawn along with an estimate on the standard error. As this concept is discussed later (Chapter 11), we suppress its rendering in the following by specifying se=FALSE.

The method argument determines which trend lines to use. Options are "lm", "glm", "gam", "loess", and "rlm". The default depends on the size of the data, with "loess" being used for smaller values of n. In the following we fit a regression line instead (middle graphic of Figure 5.13).

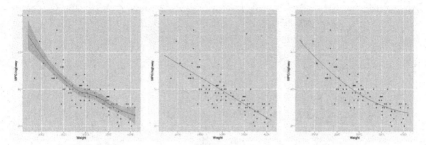

Figure 5.13: For the same scatterplot, three different trend lines are added. The first the default computed by loess and showing an estimated confidence band, the second a linear regression line, and the third a quadratic polynomial fit.

```
p + geom_smooth(method="lm", se=FALSE)
```

The formula argument allows for different specification of the formula used to compute the trend line. This is specified in terms of the aesthetic variables, x and y. The default is simply y ~ x. The trend in the data of Figure 5.13 looks less linear, and more like a polynomial. The specification poly(x, 2) will force a second-degree polynomial to be fit. Other transformations of the x data are possible, but recall the typical mathematical operations are overloaded, so they must be wrapped with I.

The following command produces the rightmost graphic in Figure 5.13. The resulting curve appears to do a better job of tracking the relationship implied by the data than the straight line of the middle graphic.

```
p + geom_smooth(method="lm", formula=y ~ poly(x,2), se=FALSE)
```

Faceting

The use of different panels in lattice makes comparisons across groups much easier. This feature is introduced into ggplot2 through *facets*. There are two faceting functions: facet_grid and facet_wrap. The former, like a matrix, has rows and columns representing something; the latter is more like a vector with new facets wrapping around to utilize horizontal and vertical space.

The faceting specification is done using R's model formula syntax. To get faceting by a single variable, say f, the formula . ~ f or f ~ . can be used. The difference being the left-hand specification facets the rows, the right-hand specification the columns, so the latter will display the data grouped by f in rows. For a two-dimensional grid, the specification is f ~ g.

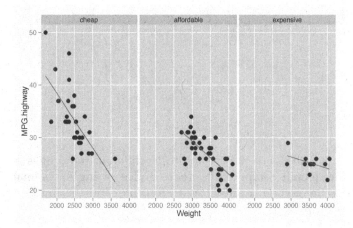

Figure 5.14: Plots of highway mileage by weight for ranges of car prices using ggplot2.

The following command revisits the bend found in Galton's data by [58]. The loess line for a fit of child's height by parent's height suggested two distinct regression lines, likely due to differences in gender. The PearsonLee (HistData) works with disaggregated data, whereas Galton worked with aggregated data. Using PearsonLee, we can facet by the gender of the parent and the child to see if the relationships differ.

```
p <- ggplot(PearsonLee, aes(y=child,x=parent))
p + geom_point(alpha=0.5) + geom_smooth(method="loess") +
   facet_grid(par ~ chl)
```

The left graphic in Figure 5.15 shows a bend for the Mother level.

• **Example 5.1: Redoing Figure 5.3**
We redo the example of plotting highway mileage modeled by car weight based on the price range of the car using ggplot2 commands.

```
Cars93 <- transform(Cars93, price=cut(Price, c(0, 15, 30, 75),
                labels=c("cheap", "affordable", "expensive")))
p <- ggplot(Cars93) + aes(x=Weight, y = MPG.highway) +
   geom_point(cex=3) + geom_smooth(method="lm", se=FALSE)  +
   facet_grid( ~ price)
p
```

The result can be seen in Figure 5.14.   ••

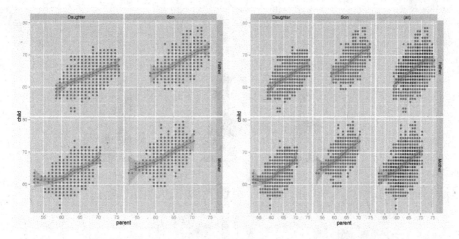

Figure 5.15: Samples of `facet_grid`. The right-hand graphic adds marginal information through `margins=TRUE`.

Margins A grid display is like a contingency table, and, as with contingency tables, the margins may be of secondary interest. Adding margins is done by specifying a value to the `margins` argument of `facet_grid`. In the following, we want to take margins over the children. This can be specified by the variable name for which the faceting is achieved through:

```
p <- ggplot(PearsonLee, aes(y=child,x=parent))
p + geom_point(alpha=0.5) +
  geom_smooth(method="loess") +
  facet_grid(par ~ chl, margins="chl")  # or margins=TRUE for both
```

The right graphic of Figure 5.15 shows the result. The measurements based on fathers show a slight bend for taller fathers, the measurements based on mothers show a kink in the middle, suggestive of two different regimes.

Wrap The `facet_wrap` function is useful to organize the layout when faceting by a factor (or interaction of factors). The use is similar to how `facet_grid` is employed with one-dimensional faceting, where a one-sided formula is specified. There are arguments to adjust how the values are laid out, but we do not illustrate. The following code will produce five histograms—one for each experiment—arranged in a grid (Figure 5.16):

```
p <- ggplot(morley, aes(x=Speed)) + geom_histogram(binwidth=50)
p + facet_wrap( ~ Expt)
```

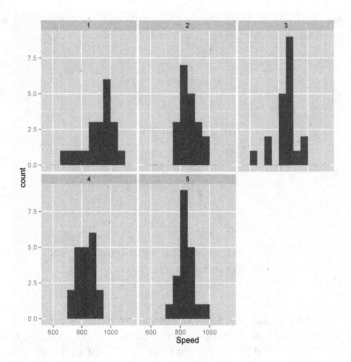

Figure 5.16: Scatterplots of speed versus run for the morley data. The facet_wrap function is applied to wrap the one-dimensional display into two dimensions.

As with lattice graphics, the panels have the same scales. This makes cross-panel comparisons easier. However, there may be occasions where this is not desirable. In that case, the facet_wrap has the scales option where we can specify "free", "free_x", or "free_y" to allow those scales to depend just on the panel data.

Problems

5.7 For the mtcars data set, produce graphics of the following using ggplot2:

1. Boxplots of miles per gallon (mpg) for groups defined by the number of gears (gear).

2. A scatterplot of mpg modeled by weight (wt) with a trend line added.

3. A scatterplot of mpg modeled by horsepower (hp). Create facets by the number of cylinders (cyl) and gear.

Populations

Statistical inference is the process of forming judgments about a population based on a sample from the population. In this chapter we describe populations and samples from a population (our data sets) using the language of probability.

6.1 Populations

To make statistical inferences based on data we use a probability model for the data. Consider a univariate data set consisting of measurements of some variable. A single data point is just one of a possible range of values. We call the *population* of the variable a description of the range of possible values. We use the term *random variable* to be a random number drawn from a population. A data point then will be a realization of some random variable. We make a distinction between whether or not we have observed or realized a random variable. Once observed, the value of the random variable is known. Prior to being observed, it is full of potential—it can be any value in the population it comes from. For most cases, not all values or ranges of values of a population are equally likely, so to fully describe a random variable prior to observing it, we need to indicate the probability that the random variable is some value or in a range of values. We refer to a description of the range and the probabilities as the *distribution of a random variable*, though we may call this a population as well.

By probability we mean some number between 0 and 1 that describes the likelihood of our random variable having some value. Our intuition for probabilities may come from a physical understanding of how the numbers are generated. For example, when tossing a fair coin we would think that the probability of heads would be one-half. Similarly, when a die is rolled the probability of rolling a four would be one-sixth. These are both examples in which all outcomes are equally likely and finite in number. Let an event be a set of outcomes.[1] For a situation where all outcomes are equally likely and the total number of outcomes is finite, the definition of probability is:

$$P(E) = \frac{\text{\# events in } E}{\text{\# events in total}}. \quad \text{(equally likely outcomes)}$$

---

[1]For non-finite collections of outcomes, there are technical details needed.

For this defintion, the following rules will be satisfied:

- $P(E) \geq 0$ for all events.

- The event of all possible outcomes has probability 1.

- If A and B are two disjoint events, then $P(A \text{ or } B) = P(A) + P(B)$.

A mathematical consequence of the above is that for any two events $P(A \text{ or } B) = P(A) + P(B) - P(A \text{ and } B)$.

• Example 6.1: A randomly selected person
The following contingency table shows the relationship between gender and smoking status for a cohort in the data set survey (MASS):

```
tbl <- xtabs(~ Sex + Smoke, data=survey)
tbl

##          Smoke
## Sex       Heavy Never Occas Regul
##    Female     5    99     9     5
##    Male       6    89    10    12
```

There are 237 participants in the survey, but only 235 represented above (sum(tbl)). If we select *one at random*—that is each participant is equally likely to be selected—then the probability we select a female would be the number of females divided by the total number of people represented:

```
margin.table(tbl, margin=1)

## Sex
## Female    Male
##    118     117

118 / (118 + 117)                          # sum(tbl[1,]) / sum(tbl)

## [1] 0.5021
```

Whereas the probability of selecting a heavy smoker would be $(5 + 6)/235 = 0.046$.

To illustrate the last rule above, the probability of selecting a female *or* a heavy smoker would be $((5 + 99 + 9 + 5) + 6)/235$. We could rewrite this in the form $P(\text{female}) + P(\text{heavy smoker}) - P(\text{female } and \text{ heavy smoker})$ as $(5 + 99 + 9 + 5)/235 + (5 + 6)/235 - 5/235$.   ••

For situations where our intuition comes about by performing the same action over and over again, our idea of the probability of some event comes from a proportion of times that event occurs. For example, the batting average of a baseball player is a running proportion of a player's success at bat. Over the course of a season, we expect this number to get closer to the probability that an official at bat will be a success. This is an example in which a *long-term frequency* is used to give a probability.

For other populations, the probabilities are simply assigned or postulated, and our model is accurate as far as it matches the reality of the data collected. An example of this may be our assumption that the distribution of heights of adult males follows a bell-shaped distribution. This may not be precisely true, but is a very good approximation for most uses.

For all these cases, the rules above[2] serve to define a probability.

We indicate probabilities using a $P()$ and random variables with letters such as X.[3] For example, $P(X \leq 5)$ would mean the probability, the random variable X, is less than or equal to 5.

Discrete random variables

We have seen that numeric data can be discrete or continuous. Our language of probability will also come in these two flavors.

Let X be a discrete random variable. That is, a random variable whose possible outcomes are some discrete set such as {yes, no} or {0,1,2,...}. The *range* of X is the set of all k where $P(X = k) > 0$. The distribution of X is a specification of these probabilities. The rules of probability imply that distributions are not arbitrary, as for each k in the range, $P(X = k) > 0$ and $P(X = k) \leq 1$. Furthermore, as X has some value, we have $\sum_k P(X = k) = 1$.

Here are a few examples for which the distribution can be calculated.

- **Example 6.2: Number of heads in two coin tosses**

If a coin is tossed two times we can keep track of the outcome as a pair. For example, (H, T) could denote "heads" then "tails." The set of all possible outcomes is $\{(H,H), (H,T), (T,H), (T,T)\}$. If X is the number of heads, then X is either 0, 1, or 2. Intuitively, we know that for a fair coin all the outcomes have the same probability, so we have $P(X = 0) = 1/4$, $P(X = 1) = 1/2$, and $P(X = 2) = 1/4$. ••

- **Example 6.3: Picking balls from a bag**

Imagine a bag with N balls, of which R are red and $N - R$ are green. We pick a ball, note its color, replace the ball, and repeat. Let X be the number of red balls chosen. As in the previous example, X is 0, 1, or 2. The probability that

---

[2]With a generalization of the last rule to the probability of countably many *disjoint events* is the sum of the individual probabilities.

[3]When referring to a single random variable, to emphasize the distinction, we use a capital letter for the random variable and a lowercase letter for the possible outcome. In later chapters, we utilize lower case letters for each.

$X = 2$ is intuitively $(R/N) \cdot (R/N)$ as R/N is the probability of picking a red ball on any one pick. The probability that $X = 0$ is $((N - R)/N)^2$ by the same reasoning, and as all probabilities add to 1, $P(X = 1) = 2(R/N)((N - R)/N)$. This specifies the distribution of X.

The binomial distribution describes in general the result of selecting n balls in this manner, not simply two.   ••

The intuition that leads us to multiply two probabilities together is due to the two events being independent. Two events are *independent* if knowledge that one occurs doesn't change the probability of the other occurring.[4] Two events are disjoint if they can't both occur for a given outcome, as they share no outcome in common. Probabilities add with disjoint events.

• Example 6.4: Specifying a distribution

We can specify the distribution of a discrete random variable by first specifying the range of values and then assigning to each k a number $p_k = P(X = k)$ such that $\sum p_k = 1$ and $p_k \geq 0$. To visualize a physical model where this can be realized, imagine making a pie chart with areas proportional to p_k, placing a spinner in the middle, and spinning. The ending position determines the value of the random variable.   ••

Figure 6.1 shows a *spike plot* of a distribution and a spinner model to realize values of X. (The chance a well-spun spinner will end in a numbered region is proportional to the number's area.) A spike plot shows the probabilities for each value in the range of X as spikes, emphasizing the discreteness of the distribution. The spike plot is made with the following commands:

```
k <- 0:4
p <- c(1, 2, 3, 2, 1); p <- p/sum(p)
plot(k, p, type="h", xlab="k", ylab="probability", ylim=c(0,max(p)))
points(k, p, pch=16, cex=2)        # add the balls to top of spike
```

The argument type="h" plots the vertical lines of the spike plot.

Using sample to generate random values

R does not have a "spinner" function, but it does have the sample function to generate observations of a discrete random variable with a specific distribution. If the data vector k contains the values we are sampling from, and p contains the probabilities of each value being selected, then the command sample(ks, size=1, prob=p) will select one of the values of ks with probabilities specified by p.

---

[4]Technically, two events are independent if $P(A \text{ and } B) = P(A) \cdot P(B)$.

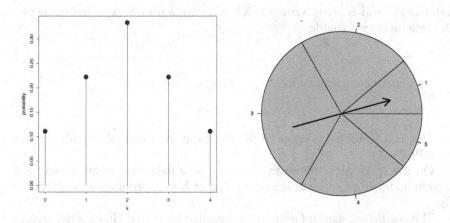

Figure 6.1: Spike plot of distribution of X and a spinner model to realize values of X with the specified probabilities.

For example, the number of heads for two coin tosses can be simulated as follows:

```
k <- 0:2
p <- c(1, 2, 1); p <- p/sum(p)                # add to 1
sample(k, size=1, prob=p)

## [1] 2

sample(k, size=1, prob=p)                      # likely different

## [1] 1
```

The default probabilities for prob make each value of k equally likely. We can use this to simulate rolling a pair of dice and adding their values:

```
sample(1:6, size=1) + sample(1:6, size=1)

## [1] 4
```

The mean and standard deviation

For a data set, the mean and standard deviation are summaries of center and spread. For random variables these concepts carry over, though the definitions are different.

The *population mean* is denoted by μ (the Greek letter *mu*). If X is a random variable with this population, then the mean is also called the "the expected value of X" and is often written $E(X)$. A formula for the expected value of a discrete random variable is

$$\text{mean of random variable} \quad \mu = E(X) = \sum k P(X = k).$$

This is a weighted average of the values in the range of X with weights $p_k = P(X = k)$.

On the spike plot, the mean is viewed as a balancing point if there is a weight assigned to each value in the range of X proportional to the probability.

The *population standard deviation* is denoted by σ (the Greek letter *sigma*). The standard deviation is the square root of the variance. If X is a discrete random variable, then its variance is defined by $\sigma^2 = \text{VAR}(X) = E((X - \mu)^2)$. This is the expected value of the random variable $(X - \mu)^2$. That is, the population variance measures spread in terms of the expected squared distance from the mean.

• **Example 6.5: Picking balls from a bag**

Again consider the experiment with a bag with N balls, of which R are red and $N - R$ are green. Let X be the number of red balls chosen when we pick two balls with replacement. Then

$$P(X = 0) = \frac{(N - R)^2}{N^2}, P(X = 1) = 2\frac{R(N - R)}{N^2}, P(X = 2) = \frac{R^2}{N^2}.$$

More succinctly, if we set $p = R/N, q = 1 - p$, then these values are q^2, $2pq$, and p^2. The mean then becomes:

$$E(X) = 0 \cdot q^2 + 1 \cdot 2pq + 2 \cdot p^2 = 2p(1 - p) + 2p^2 = 2p,$$

and

$$\text{VAR}(X) = E((X - 2p)^2) = (0 - 2p)^2 \cdot q^2 + (1 - 2p)^2 \cdot 2pq + (2 - 2p)^2 \cdot p^2 = 2pq.$$

That these formulas for more than one selection can be written so simply in terms of p and q—values which summarize a single selection—is no coincidence.   ••

Continuous random variables

Continuous data is modeled by continuous random variables. Due to the continuum of possible values, a new means of defining probabilities must

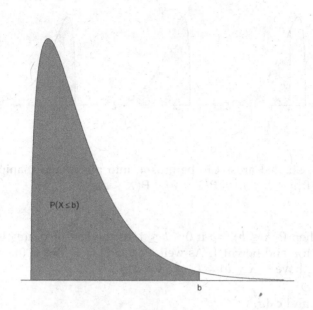

P(X≤b)

Figure 6.2: $P(X \leq b)$ is defined by the area to left of b under the density of X.

be used. Rather than try the impossible task of specifying $P(X = k)$ for all possible k, instead the probability of ranges, $P(a < X \leq b)$ are specified. This can be done through a function, $F(b) = P(X \leq b)$ or through a related function $f(x)$ which has $P(a < X \leq b)$ equalling the area under the graph of $f(x)$ between a and b.

For a given random variable X, the function $f(x)$ is referred to as the *density* of X. The relation between $f(x)$ and $F(b)$ above is defined through calculus. However, we shall see that in most cases $f(x)$ helps us visualize the value and we shall use a suitable R function for computing $F(b)$.

Figure 6.2 shows an illustration of some $f(x)$ and the computation of $F(b)$ for that f. The shaded area represents $F(b)$. The rules of probability put some restriction on potential densities: they must be non-negative and have total area equal to 1.

Areas can also be broken up into pieces, as Figure 6.3 illustrates, showing $P(a < X \leq b) = P(X \leq b) - P(X \leq a)$. Rearranging, this says $P(X \leq a) + P(a < X \leq b) = P(X \leq b)$, a special case of the rule of probability for disjoint sets. A similar reasoning yields the useful complement rule: for any b, $P(X \leq b) = 1 - P(X > b)$.

For example, the uniform distribution on $[0,1]$ has density $f(x) = 1$ on the interval $[0,1]$ and is 0 otherwise. Let X be a random variable with this

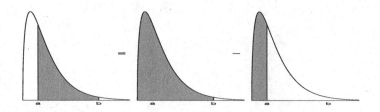

Figure 6.3: Shaded areas can be broken into pieces and manipulated. This illustrates $P(a < X \leq b) = P(X \leq b) - P(X \leq a)$.

density. Then $P(X \leq b) = b$ if $0 \leq b \leq 1$, as the specified area is a rectangle with length b and height 1. As well, $P(X > b) = 1 - b$ for the same reason. Clearly, we have $P(X \leq b) = 1 - P(X > b)$.

The p.d.f. and c.d.f.

For a discrete random variable it is common to define a function $f(k)$ by $f(k) = P(X = k)$. Similarly, for a continuous random variable X, it is common to denote the density of X by $f(x)$. Both usages are called p.d.f.'s. For the discrete case, p.d.f. stands for probability *distribution* function, and for the continuous case, probability *density* function. The cumulative distribution function, c.d.f., is $F(b) = P(X \leq b)$. In the discrete case this is given by $\sum_{k \leq b} P(X = k)$, and in the continuous case it is the area to the left of b under the density $f(x)$.

The mean and standard deviation

The concepts of the mean and standard deviation also apply for continuous random variables, although their definitions require calculus. The intuitive notion for the mean of X is that it is the balancing point for the density of X. The notation μ or $E(X)$ is used for the mean, and σ or $SD(X)$ is used for the standard deviation.

If X has a uniform distribution on $[0,1]$, then the mean is $1/2$. This is clearly the balancing point of the graph of the density, which is constant on the interval. The variance can be calculated to be $1/12$, so σ is about .289.

Quantiles

The quantiles of a data set roughly split the data by proportions. Let X be a continuous random variable with positive density. Referring to Figure 6.2, we see that for any given area between 0 and 1, there is a b for which the area to the right of b under f is the desired amount. That is, for each p in $[0,1]$ there is a b such that $P(X \leq b) = p$.[5] This defines the p-quantile or $100 \cdot p$ percentile of X. The quantile function is inverse to the c.d.f., as it returns the x value for a given area, whereas the c.d.f. returns the area for a given x value.

Sampling from a population

Our probability model for a data point is that it is an observation of a random variable whose distribution is described by the parent population. To perform statistical inference about a parent population, we desire a *sample* from the population. That is, a sequence of random variables $X_1, X_2, \ldots, X_n$. A sequence is *identically distributed* if each random variable has the same distribution. A sequence is *independent* if knowing the value of some of the random variables does not give additional information about the distribution of the others. A sequence that is both independent and identically distributed (i.i.d.) is called a *random sample*.

Toss a coin n times. If we let X_i be 1 for a heads on the ith coin toss and 0 otherwise, then clearly $X_1, X_2, \ldots, X_n$ is an *i.i.d.* sequence. For the spinner analogy of generating discrete random variables, the different numbers will be *i.i.d.* if the spinner is spun so hard each time that it forgets where it started and is equally likely to stop at any angle.

If we get our random numbers by randomly selecting from a finite population, then the values will be independent if the sampling is done with replacement. This might seem counterintuitive, as there is a chance a member is selected more than once, so the values seem dependent. However, the distribution of a future observation is not changed by knowing a previous observation. Whereas, when sampling without replacement from n items, the random variables $X_1, X_2, \ldots, X_n$ will have the same distribution but will be dependent: e.g., when there are n things to choose, if you know the first $n - 1$ values, then X_n must be the value that has not been chosen.

Random samples generated by sample

The sample function will take samples of size n from a discrete distribution by specifying size=n. The sample will be taken with replacement if we specify replace=TRUE. As just mentioned, this is important if we want to produce an *i.i.d.* sample. The default is to sample without replacement.

```
sample(0:1,size=10,replace=TRUE) # toss a coin 10 times.

## [1] 1 0 1 1 1 0 1 0 0 1

sample(1:6,size=10,replace=TRUE) # roll a die 10 times
```

---

[5]In general, a distribution can be specified by $F(b)$ alone, and $F(b)$ need not be continuous, as it is when there is a density. In that case, the quantile is defined by the smallest b with $P(X \leq b) \geq p$.

```
## [1] 3 3 6 4 6 5 5 6 4 3

## sum of roll of a pair of dice roll 10 times
sample(1:6,size=10,replace=TRUE) + sample(1:6,size=10,replace=TRUE)

## [1]  7 11  6 10  7  5 10  4 10  6
```

• **Example 6.6: Public-opinion polls as random samples**
The goal of a public-opinion poll is to find the proportion of a target population that shares a given attitude. This is achieved by selecting a sample from the target population and finding the sample proportion who have the given attitude. A public-opinion poll can be thought of as a random sample from a target population if each person polled is randomly chosen from the entire population with replacement. Assume we know that the target population of 10,000 people has 6,800 that would answer "yes" to our survey question. Then a sample of size 10 could be generated by

```
sample(rep(0:1, c(3200,6800)), size=10, replace=TRUE)

## [1] 1 0 1 1 1 1 1 1 1 1
```

(The rep function produces 10,000 values: 3,200 0's and 6,800 1's.)

The target population is different from the "population," or distribution, of the random variables. For the responses, the possible values are coded with a 0 or 1 with respective probabilities $1 - p$ and p. Using this distribution, a random sample can also be produced by specifying the probabilities using prob:

```
p <- 0.68
sample(0:1, size=10, replace=TRUE, prob=c(1-p, p))

## [1] 0 1 1 1 0 0 1 1 0 0
```

••

Sampling distributions

A *statistic* is a numeric value summarizing a random sample. Examples are the sample mean ($\bar{X} = (X_1 + X_2 + \cdots + X_n)/n$) and the sample median. When a statistic depends on a random sample, it, too, is a random variable. As this is our typical usage, to emphasize this, we use a capital $\bar{X}$. The distribution of a statistic is known as its *sampling distribution*.

The sampling distribution of a statistic can be quite complicated. However, for many common statistics, properties of the sampling distribution are known and are related to the corresponding population values.

For example, the sample mean of a random sample has

$$E(\bar{X}) = \mu_{\bar{X}} = \mu \quad \text{and} \quad SD(\bar{X}) = \sigma_{\bar{X}} = \frac{\sigma}{\sqrt{n}}.$$

That is, the mean of $\bar{X}$ is the same as the mean of the parent population, and the standard deviation of $\bar{X}$ is related to the standard deviation of the parent population, but it differs as it is smaller by the factor of $1/\sqrt{n}$. These facts make it possible to use $\bar{X}$ to make inferences about the population mean.

Problems

6.1 Toss two coins. Let X be the resulting number of heads and Y be the number of tails. Find the distribution of each.

6.2 Roll a pair of dice. Let X be the largest value shown on the two dice. Use `sample` to simulate five values of X.

6.3 The National Basketball Association lottery to award the first pick in the draft is held by putting 1,000 balls into a hopper and selecting one. The teams with the worst records the previous year have a greater proportion of the balls. The data set nba.draft contains the ball allocation for the year 2002. Use `sample` with `Team` as the data vector and `prob=Balls` to simulate the draft. What team do you select? Repeat until Golden State is chosen. How long did it take?

6.4 Let $f(x) = x$ for $0 \le x \le \sqrt{2}$ be the p.d.f. of a triangular random variable X. Using geometry, find $P(X \le b)$ for $0 \le b \le \sqrt{2}$.

6.5 Let X have the uniform distribution on $[0,1]$. That is, it has density $f(x) = 1$ for $0 \le x \le 1$, and is otherwise 0. For $0 \le p \le 1$ find the quantile function that returns b, where $P(X \le b) = p$.

6.6 Repeat the previous problem for the triangular distribution with density $f(x) = x$ for $0 \le x \le \sqrt{2}$.

6.7 Toss two coins. Let X be the number of heads and Y the number of tails. Are X and Y independent?

6.2 Families of distributions

In statistics there are a number of distributions that come in families. Each
family is described by a function that has a number of parameters characterizing the distribution. For example, the uniform distribution is a continuous
distribution on the interval $[a,b]$ that assigns equal probability to equal-sized
areas in the interval. The parameters are a and b, the endpoints of the intervals.

The density and population summaries of a family of distributions are
often represented in terms of these parameters. For the uniform, if $f(x)$ is the
function which is 1 on the interval $[0,1]$ and 0 otherwise, then the density of
the general uniform is $(1/(b-a)) \cdot f((x-a)/(b-a))$; the mean is $(b+a)/2$
and the variance is $(b-a)^2/12$.

The d, p, q, and r functions

R has four types of functions for getting information about a family of distributions:

- The "d" functions return the p.d.f. of the distribution.

- The "p" functions return the c.d.f. of the distribution.

- The "q" functions return the quantiles.

- The "r" functions return random samples from a distribution.

These functions are all used similarly. Each family has a name and some
parameters. The function name is found by combining either d, p, q, or r with
the name for the family. The parameter names vary from family to family but
are consistent within a family.

For example, the uniform distribution on $[a,b]$ has two parameters. The
family name is unif. In R the parameters are named min and max.

```
dunif(x=1, min=0, max=3)            # 1/3 of area is to left of 1

## [1] 0.3333

punif(q=2, min=0, max=3)            # 1/(b-a) is 2/3

## [1] 0.6667

qunif(p=1/2, min=0, max=3)          # half way between 0 and 3

## [1] 1.5

runif(n=1, min=0, max=3)            # a random value in [0,3]

## [1] 2.416
```

The above commands are for the uniform distribution on $[0,3]$. They show that the density is $1/3$ at $x = 1$ (as it is for all $0 \leq x \leq 3$); the area to the left of 2 is $2/3$; the median or .5-quantile is 1.5; and a realization of a random variable is some randomly chosen value in $[0,1]$. This last command will vary each time it is run.

It is useful to know that the arguments to these functions are vectorized. For example, multiple quantiles can be found at once. These commands will find the quintiles:

```
ps <- seq(0, 1, by=.2)                               # probabilities
names(ps) <- as.character(seq(0, 100, by=20)) # give names
qunif(ps, min=0, max=1)

##    0  20  40  60  80 100
## 0.0 0.2 0.4 0.6 0.8 1.0
```

This command, on the other hand, will find five uniform samples from five different distributions (Uniform$(0,1)$, Uniform$(0,2)$, ..., Uniform$(0,5)$):

```
runif(5, min=0, max=1:5)                         # recycles min

## [1] 0.9718 0.6858 1.8941 1.6402 1.7285
```

• **Example 6.7: Relating the d and r functions**
For continuous distributions, the d function describes the theoretical density. We have seen the density function which estimates a density from a sample. The two are not the same, though related. Figure 6.4 shows both the d function for the uniform and the density function for a random sample produced by the r function.

```
x <- runif(100)                              # large sample, 1000 points
d <- density(x)
curve(dunif, -0.1, 1.1,
      ylim=c(0, max(d$y, 1)))                # plots function
lines(d, lty=2)                              # add density estimate
rug(x)                                       # indicates sample
```

For the uniform distribution, the discontinuities at 0 and 1 make this basic approach a bit less successful, but the relationship can be seen: a large sample from a distribution can be used to approximate that distribution. ••

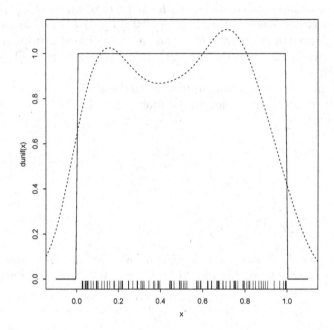

Figure 6.4: Graphic showing d function for the uniform distribution and density estimate from a random sample produced by the r function.

Binomial, normal, and some other named distributions

There are a few basic distributions that are used in many different probability models: among them are the Bernoulli, binomial, and normal distributions.

Bernoulli random variables

A *Bernoulli random variable* X is one that has only two values: 0 or 1. The distribution of X is characterized by $p = P(X = 1)$. We use Bernoulli(p) to refer to this distribution. (Jakob Bernoulli was an early figure in probability known for his "law of large numbers" which relates the frequentist's notion of a probability with that of the rules of probability.)

Often the term "success" is given to the event when $X = 1$ and "failure" to the event when $X = 0$. If we toss a coin and let X be 1 if a heads occurs, then X is a Bernoulli random variable where the value of p would be $1/2$ if the coin is fair. A sequence of coin tosses would be an *i.i.d.* sequence of Bernoulli random variables, also known as a sequence of Bernoulli trials. A Bernoulli random variable has a mean $\mu = p$ and variance $\sigma^2 = p(1 - p)$.

In R, the sample command can be used to generate random samples from this distribution. For example, to generate ten random samples when $p = 1/4$ can be done with

```
n <- 10; p <- 1/4
sample(0:1, size=n, replace=TRUE, prob=c(1-p, p))

## [1] 0 0 0 1 0 1 0 0 0 0
```

Binomial random variables

A *binomial random variable* X counts the number of successes in n Bernoulli trials. There are two parameters that describe the distribution of X: the number of trials, n, and the success probability, p. Let Binomial(n, p) denote this distribution. The possible range of values for X is $0, 1, \ldots, n$.

The distribution of X is known to be

$$P(X = k) = \binom{n}{k} p^k (1 - p)^{n-k}.$$

The term $\binom{n}{k}$ is called the binomial coefficient and is defined by

$$\binom{n}{k} = \frac{n!}{(n - k)! k!}.$$

The standard notation $n!$ is for the factorial of n, or $n \cdot (n - 1) \cdots 2 \cdot 1$. By convention, $0! = 1$.

The binomial coefficient counts the number of ways k objects can be chosen from n distinct objects and is read "n choose k." The choose function finds the binomial coefficients. (The binomial coefficients are part of Pascal's triangle and the expansion of $(a + b)^n = \sum_k \binom{n}{k} a^k b^{n-k}$.)

The mean of a Binomial(n, p) random variable is $\mu = np$, and the standard deviation is $\sigma = \sqrt{np(1 - p)}$.

In R the family name for the binomial is binom, and the parameters are labeled size for n and prob for p.

• Example 6.8: Tossing ten coins

Toss a coin ten times. Let X be the number of heads. If the coin is fair, X has a Binomial$(10, 1/2)$ distribution.

The probability that $X = 5$ can be found directly from the distribution with the choose function:

```
choose(10,5) * (1/2)^5 * (1/2)^(10-5)

## [1] 0.2461
```

This work is better done using the "d" function, dbinom:

```
dbinom(5, size=10, prob=1/2)
```

```
## [1] 0.2461
```

The probability that there are six or fewer heads, $P(X \leq 6) = \sum_{k \leq 6} P(X = k)$, can be given either of these two ways:

```
sum(dbinom(0:6, size=10, prob=1/2))
```

```
## [1] 0.8281
```

```
pbinom(6, size=10, p=1/2)
```

```
## [1] 0.8281
```

If we wanted the probability of seven or more heads, we could answer using $P(X \geq 7) = 1 - P(X \leq 6)$, or using the extra argument lower.tail=FALSE. This returns $P(X > k)$ rather than $P(X \leq k)$:

```
sum(dbinom(7:10,size=10,prob=1/2))
```

```
## [1] 0.1719
```

```
1 - pbinom(6,size=10,p=1/2)
```

```
## [1] 0.1719
```

```
pbinom(6,size=10,p=1/2, lower.tail=FALSE) # k = 6 not 7!
```

```
## [1] 0.1719
```

A spike plot (Figure 6.5) of the distribution can be produced using dbinom:

```
n <- 10; p <- 1/2
heights <- dbinom(0:10, size=n, prob=p)
plot(0:10, heights, type="h",
     main="Spike plot of X", xlab="k", ylab="p.d.f.")
points(0:10, heights, pch=16, cex=2)
```

• •

• **Example 6.9: Binomial model for a public-opinion poll**
In a public-opinion poll, the proportion of "yes" respondents is used to make inferences about the population proportion. If the respondents are chosen by

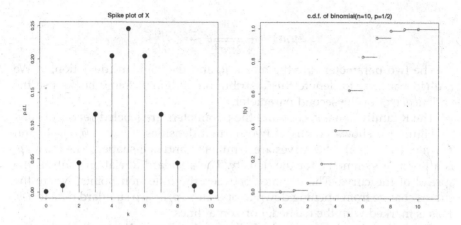

Figure 6.5: Spike plot of Binomial(10,1/2) distribution, and plot of its cumulative distribution function.

sampling with replacement from the population, and the "yes" responses are coded by a 1 and the "no" responses by a 0, then the sequence of responses is an *i.i.d.* Bernoulli sequence with parameter p, the population proportion. The number of "yes" respondents is then a Binomial(n, p) random variable where n is the size of the sample. (Of course, in practice it is very difficult to sample a population at random. Non-response, or even misleading responses, can further complicate the matter.)

For instance, if it is known that 62% of the population would respond favorably to the question were they asked, and a sample of size 100 is asked, what is the probability that 60% or less of the sample responded favorably?

```
pbinom(60, size=100, prob=0.62)

## [1] 0.3759
```

●●

Normal random variables

The normal distribution is a continuous distribution giving concrete meaning to the term "bell-shaped." It is used to describe many different populations in nature, such as a distribution of heights, and additionally describes the sampling distribution of many different statistics.

The normal distribution is a family of distributions with density given by:

$$f(x|\mu,\sigma) = \frac{1}{\sqrt{2\pi\sigma^2}} e^{-\frac{1}{2\sigma^2}(x-\mu)^2}.$$

The two parameters are the mean, μ, and the standard deviation, σ. We use Normal(μ,σ) to denote this distribution, although many books use the variance, σ^2, for the second parameter.

The R family name is norm and the parameters are labeled mean and sd.

Figure 6.6 shows graphs of two normal densities, $f(x|\mu = 0, \sigma = 1)$ and $f(x|\mu = 4, \sigma = 1/2)$. The curves are symmetric and bell-shaped. The mean, μ, is a point of symmetry for the density. The standard deviation controls the spread of the curve. The distance between the inflection points, where the curves change from opening down to opening up, is two standard deviations. This is marked with the dashed, horizontal lines.

The figure also shows two shaded areas. Let Z have Normal$(0,1)$ distribution and X have Normal$(4,1/2)$ distribution. Then the left shaded region is $P(Z \leq 1)$ and the right one is $P(X \leq 4.5)$. The random variable Z is called a *standard normal*, as it has mean 0 and variance 1. A key property of the normal distribution is that for any normal random variable the z-score, $(X - \mu)/\sigma$, is a standard normal. This implies that areas are determined by z-scores. In Figure 6.6 the two shaded areas are the same, as each represents the area to the left of 1 standard deviation above the mean.

We can verify this with the "p" function:

```
pnorm(1, mean=0, sd=1)

## [1] 0.8413

pnorm(4.5, mean=4, sd=1/2)               # same z-score as above

## [1] 0.8413
```

It is useful to know some areas for the normal distribution based on z-scores. For example, the IQR is the range of the middle 50%. We can find this for the standard normal by breaking the total area into quarters:

```
qnorm(c(0.25, 0.5, 0.75))

## [1] -0.6745  0.0000  0.6745
```

We use qnorm to specify the area we want. The mean and standard deviation are taken from the defaults of 0 and 1. For any normal random variable, this says the IQR is about 1.35σ.

To answer how much area is no more than one standard deviation from the mean, we use pnorm:

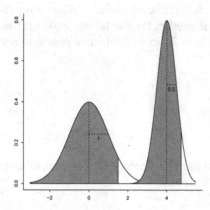

Figure 6.6: Two normal densities: the standard normal, $f(x;0,1)$, and $f(x;4,1/2)$. For each, the shaded area is the same, and amounts to area to left of a z-score of 1.

```
pnorm(1) - pnorm(-1)

## [1] 0.6827
```

which shows that roughly 68% of the area is in this range. For two and three standard deviations the numbers are 95% and 99.7%. We illustrate two ways to find these:

```
1 - 2*pnorm(-2)              # subtract area of two tails

## [1] 0.9545

diff(pnorm(c(-3, 3)))        # use diff to subtract

## [1] 0.9973
```

This says that 95% of the time a normal random variable is within two standard deviations of the mean, and 99.7% of the time it is within three standard deviations of its mean. These three values, 68%, 95%, and 99.7%, are the rules of thumb mentioned in Chapter 2.

• **Example 6.10: What percent of men are at least 6 feet tall?**
Many distributions in nature are well approximated by the normal distribution. For example, the population of heights for adult males within an ethnic class. Assume for some group the mean is 70.2 inches, and the standard deviation is 2.89 inches. What percentage of adult males are taller than 6 feet?

What percentage are taller than 2 meters? Assuming the model applies for all males, what does it predict for the tallest male on the planet?

We convert 6 feet into 72 inches and use pnorm to see that 73% are 6 feet or shorter:

```
mu <- 70.2; sigma <- 2.89
pnorm(72, mean=mu, sd=sigma)

## [1] 0.7333
```

To answer the question for meters we convert to metric. Each inch is 2.54 centimeters, or 0.0254 meters.

```
conv <- 0.0254
pnorm(2/conv, mean=mu, sd=sigma)

## [1] 0.9984
```

That is, this model predicts that fewer than 1% of men will be 2 meters or taller.

Finally, the tallest man could be found using quantiles. There are roughly 3.5 billion males, so the tallest man would be in the top $1/(3.5$ billion) quantile:

```
p = 1 - 1 / (3.5e9)
qnorm(p, mu, sigma)/12

## [1] 7.343
```

This predicts a bit over 7 feet 4 inches, maybe not even the tallest in the NBA. Expecting a probability model with just two parameters to describe a distribution like this is completely asking too much, but for most computations it is appropriate to use.   ••

• **Example 6.11: Testing the rules of thumb**
We can test the rules of thumb using random samples from the normal distribution as provided by rnorm.

First we create 1,000 random samples and assign them to res:

```
mu <- 100; sigma <- 10
res <- rnorm(1000, mean=mu, sd=sigma)
```

```
k <- 1; sum(res > mu - k*sigma & res < mu + k*sigma)

## [1] 673

k <- 2; sum(res > mu - k*sigma & res < mu + k*sigma)

## [1] 953

k <- 3; sum(res > mu - k*sigma & res < mu + k*sigma)

## [1] 996
```

Our simulation has 67.3%, 95.3%, and 99.6% of the data within 1, 2, and 3 standard deviations of the mean. If we repeat this simulation, the answers will likely differ slightly.   ••

Popular distributions to describe populations

Many populations are well described by the normal distribution; others are not. For example, a population may be multimodal, not symmetric, or have longer tails than the normal distribution. Many other families of distributions have been defined to describe different populations. We highlight a few.

Uniform distribution

The uniform distribution on $[a, b]$ is useful to describe populations that have no preferred values over their range. For a finite range of values, the sample function can choose one with equal probabilities. The uniform distribution would be used when there is a range of values that is continuous.

The density is a constant on $[a, b]$. As the total area is 1, the height is $1/(b - a)$. The mean is in the middle of the interval, $\mu = (a + b)/2$. The variance is $(b - a)^2/12$. The distribution has short tails.

As mentioned, the family name in R is unif, and the parameters are min and max with defaults 0 and 1. We use Uniform(a, b) to denote this distribution. The left graphic in Figure 6.7 shows a histogram and boxplot of 50% random samples from Uniform$(0, 10)$. On the histogram are superimposed the empirical density and the population density. The random sample is shown using the rug function.

```
res = runif(50, min=0, max=10)
## fig= setting uses bottom 35% of diagram
par(fig=c(0,1,0,.35))
boxplot(res,horizontal=TRUE, bty="n", xlab="uniform sample")
## fig= setting uses top 75% of figure
```

```
par(fig=c(0,1,.25,1), new=TRUE)
hist(res, prob=TRUE, main="", col=gray(.9))
lines(density(res),lty=2)
curve(dunif(x, min=0, max=10), lwd=2, add=TRUE)
rug(res)
```

(We overlaid two graphics by using the fig argument to par. This parameter sets the portion of the graphic device to draw on. You may manually specify the range on the *x*-axis in the histogram using xlim to get the axes to match. Other layouts are possible, as detailed in the help page ?layout.)

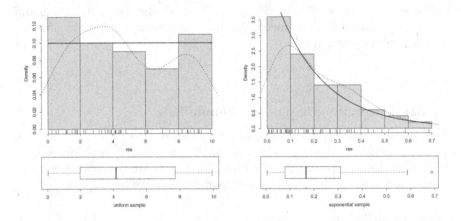

Figure 6.7: Histogram and boxplot of 50 samples from the Uniform$(0, 10)$ distribution and the Exponential$(1/5)$ distribution. Both empirical densities and population densities are drawn.

Exponential distribution

The exponential distribution is an example of a skewed distribution. It is a popular model for populations such as the length of time a light bulb lasts. The density is $f(x|\lambda) = \lambda e^{-\lambda x}$, $x \geq 0$. The parameter λ is related to the mean by $\mu = 1/\lambda$ and to the standard deviation by $\sigma = 1/\lambda$.

In R the family name is exp and the parameter is labeled rate. We refer to this distribution as Exponential(λ).

The right graphic of Figure 6.7 shows a random sample of size 50 from the Exponential$(1/5)$ distribution.

Weibull distribution   The exponential distribution is a special case of the Weibull distribution which has two parameters: one, scale, to control the rate of decay of the density (like the exponential, but in a reciprocal relationship) and one, shape, to control an initial growth from 0. When shape=1 the distribution is identical to an exponential distribution.

Lognormal distribution

The lognormal distribution is a heavily skewed continuous distribution on the positive numbers. A lognormal random variable, X, has its name as $\log(X)$ is normally distributed. Lognormal distributions describe populations such as income distribution.

In R the family name is lnorm. The two parameters are labeled meanlog and sdlog. These are the mean and standard deviation of $\log(X)$, not of X.

Figure 6.8 shows a sample of size 50 from the lognormal distribution, with parameters meanlog=0 and sdlog=1.

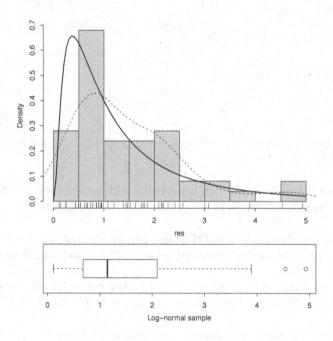

Figure 6.8: Histogram and boxplot of 50 samples from lognormal distribution with meanlog=0 and sdlog=1.

Sampling distributions

The following three distributions we will encounter in subsequent chapters as they are used to describe sampling distributions. These are the t-distribution, the F-distribution, and the chi-squared distribution (sometimes written using the Greek symbol χ).

The family names in R are t, f, and chisq. Their parameters are termed "degrees of freedom" and are related to the sample size when used as sampling distributions. For the t and chi-squared distributions, the degrees-of-freedom argument is df. For the F-distribution, as two degrees of freedom are specified, the arguments are df1 and df2.

For example, values l and r for each distribution containing 95% of the area can be found as follows:

```
qt(c(0.025, 0.975), df=10)              # 10 degrees of freedom

## [1] -2.228  2.228

qf(c(0.025, 0.975), df1=10, df2=5)      # 10 and 5 degrees of freedom

## [1] 0.2361 6.6192

qchisq(c(0.025, 0.975), df=10)          # 10 degrees of freedom

## [1]  3.247 20.483
```

Problems

6.8 A die is rolled five times. What is the probability of three or more rolls of four?

6.9 Suppose a decent bowler can get a strike with probability $p = .3$. What is the chance he gets 12 strikes in a row (assuming independence)?

6.10 A fair coin is tossed 100,000 times. The number of heads is recorded. What is the probability that there are between 49,800 and 50,200 heads?

6.11 Suppose that, on average, a baseball player gets a hit once every three times she bats. What is the probability that she gets four hits in four at bats?

6.12 Use the binomial distribution to decide which is more likely: rolling two dice twenty-four times and getting at least one double sixes, or rolling one die four times and getting at least one six?

6.13 In 1526 Cardano discussed the question of how many rolls of a pair of dice are needed so that it is even money that a pair of sixes will appear. What should he have got (he figured 18)?

6.14 A sample of 100 people is drawn from a population of 600,000. If it is known that 40% of the population has a specific attribute, what is the probability that 35 or fewer in the sample have that attribute?

6.15 If Z is Normal$(0,1)$, find the following:

1. $P(Z \leq 2.2)$.

2. $P(-1 < Z \leq 2)$.

3. $P(Z > 2.5)$.

4. b such that $P(-b < Z \leq b) = 0.90$.

6.16 Suppose that the population of adult, male black bears has weights that are approximately distributed as Normal$(350,75)$. What is the probability that a randomly observed male bear weighs more than 450 pounds?

6.17 The maximum score on the math ACT test is 36. If the average score for all high school seniors who took the exam was 20.6 with a standard deviation of 5.5, what percent received the passing mark of 22 or better? If 1,000,000 students took the test, how many more would be expected to fail if the passing mark were moved to 23 or better? Assume a normal distribution of scores.

6.18 A study found that foot lengths for Japanese women are normally distributed with mean 24.9 centimeters and standard deviation 1.05 centimeters. For this population, find the probability that a randomly chosen foot is less than 26 centimeters long. What is the 95th percentile?

6.19 Assume that the average finger length for females is 3.20 inches, with a standard deviation of 0.35 inches, and that the distribution of lengths is normal. If a glove manufacturer makes a glove that fits fingers with lengths between 3.5 and 4 inches, what percent of the population will the glove fit?

6.20 The term "six sigma" refers to an attempt to reduce errors to the point that the chance of their happening is less than the area more than six standard deviations from the mean. What is this area if the distribution is normal?

6.21 As any avid box reader knows, cereal is sold by weight not volume. This introduces variability in the volume due to settling. As such, the height to which a cereal box is filled is random. If the heights for a certain type of cereal and box have a Normal$(12,0.5)$ distribution in units of inches, what is the chance that a randomly chosen cereal box has cereal height of 10.7 inches or less?

6.22 For the `fheight` variable in the `father.son` data set, compute what percent of the data is within 1, 2, and 3 standard deviations from the mean. Compare to the percentages 68%, 95%, and 99.7%.

6.23 Find the quintiles of the standard normal distribution.

6.24 For a Uniform$(0,1)$ random variable, the mean and variance are $1/2$ and $1/12$. Find the area within 1, 2, and 3 standard deviations from the mean and compare to 68%, 95%, and 99.7%. Do the same for the Exponential$(1/5)$ distribution with mean and standard deviation of 5.

6.25 A q-q plot is an excellent way to investigate whether a distribution is approximately normal. For the symmetric distributions Uniform$(0,1)$, Normal$(0,1)$, and t with 3 degrees of freedom, take a random sample of size 100 and plot a quantile-normal plot using qqnorm. Compare the three and comment on the curve of the plot as it relates to the tail length. (The uniform is short-tailed; the t-distribution with 3 degrees of freedom is long-tailed.)

6.26 For the t-distribution, we can see that as the degrees of freedom get large the density approaches the normal. To investigate, plot the standard normal density with the command

```
curve(dnorm(x), -4, 4)
```

and add densities for the t-distribution with $k = 5, 10, 25, 50,$ and 100 degrees of freedom. These can be added as follows:

```
k <- 5; curve(dt(x, df=k), lty=k, add=TRUE)
```

6.27 The mean of a chi-squared random variable with k degrees of freedom is k. Can you guess the variance? Plot the density of the chi-squared distribution for $k = 2, 8, 18, 32, 50,$ and 72, and then try to guess. The first plot can be done with curve, as in

```
curve(dchisq(x,df=2), 0, 100)
```

Subsequent ones can be added with

```
k <- 8; curve(dchisq(x,df=k), add=TRUE)
```

6.3 The central limit theorem

It was remarked that for an *i.i.d.* sample from a population, the distribution of the sample mean had expected value μ and standard deviation $\sigma/\sqrt{n}$, where μ and σ are the population parameters. For large enough n, we see in this section that the sampling distribution of $\bar{X}$ is normal or approximately normal.

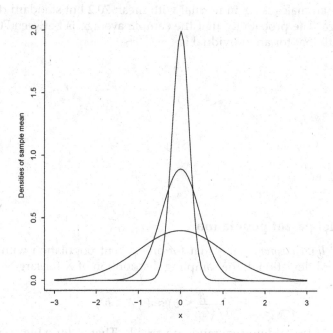

Figure 6.9: Density of $\bar{X}$ for $n = 5$ and $n = 25$ along with parent population Normal$(0,1)$. As n increases, the density concentrates on μ.

Normal parent population

When the sample $X_1, X_2, \ldots, X_n$ is drawn from a Normal(μ, σ) population, the distribution of $\bar{X}$ is precisely the normal distribution.

Figure 6.9 draws densities for the population, and the sampling distribution of $\bar{X}$ for $n = 5$ and 25 when $\mu = 0$ and $\sigma = 1$.

```
n <- 25; curve(dnorm(x, mean=0, sd=1/sqrt(n)), -3, 3,
                xlab="x", ylab="Densities of sample mean", bty="l")
n <- 5;  curve(dnorm(x, mean=0, sd=1/sqrt(n)), add=TRUE)
n <- 1;  curve(dnorm(x, mean=0, sd=1/sqrt(n)), add=TRUE)
```

The center stays the same, but as n gets bigger, the spread of $\bar{X}$ gets smaller and smaller. If the sample size increases by a factor of 4, the standard deviation is $1/2$ that population's. The density concentrates on the mean. That is, with greater and greater probability, the random value of $\bar{X}$ is close to the mean, μ, of the parent population. This phenomenon of the sample average concentrating on the mean is known as the *law of large numbers*.

For example, if adult male heights are normally distributed with mean 70.2 inches and standard deviation 2.89 inches, the average height of 25 randomly chosen males is again normal with mean 70.2 but standard deviation 1/5 as large. The probability that the sample average is between 70 and 71 compared to that for an individual is:

```
mu <- 70.2; sigma <- 2.89; n <- 25
pnorm(71, mu, sigma/sqrt(n)) - pnorm(70, mu, sigma/sqrt(n)) # average

## [1] 0.5522

pnorm(71, mu, sigma) - pnorm(70, mu, sigma)                 # person

## [1] 0.1366
```

Nonnormal parent population

The *central limit theorem* states that for any parent population with mean μ and standard deviation σ, the sampling distribution of $\bar{X}$ for large n satisfies

$$P(\frac{\bar{X} - \mu}{\sigma/\sqrt{n}} \leq b) \approx P(Z \leq b),$$

where Z is a standard normal random variable. That is, for n big enough, the standardized distribution of $\bar{X}$ is approximately a standard normal.

Figure 6.10 illustrates the central limit theorem for Exponential(1) data. This parent population and simulations of the distribution of $\bar{X}$ for $n = 5, 25,$ and 100 are drawn. As n gets bigger, the sampling distribution of $\bar{X}$ becomes more and more bell shaped.

Figure 6.10 was produced by simulating the sampling distribution of $\bar{X}$. Simulations will be discussed in the next chapter.

• Example 6.12: Average service time
The time it takes to check out at a grocery store can vary widely. A certain checker has a historic average of one-minute service time per customer, with a one-minute standard deviation. If she sees 20 customers, what is the probability that her check-out times average 0.9 minutes or less?

We assume that each service time has the unspecified parent population with $\mu = 1$ and $\sigma = 1$, and the sequence of service times is *i.i.d.* As well, we assume that n is large enough that the distribution of $\bar{X}$ is approximately Normal($\mu, \sigma/\sqrt{20}$). Then $P(\bar{X} \leq 0.9)$ is given by

```
pnorm(0.9, mean=1, sd = 1/sqrt(20))

## [1] 0.3274
```

••

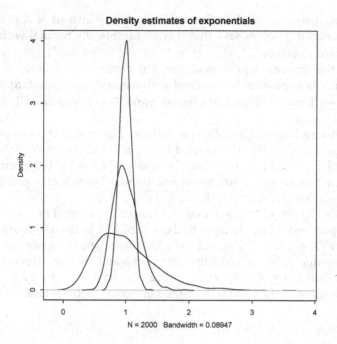

Figure 6.10: Density estimates for $\bar{X}$ when $n = 5, 25, 100$ for an Exponential(1) population. As n increases, the density becomes bell shaped and concentrates on $\mu = 1$.

There are other consequences of the central limit theorem. For example, if we replace σ with the sample standard deviation s when we standardize $\bar{X}$, we still have

$$P\left(\frac{\bar{X} - \mu}{s/\sqrt{n}} \leq b\right) \approx P(Z \leq b).$$

This fact will be behind many of the statements of statistical inference. This does not tell us what the sampling distribution is when n is not large; that will be discussed later.

In this next example, we show how the central limit theorem applies to the binomial distribution for large n.[6]

• **Example 6.13: The normal approximation to the binomial distribution**
For an *i.i.d.* sequence of Bernoulli trials $X_1, X_2, \ldots, X_n$ with success probability p, the sample mean, $\bar{X}$, is simply the number of successes divided by n, or the

---

[6]Historically, the ideas of the central limit theorem were first discussed for a case of the binomial by de Moivre in the early 1700s. Laplace, in 1812, demonstrated it for the binomial in general.

proportion of successes. We will use the notation $\hat{p}$ instead of $\bar{X}$ in this case. The central limit theorem says that $\hat{p}$ is asymptotically normal with mean p and standard deviation $\sqrt{p(1-p)/n}$.

If X is the number of successes, then X is Binomial(n, p). Since $X = n\hat{p}$, we know that X is approximately normal with mean np and standard deviation $\sqrt{np(1-p)}$. That is, a binomial random variable is approximately normal if n is large enough.

Let X have a Binomial$(30, 2/3)$ distribution. Figure 6.11 shows a plot of the distribution over $[10, 30]$. The shaded boxes above each integer k have base 1 and height $P(X = k)$, so their area is equal to $P(X = k)$. The normal curve that is added to the figure has mean and standard deviation equal to that of X: $\mu = 30 \cdot 2/3 = 20$ and $\sigma = \sqrt{30 \cdot 2/3 \cdot 1/3}$.

From the figure, we can see that the area of the shaded boxes, $P(k \le 22)$, is well approximated by the area to the left of 22.5 under the normal curve. This says $P(X \le 22) \approx P(Z \le (22.5 - \mu)/\sigma)$ for a standard normal Z. For a general binomial random variable with mean μ and standard deviation σ, the approximation $P(a \le X \le b) \approx P((a - 1/2 - \mu)/\sigma \le Z \le (b + 1/2 - \mu)/\sigma)$ is an improvement to the central limit theorem often seen in statistics text. ••

Problems

6.28 Compare the exact probability of getting 42 or fewer heads in 100 coin tosses to the probability given by the normal approximation.

6.29 Historically, a certain baseball player has averaged three hits every ten official at bats (he's a 300 hitter). Assume a binomial model for the number of hits in a 600-at-bat season. What is the probability the player has a batting average higher than 0.350? Use the normal approximation to answer.

6.30 Assume that a population is evenly divided on an issue ($p = 1/2$). A random sample of size 1,000 is taken. What is the probability the random sample will have 550 or more in favor of the issue? Answer using a normal approximation.

6.31 An elevator can safely hold 3,500 pounds. A sign in the elevator limits the passenger count to 15. If the adult population has a mean weight of 180 pounds with a 25-pound standard deviation, how unusual would it be, if the central limit theorem applied, that an elevator holding 15 people would be carrying more than 3,500 pounds?

6.32 A restaurant sells an average of 25 bottles of wine per night, with a variance of 4. Assuming the central limit theorem applies, what is the probability that the restaurant will sell more than 775 bottles in the next 30 days?

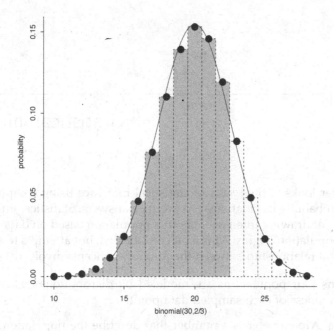

Figure 6.11: Plot of Binomial(30,2/3) distribution marked by dots. The area of the rectangle covering k is the same as the probability of k successes. The drawn density is the normal distribution with the same population mean and standard deviation as the binomial.

6.33 A traffic officer writes an average of four tickets per day, with a variance of one ticket. Assume the central limit theorem applies. What is the probability that she will write fewer than 75 tickets in a 21-day cycle?

## 7

Statistical inference

This chapter looks at the basics of statistical inference using computation—to avoid probability calculations—to produce answers. Statistical inference is the process of drawing inferences about a population based on data sampled from the population. Figure 7.1 is a bit complicated, but attempts to show the fundamental relationship between the four key concepts involved:

populations Our populations are modeled by distributions describing the randomness of each sampled data point.

parameters A parameter is a number that describe the population, such as the mean (μ) or the standard deviation (σ). Parameters are typically denoted by Greek letters and typically are related to the description of the population as a family of related functions.

samples A sample is a collection of observations from the population, where an observation is a realization of a random variable with the population's distribution. For our purposes, a sample is usually assumed to be a random sample from the population (implying independence). A sample has a certain size, n.

statistics A numeric summary of a sample. For example, the *sample* mean $\bar{x}$ or the *sample* standard deviation s.

The figure shows a population, the normal distribution, summarized by the parameters μ and σ. The population is drawn with asolid line, the parameters represented with dashed lines. From this population the graphic shows 10 samples, each of size 16. The values in a given sample appear along a horizontal line as gray dots. For each sample, the sample mean, $\bar{x}$, is indicated with a square. The population is described by the parameters, and is represented by the samples. As the statistic(s) summarize the sample, the key statistical question is how can we infer information about the parameters from the statistic?

Comparing the inset density estimate for the distribution of $\bar{x}$ with the theoretical distribution for a single value, we would be led to say that the distribution of $\bar{x}$ is centered in the same place as the population—that is randomness of the statistic is centered at the randomness of the population, and

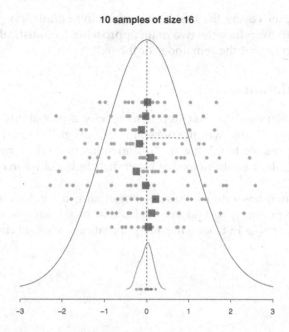

Figure 7.1: A population from which 10 samples of size 16 are chosen. Each sample is summarized by its mean (the boxes). These 10 means are summarized with the scaled, inset density estimate. This density is bell shaped, centered on the population mean, but has smaller spread.

the distribution is bell shaped. However, it is apparent that the distribution of $\bar{x}$ has smaller variability than the population.

In general, for a simulation of $\bar{x}$ these observations can be made explicit:

- For any random sample, the sampling distribution of $\bar{x}$ is centered at the population mean, μ.

- For any random sample, the sampling distribution of $\bar{x}$ has a standard deviation given by $\sigma/\sqrt{n}$ where σ is the standard deviation of the population.

- Further, if the population is normally distributed (as in the figure) the distribution of $\bar{x}$ is normally distributed.

The point is, the simulated data can be used to investigate relationships producing answers that while not always as precise, can give excellent insight.

This chapter covers the basics of performing simulations in R and then applies this to introduce the two main approaches to statistical inference that are used throughout the remainder of the text.

7.1 Simulation

For our simulations, the first step is to specify a probability model for the data. In Figure 7.1 this was the normal distribution. Other families of distributions are possible too. R has its "r" functions to produce random samples from the popular distributions. For other distributions, we may need to work harder.

Our simulations often involve summarizing a random sample with a statistic. For example, to find one realization of the sample mean of a random sample of size 16 from a normal population we might have:

```
mu <- 100; sigma <- 16              # population parameters
x <- rnorm(16, mean=mu, sd=sigma)   # our sample
mean(x)                             # xbar

## [1] 98.85
```

The model is specified in this example by a choice of distributional family (rnorm) and a choice of parameters ($\mu = 100$, $\sigma = 16$). With this, R can be used to produce a random sample of a given size. The value $mean(x)$ is the one produced. Running this again would produce a different value each time.[1] In real life, creating a random sample can be very expensive either in time, resources, or both. This is not so with the computer, where running a simulation again is just a matter of re-executing a line above. Though one informal description of insanity is "repeating the same action while expecting a different result," in simulation, due to the randomness, repeating a simulation can give us great insight into the distribution's shape, its tails, its mean and variance, related probabilities, etc.

Repeating a simulation easily

Repeating a simulation is the key to gaining insight from a simulation, as we can easily use graphical summaries of the simulated numbers to read into the mechanism that produces them, in the same way the inset of Figure 7.1 informed us about the nature of the sampling distribution of $\bar{x}$. The are many other alternatives. Figure 7.2 shows boxplots for many different simulations from a normal population with $50, 250$, and $1,000$ samples taken, respectively. By layering the boxplots, one can see the underlying variability.[2]

---

[1] In theory anyways, there is the *very* remote chance that due to approximations and round off, the same value could be achieved.

[2] This idea is a modification of one by Chris Wild who introduced animated boxplots to indicate variability due to sampling.

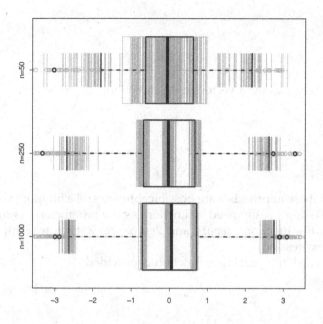

Figure 7.2: Boxplots showing different simulated data. For each sample size, several random samples from a normal population are taken and summarized with boxplots. For smaller samples, there is more variability in the estimates for the median, Q_1, Q_3, ..., than for the large sample. The heavier line summarizes just one of the several samples.

There are several ways to repeat an expression in R. In Section 4.3 we mentioned briefly for loops to repeat a process, the method used below:

```
mu <- 100; sigma <- 16
M <- 4; n <- 16                          # change M for more
res <- numeric(M)
for (i in 1:M) {
  res[i] <- mean(rnorm(n, mean=mu, sd=sigma))
}
res

## [1] 102.12  96.93  95.21 101.74
```

This is fairly direct, but has some bookkeeping details (e.g., creating a container for the output, res) that are best left to the computer. We also saw in Section 4.3 some alternatives. Here are a few. First we create a function to call our expression to create a single $\bar{x}$ to use with sapply:

```
xbar <- function(i)  mean(rnorm(n, mean=mu, sd=sigma)) #  function
sapply(1:M, xbar)
```

```
## [1]  99.25  89.83 109.48 103.69
```

Alternately, we could vectorize this function:

```
Xbar <- Vectorize(xbar)
Xbar(1:M)
```

```
## [1]  98.16 104.84 105.93 102.08
```

Both of these approaches are possible, but are still a bit more verbose than desired. We don't really need a function, as the parameter i is there just to work with the functions sapply and Vectorize. Rather, we simply need to repeat an expression.

For this task the replicate function is provided:

```
replicate(M, mean(rnorm(n, mean=mu, sd=sigma))) # replicate(n, expr)
```

```
## [1] 100.3 100.2 105.2 103.9
```

Without random number generation, replicate would simply produce M identical outputs, but with the expression above it produces M different realizations of $\bar{x}$ for a normal population.

The expression above is a single command. As with other R functions, multi-step expressions can be used if enclosed in matching braces.

Using apply For some simulations it can be faster to first generate all the numbers, then take the statistic. For this task, apply is useful. The basic example could be done through:

```
x <- matrix(rnorm(M*n, mean=mu, sd=sigma), nrow=n)
dim(x)                                       # M columns, n rows
```

```
## [1] 16  4
```

```
apply(x, 2, mean)
```

```
## [1] 102.50 103.15 100.99  97.84
```

• **Example 7.1: The central limit theorem through approximation**
For a normal population, the sampling distribution of $\bar{x}$ is also normal distributed. We will verify this through a simulation.

We begin by making a function to find the z-score of a sample from $\bar{x}$ where we center by the population mean and scale by the population standard deviation:

```
zstat <- function(x, mu, sigma) {
  (mean(x) - mu) / (sigma/sqrt(length(x)))
}
```

Simulating the sampling distribution of this statistic is no different from before. In the leftmost graphic of Figure 7.3 we summarize the graphic with a qqnorm plot, which indicates normality of the sample.

```
M <- 2000; n <- 7
mu <- 100; sigma <- 16
res <- replicate(M, {
  x <- rnorm(n, mean=mu, sd=sigma)
  zstat(x, mu, sigma)
})
qqnorm(res, main="Normal, n=7")
```

If we use a different population, the sampling distribution of $\bar{x}$ is no longer normal, though the central limit theorem says that as n gets large it will become normal.

A natural question is: what n is large enough? The answer depends on the population—populations which are either skewed or long-tailed will need larger values of n.

Let's look at the right-skewed exponential distribution. We need to know that it has population mean and population standard deviation are both given by 1/rate:

```
M <- 2000; n <- 7
rate <- 2; mu <- sigma <- 1/rate
res <- replicate(M, {
  x <- rexp(n, rate=rate)
  zstat(x, mu, sigma)
})
qqnorm(res, main="Exponential, n=7")
```

The middle graphic of Figure 7.3 shows an upward curve in the quantile-normal plot. This is indicative of a right-tailed distribution. The central limit phenomenon has not removed the initial population asymmetry when $n = 7$.

Now we repeat with $n = 150$. The right graphic in Figure 7.3 shows that the sampling distribution is now approximately normal.

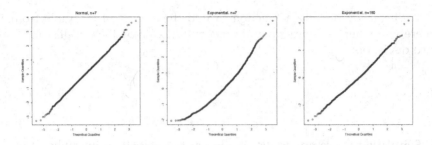

Figure 7.3: The left graphic shows a simulated sampling distribution of $\bar{x}$ for a normal population. This is known to be normally distributed, as supported by the graphic. The other two graphics show the sampling distribution of $\bar{x}$ for an exponential population with rate 2 for samples of size 7 and 150. The middle graphic shows that for $n = 7$ the central limit effect has not dominated, as it does for the $n = 150$ case.

```
n <- 150
res <- replicate(M, {
  x <- rexp(n, rate=rate)
  zstat(x, mu, sigma)
})
qqnorm(res, main="Exponential, n=150")
```

●●

● **Example 7.2: The *t*-statistic**
The z-statistic defined above standardizes by the standard deviation of $\bar{x}$, known to be $\sigma/\sqrt{n}$. In applications, the knowledge of σ is not generally known, so is estimated by the sample standard deviation s. This gives the t-statistic:

```
tstat <- function(x, mu) (mean(x) - mu) / (sd(x) / sqrt(length(x)))
```

As s is random, the distribution of the t-statistic should differ from that of the z-statistic: we would expect it to have longer tails as larger values are produced when $s < \sigma$, but also that this difference decreases for larger values of n.

Gosset [26] simulated the distribution of the above statistic using data published in 1902 on the finger lengths of 3,000 male criminals. The population of the finger lengths was approximately normal, and Gosset took samples of size 4 to simulate the t-statistic. We are more fortunate than having to

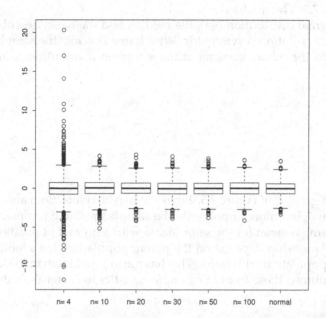

Figure 7.4: Simulations of the sampling distribution of the *t*-statistic for different values of *n*. As *n* gets bigger the boxplots show the sampling distribution becomes more like the normal distribution on the right.

compute 750 sample standard deviations by hand. To perform a simulation with $n = 4$ and $M = 750$ we have:[3]

```
mu <- 0; sigma <- 1                           # use defaults
M <- 750; n <- 4
res <- replicate(M, tstat(rnorm(n, mu, sigma), mu))
boxplot(res)
```

Figure 7.4 shows several such simulations for differing values of *n*. The boxplots show that the sampling distribution is symmetric and centered at 0 (as expected because the mean is subtracted). For smaller *n* the values of *s* less than σ influence the length of the tails and the variability. As *n* increases, it appears the sampling distribution of the *t*-statistic becomes normal (the reference boxplot on the right side of the graphic). ••

---

[3]If we knew the distribution was the *t*-distribution with $n - 1$ degrees of freedom, we could just do the simulation with rt(n, df=4-1).

• **Example 7.3: The median**

For the normal distribution both the median and the mean describe the center, as the distribution is symmetric. Why is one statistic (the sampling mean) preferred to the other? Looking at the sampling distributions can be informative:

```
M <- 1000; n <- 35
res_mean <- replicate(M, mean(rnorm(n)))
res_median <- replicate(M, median(rnorm(n)))
boxplot(list("sample mean"=res_mean, "sample median"=res_median),
        main="Normal population")
```

The left graphic of Figure 7.5 shows the two boxplots. Both are centered at 0 as expected, but more importantly the sample median has more variability than the sample mean for the same size n with a normal population.

This relationship depends on the parent population. Let's look at the exponential population with rate 1. This has mean 1 and median $\log(2) = 0.693$. Here we subtract these from our sample statistics to center the values:

```
M <- 1000; n <- 35
res_mean <- replicate(M, mean(rexp(n))) - 1
res_median <- replicate(M, median(rexp(n))) - log(2)
boxplot(list("sample mean"=res_mean, "sample median"=res_median),
        main="Exponential population")
```

The right graphic shows the sample mean and median having comparable spreads when the population has an exponential distribution.   ••

• **Example 7.4: Estimating probabilities**

There are many different types of lotteries, but a common one is when there are N distinct numbers of which a player selects k. From the N numbers, j are chosen at random without replacement. A player wins if 1 or more numbers are matched. (Of course, they generally win more for matching more balls.) While often $k = j$, this is not always the case. Take for example, the case where $N = 80$, $k = 20$, and $j = 10$. Which is more likely: matching 1 ball or matching 0 balls?

This is a fun problem to compute by hand with the basic rules of probability, but here we will simulate the answer. Without any loss of generality, we can assume the numbers are 1 through N and the player chooses the numbers 1 through k. Here the simulation would be done using sample. A single sample is:

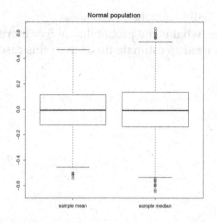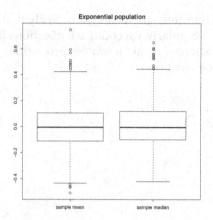

Figure 7.5: Illustration that the ratio of the variability of the sample mean and sample median depend on the population. In this graphic, the exponential distribution is the population and the ratio appears to be about 1 (the actual value by theory).

```
N <- 80; k <- 20; j <- 10
x <- sample(1:N, j, replace=FALSE)
x
```

```
## [1] 26 70 59 80  2 14 17  4  9 69
```

Here is how we can count the number of matches:

```
sum(x %in% 1:k)
```

```
## [1] 5
```

Now we can simulate this, say $M = 10,000$ times:

```
res <- replicate(10000, {
  x <- sample(1:N, j, replace=FALSE)
  sum(x %in% 1:k)
})
```

The proportion of values gives an estimate to the underlying probabilities:

```
c("zero matches"= sum(res == 0)/length(res), # prop.table to see all
  "one match"   = sum(res == 1)/length(res))
```

```
## zero matches   one match
##       0.0474      0.1835
```

From this, it appears more likely that 1 ball is matched than none.

Similarly we could ask questions like, what is the probability of 5 or more matches? Again a sample proportion is used to estimate this, but in this case the condition would be:

```
sum(res >= 5) / length(res)              # 5 or more matches

## [1] 0.0643
```

••

Problems

7.1 Simulate 1,000 rolls of a pair of die. Which is more common, a roll of 7 or 8?

7.2 For the `rivers` data set, take a 1,000 random samples of size 10. Compare the mean of the sample means, with the sample mean of the data in `rivers`.

7.2 Significance tests

Imagine a study of some new treatment to improve performance. For example, does consuming an amount of honey during exercise increase performance? To test such a treatment, a cohort of 7 conveniently selected people is gathered. The investigator randomly assigns these 7 into a control group of 3 and a treatment group of 4. The control group should give a baseline to compare the treatment group with. The data collected is some measurement where smaller values are better:[4]

```
Control:   23, 33, 40
Treatment: 19, 22, 25, 26
```

```
ctrl <- c(23, 33, 40)
treatment <- c(19, 22, 25, 26)
the_data <- stack(list(ctrl=ctrl, treatment=treatment))
aggregate(values ~ ind, the_data, mean)

##          ind values
## 1       ctrl     32
## 2  treatment     23
```

---

[4]This data comes from an influential article of Cobb [11] who attributes Ernst [19]. There is a slight difference, as Ernst has a value of 23, not 22.

A difference of 9 is noted. The researcher is thrilled. It seems something as simple as a bit of honey can improve performance.

The skeptic might be a bit more cautious. Perhaps the difference is simply due to the randomization chosen. Maybe the selected cohort put the better performers into the treatment group and the honey made no difference at all.

Well, with simulation this can be investigated. In this case we can enumerate all the possible different randomizations. The combn function will list them through their indices:

```
cmbs <- combn(7, 3)                                # 35 possibilities
cmbs[, 1:6]                                        # first 6 columns

##      [,1] [,2] [,3] [,4] [,5] [,6]
## [1,]   1    1    1    1    1    1
## [2,]   2    2    2    2    2    3
## [3,]   3    4    5    6    7    4
```

The first would just be the observed randomization. To see, we have

```
i <- 1
ind <- cmbs[,1]                                    # in columns
obs <- mean(the_data$value[ind]) - mean(the_data$value[-ind])
obs

## [1] 9
```

The use of negative indexing is very convenient in this case, as we are partitioning our set of 7 scores into groups of 3 and 4.

To compute the above for all 35 randomizations can be done efficiently with apply:

```
res <- apply(cmbs, 2, function(ind) {
  mean(the_data$value[ind]) - mean(the_data$value[-ind])
})
```

The values in res represent the *randomization distribution* for the difference of the group means. Okay, we have the 35 different values. To see how unusual, or extreme, 9 is we can count how many of the values are larger:

```
sum(res >= obs)

## [1] 3

sum(res >= obs) / length(res)

## [1] 0.08571
```

So 3 of 35 random assignments or about 8.57% will produce a difference this large or larger. This value seems unlikely, somewhere between 1 in 10 and 1 in 20, but not "impossibly" small.

Whether this is a true difference or not is ambiguous, but what the above illustrates is a process for investigating differences from random samples. Had the process produced a less ambiguous answer then an answer on whether honey intake improved performance (all other things being controlled for) could be made.

What is not possible from this data scenario would be to try and generalize to a wider population. This is due to the 7 recruits not being representative of some larger population. For that, random sampling is often employed to select a sample likely to be representative.

• **Example 7.5: Simulating the randomization distribution**
Does caffeine make you jittery? The popular impression of caffeine consumption is that it makes the consumer more jittery. Is that really so?[5]

A student decides to test the impression. She recruits 20 classmates to participate and randomly divides them into two cohorts of size 10. She serves all a large cup of coffee, but one cohort has decaffeinated and one caffeinated. The number of finger taps the students made is then secretly counted with the aid of videotape. The data is then entered into R through these commands:

```
caf <- c(245, 246, 246, 248, 248, 248, 250, 250, 250, 252)
no_caf <- c(242, 242, 242, 244, 244, 245, 246, 247, 248, 248)
the_data <- stack(list(caffeine=caf, no_caffeine=no_caf))
```

Is there a difference in the mean number of taps? Well, the sample means have this difference:

```
obs <- mean(caf) - mean(no_caf)
obs
```

```
## [1] 3.5
```

So, there is a difference of 3.5 taps between the groups. The natural question is: does this indicate coffee makes students more jittery? Of course, as in the last example, one could say the difference is attributable to the random allocation. In that example, the presence of a difference was left for debate and there the difference was more than twice as large (9).

If we assume the skeptical approach to be the case, then we can ask how unusual this difference would be in the distribution of all possible random assignments of the 20 into two groups of 10. There are choose(20, 10) =

---

[5]This data comes from http://lock5stat.com/statkey/ where a set of applets is hosted to illustrate randomization techniques discussed in the text [39].

184,756 different randomizations. Clearly looking at each is not doable by hand, though still possible by R. However, numbers like 40 and 20 would not be. So if this approach is to be widely useful, a different means is needed.

We follow Dwass [17] who noted that a shortcoming of this line of investigation was the issue of computing a statistic for all the possible permutations and suggested taking random samples (simulating) from the randomization distribution instead.

To find a single random assignment is simple with the sample function, this command produces one taking 10 numbers at random from the values 1 through 20 without replacement:

```
sample(1:20, 10, replace=FALSE)

## [1] 12 15 10 19  8 20  2  9  5  1
```

A simulation then can be done by repetition:

```
res <- replicate(2000, {
  ind <- sample(1:20, 10, replace=FALSE)
  mean(the_data$value[ind]) - mean(the_data$value[-ind])
})
```

The probability of a randomization producing a value as or more extreme than the observed one is found with:

```
sum(res > obs) / length(res)

## [1] 0.0025
```

This value is very small. In this case, there is not much ambiguity implying that caffeine increases tapping with this group of 20 students.

So, on one hand a difference of 9 was debatable, and here a smaller difference of 3.5 is pretty conclusive. How can that be? We will see the key to this seeming paradox is understanding that differences must be measured on scales suitable for the data. ●●

7.3 Estimation, confidence intervals

The last section looked at two questions where we were curious if some treatment induced a notable effect. This section has a different question. How well does a sample statistic estimate a parameter?

A common news item is a report on a proportion in a sample. For example, in mid-2013, following a U.S. Supreme Court decision on the defense against marriage act, a poll was taken by Princeton Survey Research Associates asking a random sample of 1,003 Americans whether marriages between same-sex couples should be recognized by law as valid, with the same rights of traditional marriage. It was reported that 55% said yes, a reported high.

The value 55% represents the sample, so is a statistic. The implication is it somehow represents a parameter—the proportion of all Americans who would answer yes. How does a statistic estimate an unknown parameter?

Again, we start with some probability model for the data. In the scenario above, a simple model would be that each randomly chosen person has the same probability, p, of responding yes. That is the random variables have the Bernoulli(p) distribution. Other scenarios might assume a normal distribution for the random variables producing the sample. The models we consider mostly are defined through a family and one or more parameters (e.g., p or both μ and σ).

The question of a statistic estimating a parameter is then akin to looking at values such as $\bar{x}$ in relation to μ, or $\hat{p}$ (the sample proportion from a survey) in relation to p. There are various ways to compare a relation, a simple one is to look at their difference (e.g., $\bar{x} - \mu$ or $\hat{p} - p$) or their ratio (e.g., s^2/σ^2).

Let's focus for now on the difference. How much can we say? If we have a large sample, intuition should tell us that the sample mean is a better estimate for μ than a single value. Why? Both have some expected value—μ. The key is variability. For a random sample of size n, VAR($\bar{x}$) $= \sigma^2/n$, where σ^2 is the variance of the population, or a single one of the random variables.

Let θ be some parameter and $\hat{\theta}$ be some statistic to estimate θ. (It is very common to use a "hat" to indicate an estimator for a parameter, as we did with $\hat{p}$ above.) Looking at the variability, or $E((\hat{\theta} - \theta)^2)$ is reasonable. For a random sample, this can be written in two terms:

$$E((\hat{\theta} - \theta)^2) = \text{VAR}(\hat{\theta}) + [E(\hat{\theta} - \theta)]^2 = \text{variance} + \text{bias}^2.$$

The bias is simply the difference between the expected value of the estimator and the parameter. Most statistics we encounter are *unbiased* meaning this difference is 0. Examples are $E(\bar{x}) = \mu$, $E(\hat{p}) = p$, and $E(s^2) = \sigma^2$. (It is this last equality that drives the division by $n - 1$, as opposed to n, in the formula for the variance.) Not all are unbiased, for example $E(s) \neq \sigma$, though the difference gets negligible as n gets larger. The left-hand side of the above formula is the key, the right-hand side can manipulated to exchange biasness for reduced variability and an overall smaller sum.

A finer question than an expectation looks at the sampling distribution—as this contains more information. For the normal-population model we've seen that we can simulate the distribution of $\bar{x} - \mu$:

```
mu <- 100; sigma <- 16
M <- 1000; n <- 4
```

```
res <- replicate(4, mean(rnorm(n, mu, sigma)) - mu)
```

In previous examples we have standardized this by $\sigma/\sqrt{n}$ and by $s/\sqrt{n}$. The former is the standard deviation of $\bar{x}$, the latter the so-called *standard error*, where estimators replace unknown parameters.

The distribution depends on n, but with a large enough sample we can ask questions like: where is most of the data? Being precise, we might ask what range about 0 contains 95% of the data?

Here we simulate the *t*-statistic again:

```
mu <- 100; sigma <- 16
M <- 1000; n <- 4

res <- replicate(M, {
  x <- rnorm(n, mu, sigma)
  SE <- sd(x)/sqrt(n)                    # standard error
  (mean(x) - mu) / SE
})
```

To find a range, we use the quantile function:

```
quantile(res, c(0.025, 0.975))

##    2.5%   97.5%
## -3.192   3.329
```

So basically we have with probability roughly 0.95 that

$$-3.19 \cdot \text{SE} < \bar{x} - \mu < 3.33 \cdot \text{SE}.$$

Or in other words, the value of $\bar{x}$ is not too many standard errors away for the mean, μ, with high probability. The key to the statement is to compare distance using a scale defined by a standard error.

When we run a simulation, we know the parameters like μ as part of our specification of the model, and generate repeated values of some statistic like $\bar{x}$. In real life, there is a different perspective: we have a single value of the statistic and don't know the parameter (though we will usually assume a family for the population). So what can we say about a parameter based on a single value of $\bar{x}$?

We can simply invert the algebra and solve for μ above to say that with probability 0.95:

$$\bar{x} - 3.33 \cdot \text{SE} < \mu < \bar{x} + 3.19 \cdot \text{SE}.$$

This formula is describing the relationship as random variables. We have a single realization. For that we have to be careful with the word "probability," so the phrasing becomes we are 95% confident that the interval $(\bar{x} - 3.33 \cdot SE, \bar{x} + 3.19 \cdot SE)$ contains the unknown μ.

So, if we know the sampling distribution of $\bar{x} - \mu$, then from a single value we can produce an interval describing μ. It isn't a guarantee, but you can have confidence in the process to be correct roughly 95% of the time.

The basic bootstrap

The rub is we need to know μ to actually simulate the sampling distribution. For this particular question and others we will encounter in the upcoming chapters we will know the sampling distribution from theoretical means. So this issue will not be an impediment.

This section briefly touches on a computer-intensive means to work around the absence of a known sampling distribution, the bootstrap method [18], [16]. In its most basic application, we will see how the bootstrap method can be used to estimate the sampling distribution.

How? A basic assumption in simulation is that if we have enough random samples from a population, that the random samples can describe that population very well. This is the starting point for the bootstrap.

First a random sample $x_1, x_2, \ldots, x_n$ is used to produce an *empirical cumulative distribution function* $\hat{F}(a) = \#\{i : x_i \leq a\}/n$. The assumption is that this approximates well the actual cumulative distribution function $F(a) = P(X \leq a)$. As such, we assume we know a reasonable proxy for the population and use that as our model.

Next, the parameter μ describing the population is replaced by the mean of the empirical c.d.f. which turns out to be just the sample mean, $\bar{x}$.

We have sample data on the difference between the charge and the Medicare payment for a diabetes diagnosis in the Medicare (UsingR) data set:

```
diabetes <- subset(Medicare,
            subset= DRG.Definition =="638 - DIABETES W CC")
gap <- with(diabetes, Average.Covered.Charges-Average.Total.Payments)
```

There is an enormous range in this data:

```
range(gap)
```

```
## [1]    -465.4 119472.4
```

Our goal is to find an interval for which we are reasonably confident the unknown mean resides.

Using the data in gap we have the value of $\bar{x}$ is:

```
xbar <- mean(gap)
xbar
```

```
## [1] 15446
```

A single random sample from our "population" is produced with sample:

```
xstar <- sample(gap, length(gap), replace=TRUE)
head(xstar)

## [1] 10297 26219  4805  4033  3785 11600
```

We took a random sample the same size as our original sample from the empirical cumulative distribution. This just means sampling at random from the original sample. The key to the basic bootstrap method is to treat these new samples as simulated random samples from the actual population. That means, finding repeated values of

```
mean(xstar) - xbar

## [1] 841.3
```

This is done as before through replicate:[6]

```
M <- 2000
res <- replicate(M, {
  xstar <- sample(gap, length(gap), replace=TRUE)
  mean(xstar) - xbar
})
```

From this simulation we can get the basic bootstrap confidence interval for μ using the quantile function:

```
alpha <- 0.05
xbar + quantile(res, c(alpha/2, 1-alpha/2))

##  2.5% 97.5%
## 13093 18119
```

7.4 Bayesian analysis

One common mistake made when describing a confidence interval is to say with probability 0.95, say, the unknown parameter resides in the produced interval. This is incorrect, as neither the interval (once produced) or parameter is presumed random, so the probabilistic terms don't apply.

---

[6]The proper R framework for bootstrapping is provided by the boot package [7].

There is another approach where this statement can make sense, the Bayesian approach. In this section we look briefly at only the simplest of cases.

Suppose we have a sequence of n Bernoulli trials. These are characterized by a success probability p. The goal is to estimate p after looking at the data. One estimator is the sample proportion $\hat{p}$ defined to be the number of successes divided by the number of trials. With this estimate, we can look at the sampling distribution of $\hat{p} - p$ and proceed in a similar manner as before.

The Bayesian approach is different. Bayes' formula makes explicit the relationship between two conditional probabilities, $P(B|A)$ and $P(A|B)$. That is what is the probability of some event B given event A occurred, and the similar-sounding, but fundamentally different, probability of an event A given event B occurred. Bayesian analysis uses this formula and a different starting point to approach estimation.

Rather than assume the parameter p is some unknown number, the Bayesian approach starts with the assumption that p is random and prior to looking at the data can be described by some probability distribution. After looking at the data, the distribution is updated to reflect the additional information. That is, there is a *prior distribution* and a *posterior distribution*. The Bayes' formula gets operationalized into the following updating rule:

$$\text{posterior distribution} \propto \text{likelihood} \cdot \text{prior distribution}.$$

The likelihood describes how likely the data is given a specific value for the parameter, and in many cases can be computed from the model. In this case, it takes a simple form. If the parameter value is given to be p and the data has s success in n trials, then each success contributes a factor of p and each failure a factor of $1 - p$ leaving:

$$\text{likelihood} = p^s(1 - p)^{n-s}.$$

The likelihood above is not a probability distribution as it does not add to 1, so when using it below we will normalize our answer.

To use this, we follow an example from Albert [3] where we send the interested reader for much more on the topic. Let p be the proportion of college students getting less than 8 hours of sleep. A researcher wants to know more about p. Prior to collecting any data she believes p is likely around 0.3 but may realistically be other values. She ascribes a distribution of values as realistic prior to any data collection:

```
p <- seq(0.05, 0.95, by=0.1)              # possible values
prior <- c(2, 4, 8, 8, 4, 2, 1, 1, 1, 1) # how likely
prior <- prior / sum(prior)               # as a probability
prior

## [1] 0.06250 0.12500 0.25000 0.25000 0.12500 0.06250 0.03125
## [8] 0.03125 0.03125 0.03125
```

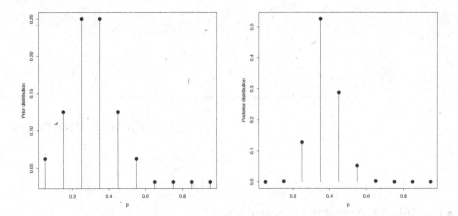

Figure 7.6: Spike graphs of prior and posterior distributions. The posterior is more concentrated due to the inclusion of the sample data.

Figure 7.6 shows a graphic of the prior distribution in the left panel.
The collected data consisted of 27 students of which 11 record that they sleep 8 or more hours.
The likelihood function is then easily computed given this data:

```
like <- p^11 * (1-p)^(27 - 11)
```

Bayes' theorem then allows us to update the prior probability simply by multiplying and normalizing:

```
posterior <- like * prior
posterior <- posterior / sum(posterior)
```

The right panel of Figure 7.6 shows the posterior distribution. We can see it is more concentrated. In particular, the initial prior only has p in $[0,0.5]$ with an 81% chance:

```
ind <- 0 <= p & p <= 0.5
sum(prior[ind])

## [1] 0.8125
```

Whereas, the posterior distribution gives a 94.4% chance of this event:

```
sum(posterior[ind])

## [1] 0.9441
```

## 8

Confidence intervals

• Example 8.1: Age of the universe

How old is the universe? As a modern question, it is now pretty well settled, roughly 13.772 billion years old with a margin of error of 55.9 million years. From a historical perspective, this level of certainty has not been the case. The age.universe data set contains estimates for the age of the universe, dating back to some early estimates based on the age of the earth.

The current best estimate is computed by the Wilkinson microwave anisotropy probe (http://map.gsfc.nasa.gov). The reported answer is not a single number, rather an interval, in this case 13.772 ± 0.059 Gyr. The precision of any measurement is limited due to measurement errors, for this delicate task effects as small as thermal distortions need to be accounted for. As well, errors can be induced by inexact models, such as a linear model being used when a curvilinear one may be more appropriate. The precision in this estimate is truly incredible.

Figure 8.1 shows other such intervals given by various people over time. Most, but not all, of the modern estimates contain the value of 13.7 billion years. Were those experiments "wrong?" Not necessarily. In this chapter we look at the process of estimating unknown values using random data and will see that a process can be "right," yet still produce values that are "wrong," in the way these are. This will be a subtlety in the framework. In this particular example though, the errors may be due to an incorrect model specification for the data—certainly in 1852 the big bang model was not used.   ••

• Example 8.2: Age of universe (redux)

In 2012, a Gallup poll[1] is summarized by "Forty-six percent of Americans believe in the creationist view that God created humans in their present form at one time within the last 10,000 years." (Clearly not all Americans believe in the science of the U.S.-funded Wilkinson probe.)

Reading further, we find that not all Americans were asked, but rather:

---

[1]http://www.gallup.com/poll/155003/Hold-Creationist-View-Human-Origins.aspx.

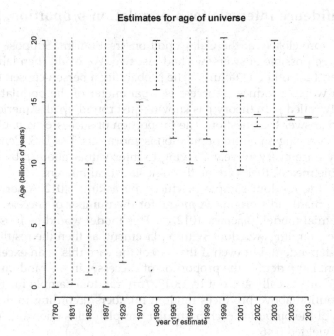

Figure 8.1: Various estimates for the age of universe, some historic, some modern. Ranges are represented by dotted lines. When the estimate is a lower bound, only a bottom bar is drawn. The current best estimate of 13.772 billion years old is drawn with a horizontal line. This estimate has a margin of error of about 0.1%.

> Results for this USA Today/Gallup poll are based on telephone interviews conducted May 10–13, 2012, with a random sample of 1,012 adults, aged 18 and older, living in all 50 U.S. states and the District of Columbia.

Of these 1,012, 46% responded as above, and only 15% indicated their belief that humans evolved without a God playing a part in the process. ••

How does a sample of size 1,012 get used to generalize to the entire population of Americans? This is the focus of this chapter. The key is that random samples are not "random" in a casual sense, but rather obey laws of probability. This introduces a language to make statements such as confidence intervals mentioned in the previous chapter. In this chapter we will see that if certain assumptions on the population that a random sample is drawn from are valid, then a certain precision can be made as to how a sample statistic can be used to estimate or make inferences about a population parameter.

8.1 Confidence intervals for a population proportion, p

Let's look more closely at the Gallup poll on creationism. Suppose, that instead of three possible answers, we had just two. We could then label one a "success" and the other a "failure." The probability a person chosen from the population would produce a success is a parameter of the population, and traditionally called p. In theory p is known – just round up all Americans and get them to answer the question. The proportion of successes is p. Of course the task of querying an entire population is enormous. The U.S. Census only tries it only once every 10 years. Trying to infer values about p by cheaper, more feasible means, then is done through statistical inference.

Imagine the random sample produced by asking 1,012 Americans the question on mankind's origins. A model for the number of successes would be the binomial model, Binomial$(1012, p)$. This model would be most appropriate if the sampling was done with replacement, as then the results would be truly independent, but even if that is not the case this is an excellent approximation. Let $\hat{p}$ denote the proportion of successes in the random sample. This is 0.46, or basically 466 out of 1,012. Any random sample for the binomial distribution above produces random numbers according to the probability rules of the binomial, each particular sample yields one value of $\hat{p}$ and this choice yielded 0.46.

As in the last chapter, it is natural to estimate the unknown p by the value $\hat{p}$, as is done in the Gallup report. But what is the accuracy? Let $\hat{p}$ denote a random variable with the Binomial$(1012, p)$ distribution. Then $\hat{p}$ has a distribution depending on p, and so would the error $\hat{p} - p$.

In the last chapter we simulated values from an assumed p to get the sampling distribution of the error. Here we do something different. The rules of probability tell us that

$$\mathsf{SD}(\hat{p}) = \sqrt{\frac{p(1-p)}{n}},$$

and the central limit theorem for binomial random variables tells us that for large enough n the distribution of $(\hat{p} - p)/\mathsf{SD}(\hat{p})$ is a standard normal.

With this information, we could use qnorm(0.025) to inform us that with probability 0.95 we have

$$-1.96 \cdot \mathsf{SD}(\hat{p}) < \hat{p} - p < 1.96 \cdot \mathsf{SD}(\hat{p}).$$

The above applies to the random variable $\hat{p}$, but we have a realization of that random variable, $\hat{p} = 0.46$. We are only 95% confident in our process that the above is the case, as once we observe our random variable $\hat{p}$, our probability language isn't appropriate.

With that small but important result made, the next task would be to solve for pair of inequalities involving p. The above can be solved to produce that, but in this case we choose to first make one more approximation.

Recall, we assume n is large enough for the central limit theorem to hold. For large enough n a different version applies as well. The standard error for $\hat{p}$ is the standard deviation with unknown parameters replaced by sample estimates. In this case it becomes:

$$SE(\hat{p}) = \sqrt{\frac{\hat{p}(1-\hat{p})}{n}}.$$

Then the following random variable also has a standard normal distribution for sufficiently large n:

$$\frac{\hat{p}-p}{SE(\hat{p})} = \frac{\text{observed} - \text{expected}}{SE}.$$

From this, we can say with 95% confidence that:

$$-1.96 \cdot SE(\hat{p}) < \hat{p} - p < 1.96 \cdot SE(\hat{p}).$$

This can now be inverted for p to produce an interval containing p based on the sample value of $\hat{p}$

$$\hat{p} - 1.96 \cdot SE(\hat{p}) < p < \hat{p} + 1.96 \cdot SE(\hat{p}),$$

or $\hat{p} \pm 1.96 \cdot SE$.

For our Gallup example, $\hat{p} = 0.46$ and $n = 1,012$. This gives

```
phat <- 0.46
n <- 1012
SE <- sqrt(phat*(1-phat)/n)
c(lower = phat - 1.96 * SE, upper = phat + 1.96 * SE)

##  lower  upper
## 0.4293 0.4907
```

(The values from the last command can also be produced with phat + c(-1,1) * 1.96 * SE.)

Margin of error The value $1.96 \cdot SE$ is known as the *margin of error*. The multiplying factor 1.96 is related to the confidence level. In the above, we used 95%. Let $\alpha = 1 - 0.95$ measure the level of confidence. (The 0.95 could be replaced by other values.) Here is how we can find our multiplying factor from α (cf. Figure 8.2):

```
conf.level <- c(0.80, 0.90, 0.95, 0.99)
alpha <- 1 - conf.level
multipliers <- qnorm(1 - alpha/2)
setNames(multipliers, conf.level)        # give some names

##   0.8   0.9  0.95  0.99
## 1.282 1.645 1.960 2.576
```

The margin of error is composed from two factors. The multiplier depends on the level of confidence—the more confidence, the bigger the multiplier, hence the bigger the margin of error. This should make sense, if there were more room for error then there should be more confidence. When using a symbol for a multiplier, we will use z^* (where z refers to the normal distribution).

The other factor is the standard error. This depends on $\hat{p}$ to some degree. (It is smaller if $\hat{p}$ is close to 0 or 1.) More importantly, is the presence of $\sqrt{n}$ in the denominator. If n is larger, the standard error will be smaller, hence the margin of error is smaller. This too should make sense—larger samples give more confidence in their results. The square root can cause frustration. To *halve* the margin of error for the same $\hat{p}$ requires *four* times the sample size.

Surprisingly, what isn't involved is the size of the population. That means a random sample of size 1,000 from a population of 300, 300 thousand, or 300 million would have the same margin of error, depending only on $\hat{p}$, n, and α. This is great when sampling at random from a population like the United States—just 1,012 people can be used to make an inference about all Americans with a margin of error of just over 3%.

For reference, we have the following.

Confidence interval for a proportion

Let $\hat{p}$ be the sample proportion from a random sample and suppose n is so large that the random variable $(\hat{p} - p)/\text{SE}$ is approximately normal. Then a $(1 - \alpha) \cdot 100\%$ confidence interval for p based on $\hat{p}$ is given by

$$\hat{p} \pm z^* \sqrt{\frac{p(1-p)}{n}}.$$

The function prop.test will compute a confidence interval based on values x and n, though through a slightly different formula. The basic call is prop.test(x, n, conf.level = 0.95) where x is the number of success, n the number of trials, and conf.level the confidence level, with default of 0.95. The similarly used function binom.test employs the binomial distribution to compute an exact confidence interval.

• Example 8.3: Physical activity for college students
A biology student reads that moderate physical activity is defined as an activity that burns about 150 calories of energy per day. She wants to know what percent of her college's 1,812 students achieve this level at least 5 times per week. To do so, with the aid of some friends, she finds a random sample of 125 students of which 80 said they engaged in moderate exercise 5 or more

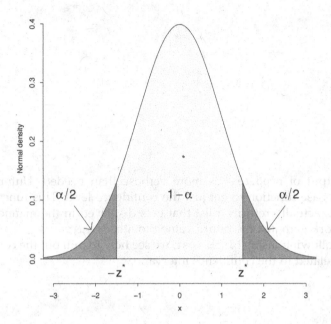

Figure 8.2: Relationship between multiplying factor z^* and confidence level $(1 - \alpha) \cdot 100\%$. Similar pictures could be made for distributions other than the normal.

times per week. What is a 90% confidence level for the proportion of all 1,812 students?

This can be answered many ways. We begin with the basic R commands:

```
x <- 80; n <- 125
phat <- x/n
alpha <- 1 - 0.90
zstar <- qnorm(1 - alpha/2)
SE <- sqrt(phat * (1 - phat) / n)
MOE <- zstar * SE
phat + c(-1, 1) * MOE

## [1] 0.5694 0.7106
```

We can compare to the value returned by prop.test:

```
prop.test(x, n)

##
```

```
##  1-sample proportions test with continuity correction
##
## data:  x out of n, null probability 0.5
## X-squared = 9.248, df = 1, p-value = 0.002358
## alternative hypothesis: true p is not equal to 0.5
## 95 percent confidence interval:
##  0.5488 0.7225
## sample estimates:
##    p
## 0.64
```

The output of prop.test is more verbose than needed. Unfortunately there is no base function to get just the confidence level. This function, like binom.test, actually returns a list that gets displayed in the manner above. One can work with just the return value though.

In the following, using binom.test, we see how to pull out the component of the list related to the confidence interval:

```
binom.test(x, n)$conf.int
```

```
## [1] 0.5493 0.7239
## attr(,"conf.level")
## [1] 0.95
```

The UsingR package provides a confint convenience method: [2]

```
confint(binom.test(x, n))
```

```
## (0.55, 0.72) with 95 percent confidence
```

••

• **Example 8.4: The missing confidence level**
In United States' newspapers the results of a survey are often printed with the sample proportion, the sample size, and a margin of error. The confidence level is almost always missing but can be inferred from the three pieces of information. If a survey has $\widehat{p} = .57$, $n = 1{,}000$, and a margin of error of 3 percentage points, what is α?

Assuming the survey is done with a random sample, we are given that $z^{*}\text{SE}(\widehat{p}) = .03$ and $\text{SE} = \sqrt{.57 \cdot (1 - .57)/1000}$. Solve for z^{*} and then $1 - \alpha$ as follows:

---

[2]The confint function is typically used with R's model formula. It is added to the UsingR package just to streamline the exposition.

```
zstar <- 0.03 / sqrt(.57*(1-.57)/1000)
alpha <- 2* pnorm(-zstar)
(1-alpha) * 100

## [1] 94.47
```

There is an implied 95% confidence level.   ••

Problems

8.1 Find an example in the media in which the results of a poll are presented. Identify the population, the size of the sample, the confidence interval, the margin of error, and the confidence level.

8.2 A random sample from the United States population is taken by using listed residential phone numbers. Which segments of the population would be missed by this sampling method?

8.3 The web site http://www.cnn.com conducts daily polls. Explain why this disclaimer is needed:

> This QuickVote is not scientific and reflects the opinions of only those Internet users who have chosen to participate. The results cannot be assumed to represent the opinions of Internet users in general, nor the public as a whole.

8.4 Suppose a Zogby poll with 1,013 randomly selected participants and the http://www.cnn.com poll (see the previous problem) with 80,000 respondents ask the same question. If there is a discrepancy in the sample proportion, which would you believe is closer to the unknown population parameter?

8.5 Find 80% and 90% confidence intervals for a survey with $n = 100$ and $\hat{p} = 0.45$.

8.6 A student wishes to find the proportion of left-handed people at her college. She surveys 100 fellow students and finds that only 5 are left-handed. If she computed a 95% confidence interval would it contain the value of $p = 1/10$?

8.7 Of the last ten times you've dropped your toast, it has landed sticky-side down nine times. If these are a random sample from the Bernoulli(p) distribution, find an 80% confidence interval for p, the probability of the sticky side landing down.

8.8 A *New York Times* article from October 9, 2003, contains this explanation about an exit survey for a California recall election:

> In theory, in 19 cases out of 20, the results from such polls should differ by no more than plus or minus two percentage points from what would have been obtained by seeking to interview everyone who cast a ballot in the recall election.

Assume a simple random sample and $\hat{p} = .54$. How big was n?

8.9 An erstwhile commercial claimed that "Four out of five dentists surveyed would recommend Trident for their patients who chew gum."
    Assume the results were based on a random sample from the population of all dentists. Find a 90% confidence interval for the true proportion if the sample size was $n = 5$. Repeat with $n = 100$ and $n = 1,000$.

8.10 A survey is taken of 250 students, and a $\hat{p}$ of 0.45 is found. The same survey is repeated with 1,000 students, and the same $\hat{p}$ value is found. Compare the two 95% confidence intervals. What is the relationship? Is the margin of error for the second one four times smaller? If not, how much smaller is it?

8.11 How big a survey is needed to be certain that a 95% confidence interval has a margin of error no bigger than 0.01? How does this change if you are asked for an 80% confidence interval?

8.12 The phrasing, "The true value, p, is in the confidence interval with 95% probability" requires some care. Either p is or isn't in a given interval. What it means is that if we *repeated* the sampling, there is a 95% chance the true value is in the random interval. We can investigate this with a simulation. The commands below will find several confidence intervals at once.

```
M <- 50; n <- 20; p <- .5;     # toss 20 coins 50 times,
alpha <- 0.10;
zstar <- qnorm(1-alpha/2)
phat <- rbinom(M, n, p)/n       # divide by n for proportions
SE <- sqrt(phat*(1-phat)/n)    # compute SE
```

We can find the proportion that contains p using

```
sum(phat - zstar*SE < p & p < phat + zstar * SE)/n

## [1] 2.25
```

and draw a nice graphic with

```
matplot(rbind(phat - zstar*SE, phat + zstar*SE),
        rbind(1:m,1:m),type="l",lty=1)
abline(v=p)                        # indicate parameter value
```

Do the simulation above. What percentage of the 50 confidence intervals contain $p = 0.5$?

8.2 Confidence intervals for the population mean

In this section, we follow the pattern of the last section to show how to create a confidence interval based on $\bar{x}$ to make statistical inference about a population mean, μ.

Let $x_1, x_2, \ldots, x_n$ be a random sample from a normal population with mean μ and standard deviation σ. Then this statistic

$$Z = \frac{\bar{x} - \mu}{\sigma/\sqrt{n}} = \frac{\bar{x} - \mu}{\text{SD}(\bar{x})}$$

will have a standard normal distribution. This implies, for example, that roughly 95% of the time Z is no larger than 2 in absolute value. Were σ known, we could use this to create a confidence interval for μ based on $\bar{x}$.

Unfortunately, it is a rare example where the standard deviation is known, but the mean is not.

As in the last section, we replace the standard deviation in the above with the standard error, simply by estimating σ with the sample standard deviation s. $(\text{SE}(\bar{x}) = s/\sqrt{n}.)$ This gives:

$$T = \frac{\bar{x} - \mu}{s/\sqrt{n}} = \frac{\bar{x} - \mu}{\text{SE}(\bar{x})} = \frac{\text{observed} - \text{expected}}{\text{SE}}.$$

In the past section, when encountering a similar statistic, it was assumed that n was large enough so that the central limit theorem implies the sampling distribution was approximately the standard normal. For this statistic, this will also be the case for large values of n, as can be seen empirically in Figure 7.4. This will also be the case for populations which are non-normal, though "large enough" will depend on the population's characteristics.

However, for small values of n the distribution of T is not normal. It was Gosset [26] who computed the distribution of the above statistic for a normally distributed population. It has the t-distribution with $n - 1$ degrees of freedom.

The t-distribution is a symmetric, bell-shaped distribution that asymptotically approaches the standard normal distribution but for small n has fatter tails. The degrees of freedom is a parameter for this distribution the way the mean and standard deviation are for the normal distribution.

As with the normal distribution, the t-distribution allows us to solve for either t^* or α in the following equation, assuming one or the other is known:

$$P\left(-t^* < \frac{\bar{x} - \mu}{SE(\bar{x})} < t^*\right) = 1 - \alpha.$$

This can be rewritten algebraically as

$$P(\bar{x} - t^* \cdot SE < \mu < \bar{x} + t^* \cdot SE) = 1 - \alpha,$$

a form which leads easily to a confidence interval.

Confidence intervals for the mean

Let $x_1, x_2, \ldots, x_n$ be a random sample from a normal population with mean μ and variance σ^2. A $(1 - \alpha)100\%$ confidence
interval for μ is given by

$$\bar{x} \pm t^* SE(\bar{x}),$$

where the multiplier t^* is related to α through the t-distribution
with $n - 1$ degrees of freedom.

For unsummarized data, the function t.test will compute confidence
intervals. A template for its usage is

```
t.test(x, conf.level=0.95)
```

The data is stored in a data vector (named x above) and the confidence level
is specified with conf.level.

Finding t^* with R   Computing the value of t^* for a given α and vice versa
is done in a manner similar to finding z^*, except that a different density
is used. Changing to a new density requires nothing more than using the
proper family name—t, for the t-distribution, and norm for the normal—and
specifying the parameter values. In particular, if n is the sample size, then the
two are related as follows:

```
tstar <- qt(1 - alpha / 2, df=n - 1)
alpha <- 2 * pt(-tstar, df=n - 1)
```

By way of contrast, for z^* the corresponding commands are

```
zstar <- qnorm(1 - alpha / 2)
alpha <- 2 * pnorm(-zstar)
```

• Example 8.5: Average height
Students in a class of 30 have an average height of 66 inches, with a standard

deviation of 4 inches. Assume that these heights are normally distributed, and that the class can be considered a random sample from the entire college population. What is an 80% confidence interval for the mean height of all the college students?

Our assumptions allow us to apply the confidence interval for the mean, so the answer is $\bar{x} \pm t^*\text{SE}$. Computing gives

```
xbar <- 66; s <- 4; n <- 30
alpha <- 1 - 0.80
tstar <- qt(1 - alpha/2, df = n-1)        # 1.311
SE <- s/sqrt(n)
MOE <- tstar * SE
xbar + c(-1,1) * MOE

## [1] 65.04 66.96
```

••

• **Example 8.6: Making coffee**
A barista at "*t*-test espresso" has been trained to set the bean grinder so that a 25-second espresso shot results in 2 ounces of espresso. Knowing that variations are the norm, he pours eight shots and measures the amounts to be 1.95, 1.80, 2.10, 1.82, 1.75, 2.01, 1.83, and 1.90 ounces. Find a 90% confidence interval for the mean shot size. Does it include 2.0?

As we have the data, we can use t.test directly. We enter in the data, verify normality (with a quantile-quantile plot that is not shown), and then call t.test:

```
ozs <- c(1.95, 1.80, 2.10, 1.82, 1.75, 2.01, 1.83, 1.90)
qqnorm(ozs)                              # approximately linear
confint(t.test(ozs, conf.level=0.80))

## (1.84, 1.95) with 80 percent confidence
```

Finding the confidence interval to be $(1.84, 1.95)$, the barista sees that 2.0 is not in the interval. The barista adjusts the grind to be less fine and switches to decaf. ••

The T-statistic is robust The confidence interval for the mean relies on the fact that the sampling distribution of $T = (\bar{x} - \mu)/\text{SE}$ is the t-distribution with $n - 1$ degrees of freedom. This is true when the x_i are a random sample from a normal population. What if that assumption in our model for the data is not appropriate?

If n is small, we can do simulations to see that the distribution of T is still approximately the t-distribution if the parent distribution of the x_i is not too far from normal. That is, the tails can't be too long, or the skew can't be too great. When n is large, the central limit theorem applies.

A statistic whose sampling distribution doesn't change dramatically for moderate changes in the population distribution is called a *robust statistic*.

• Example 8.7: Likert data
Many questionnaires are based on a 5-point scale, often called a Likert scale. If these values are numeric (e.g., one through five) can a t-test be used? Let's look at a simulation. where there are $n = 10$ questions.

```
k <- 5                                      # answers per question
n <- 10                                     # number of questions
tstat <- function(x, mu=0) (mean(x) - mu)/(sd(x)/sqrt(length(x)))
res <- replicate(2000, {
  xs <- sample(1:k, n, replace=TRUE)
  tstat(xs, (k+1)/2)
})
## look at one quantile
sum(res > qt(0.05, df=n-1, lower.tail=FALSE)) / length(res)

## [1] 0.049
```

The value of 0.049 is very close to the value 0.05, as would be the case if the sampling distribution were the t-distribution. Figure 8.3 shows quantile-quantile plots against the t-distribution for simulations with $k = 2$, 5, and 15. Clearly, when there are just two choices the distribution is not the t-distribution. Unfortunately even with 5 choices, the answers are often concentrated on just one or two. A common strategy to avoid this is to create subscores by summing related questions.   ••

One-sided confidence intervals

When finding a confidence interval for the mean for a given α, we found t^* so that $P(-t^* \leq T_{n-1} \leq t^*) = 1 - \alpha$. This method returns symmetric confidence intervals. The basic idea is that the area under the density of the sampling distribution that lies outside the confidence interval is evenly split on each side. This leaves $\alpha/2$ area in each tail. This approach is not the only one. This extra area can be allocated in any proportion to the left or right of the confidence interval. One-sided confidence intervals put the area all on one

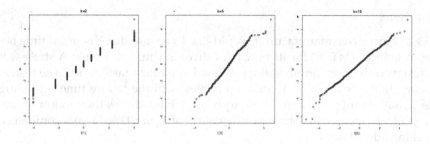

Figure 8.3: Simulations of the T-statistic for Likert data with $k = 2, 5$, and 15.

side or the other. For confidence intervals for the mean, based on the T-statistic, these would be found for a given α by finding t^* such that $P(t^* \leq T) = 1 - \alpha$ or $P(T \leq t^*) = 1 - \alpha$.

In R, the prop.test, binom.test, and t.test functions can return one-sided confidence intervals. When the argument alt="less" is used, an interval of the type $(-\infty, b]$ is printed. Similarly, when alt="greater" is used, an interval of the type $[b, \infty)$ is printed.

• **Example 8.8: Serving coffee**
The barista at "t-test espresso" is told that the optimal serving temperature for coffee is 180°F. Five temperatures are taken of the served coffee: 175, 185, 170, 184, and 175 degrees. Find a 90% confidence interval of the form $(-\infty, b]$ for the mean temperature.

Using t.test with alt="less" will give this type of one-sided confidence interval:

```
x <- c(175, 185, 170, 184, 175)
t.test(x,conf.level = 0.90, alt="less")

##
##  One Sample t-test
##
## data:  x
## t = 61.57, df = 4, p-value = 1
## alternative hypothesis: true mean is less than 0
## 90 percent confidence interval:
##    -Inf 182.2
## sample estimates:
## mean of x
##     177.8
```

The confidence interval contains the value of 180 degrees. ••

Problems

8.13 A hard-drive manufacturer would like to ensure that the mean time between failures (MTBF) for its new hard drive is 1 million hours. A stress test is designed that can simulate the workload at a much faster pace. The testers assume that a test lasting 10 days correlates with the failure time exceeding the 1-million-hour mark. In stress tests of 15 hard drives they found an average of 9.5 days, with a standard deviation of 1 day. Does a 90% confidence level include 10 days?

8.14 The stud.recs (UsingR) data set contains a sample of math SAT scores from some population in the variable sat.m. Find a 90% confidence interval for the mean math SAT score for this data.

8.15 For the homedata (UsingR) data set find 90% confidence intervals for both variables y1970 and y2000, assuming the sample represents some population. Use t.test, but first discuss whether the model assumptions are appropriate.

8.16 The variable weight in the kid.weights data set contains the weights of a random sample of children. Find a 90% confidence interval for the weight of 5-year-olds. You'll need to isolate just the 5-year-olds' data first. Here's one way:

```
yr5 <- subset(kid.weights, subset= 5*12 <= age & age < 6*12)
```

8.17 The brightness data set contains information on the brightness of stars in a sector of the sky. Find a 90% confidence interval for the mean, assuming this data is a sample from the population of all visible stars.

8.18 The data set normtemp contains measurements of 130 healthy, randomly selected individuals. The variable temperature contains normal body temperature. Does the data appear to come from a normal distribution? Is so, find a 90% confidence interval for the mean normal body temperature of the population that was sampled. Does it include 98.6°F?

8.19 The *t*-distribution is also called the Student *t*-distribution. A Guinness Brewery employee, William Gosset, derived the distribution of T to handle small samples. (Large samples of beer presumably led to less publishable science.) As Guinness did not allow publication of research results at the time, Gosset chose to publish under the pseudonym Student. The journal chosen was Pearson's journal, a colleague of Gosset and pioneer of statistics.

Gosset applied his research to a data set containing height and left-middle-finger measurements of 3,000 criminals. These values were written on cards and randomly sorted into 750 samples, each containing four criminals. (This is how simulations were done previously.) The data is available in the Macdonell (HistData) data set.

Run a simulation using 750 samples of size 4. Use quantile to find a 95% confidence interval. Compare to that found with t.test for a single sample of size 4.

As the data is tabulated, you'll need to expand it:

```
finger <- with(Macdonell, rep(finger, frequency))
```

8.20 We can investigate how robust the T-statistic is to changes in the underlying parent population from normality. In particular, we can verify that if the parent population is not too skewed or is symmetric without too heavy a tail then the T-statistic will still have the t-distribution for its sampling distribution. (This is for small samples where n is not so large that the central limit theorem is applicable.)

A simulation of the T-statistic when x_i are Normal$(0,1)$ may be done as follows:

```
n <- 10; M <- 2000
res <- replicate(M, {
  x <- rnorm(n)
  (mean(x) - 0)/(sd(x)/sqrt(n))
})
qqplot(res, rt(M, df=n-1))             # compare to t-distribution
```

The quantile-quantile plot compares the distribution of the sample with a sample from the t-distribution. If the above is executed, it should produce a graphic with points that are close to linear, as the sampling distribution is the t-distribution.

To test different parent populations you can change the line x <- rnorm(n) to some other distributions with mean 0. For example, try a short-tailed distribution with x <- runif(n)-1/2; a symmetric, long-tailed distribution with x <- rt(n, 3); a not so long-tailed, symmetric distribution with x <- rt(n, 30); and a skewed distribution with x <- rexp(n) - 1.

8.21 We can compare the relationship of the t-distribution with $n - 1$ degrees of freedom with the normal distribution in several ways. As n gets large, the t-distribution converges to the standard normal. But what happens when n is "small," and what do we mean by "large?"

A few comparative graphs can give us an idea. For $n = 10$ we can use boxplots of simulated data to examine the tails, or we can compare plots of theoretical quantiles or densities. These plots are created as follows:

```
n <- 10
## simulated data
boxplot(rt(1000, df=n - 1), rnorm(1000))
## theoretical qqplots
x <- seq(0, 1, length=150)
plot(qt(x, df=n - 1), qnorm(x))
abline(0, 1)
## compare densities
curve(dnorm(x), -3.5, 3.5)
curve(dt(x, df=n - 1), lty=2, add=TRUE)
```

Repeat the above for $n = 3, 25, 50$, and 100. What value of n seems "large" enough to say that the two distributions are essentially the same?

8.22 When the parent population is Normal(μ, σ) with known σ, then confidence intervals of the type

$$\bar{x} \pm z^* \text{SD}(\bar{x}) \text{ and } \bar{x} \pm t^* \text{SE}(\bar{x})$$

are both applicable. We have that far enough in the tail, $z^* < t^*$, but sometimes $s < \sigma$, so there is no clear winner as to which confidence interval will be smaller for a given sample.

Run a simulation 200 times in which the margin of error is calculated both ways for a sample of size 10 with $\sigma = 2$ and $\mu = 0$. Use a 90% confidence level. What percent of the time was the confidence interval using $\text{SD}(\bar{x})$ smaller?

8.3 Other confidence intervals

To form confidence intervals, we have used the key fact that certain statistics,

$$\frac{\hat{p} - p}{\text{SE}} \text{ and } \frac{\bar{x} - \mu}{\text{SE}},$$

have known sampling distributions that do not involve any population parameters. From this, we could then solve for confidence intervals for the parameter in terms of known quantities.

In general, such a statistic is called a *pivotal quantity* and can be used to generate a number of confidence intervals in various situations.

Confidence interval for σ^2

For example, if the x_i are *i.i.d.* normals, then the distribution of

$$\frac{(n-1)s^2}{\sigma^2}$$

was shown by Helmert in 1876 to be the χ^2-distribution (chi-squared) with $n - 1$ degrees of freedom. This fact allows us to solve for confidence intervals for σ^2 in terms of the sample variance s^2.

In particular, a $(1 - \alpha) \cdot 100\%$ confidence interval can be found as follows. For a given α, let l^* and r^* solve

$$P(l^* \leq \chi^2_{n-1} \leq r^*) = 1 - \alpha.$$

If we choose l^* and r^* to yield equal areas in the tails, we can find them with

```
n <- 10; alpha <- 1 - 0.90              # say
lstar <- qchisq( alpha/2, df=n-1)
rstar <- qchisq(1-alpha/2, df=n-1)
```

Then

$$P(l^* \leq \frac{(n-1)s^2}{\sigma^2} \leq r^*) = 1 - \alpha$$

can be rewritten as

$$P(\frac{(n-1)s^2}{r^*} \leq \sigma^2 \leq \frac{(n-1)s^2}{l^*}) = 1 - \alpha.$$

In other words, the interval $((n-1)s^2/r^*, (n-1)s^2/l^*)$ gives a $(1 - \alpha) \cdot 100\%$ confidence interval for σ^2.

• Example 8.9: How long is a commute?

A commuter believes her commuting times are independent and vary according to a normal distribution, with unknown mean and variance. She would like to estimate the variance to get an idea of the spread of times, knowing then it is unlikely that a commute would be two or more standard deviations longer than the average commute.

To compute the variance, she records her commute time on a daily basis. Over 10 commutes she reports a mean commute time of 25 minutes, with a sample variance of 12 minutes. What is a 95% confidence interval for the population variance?

We are given $s^2 = 12$ and $n = 10$, and we assume each x_i is normal and i.i.d. From this we find

```
s2 <- 12; n <- 10
alpha <- 1 - 0.95
lstar = qchisq(alpha/2, df=n - 1)
rstar = qchisq(1 - alpha/2, df=n - 1)
(n-1) * s2 * c(1/rstar, 1/lstar)       # CI for sigma squared

## [1]  5.677 39.994
```

After taking square roots, we get a 95% confidence interval for σ, which is $(2.383, 6.324)$.   ••

Problems

8.23 Let $x_1, x_2, \ldots, x_n$ and $y_1, y_2, \ldots, y_m$ be two *i.i.d.* samples with sample variances s_x and s_y respectively. A confidence interval for the equivalence of sample variances can be given from the following statistic:

$$F = \frac{(s_x^2/\sigma_x^2)}{(s_y^2/\sigma_y^2)}.$$

If the underlying x_i and y_i are normally distributed, then the distribution of F is known to be the F-distribution with $n-1$ and $m-1$ degrees of freedom. That is, F is a pivotal quantity, so probability statements such as $P(a \leq (s_x^2/\sigma_x^2)/(s_y^2/\sigma_y^2) \leq b)$ can be answered with the known quantiles of the F-distribution. For example,

```
n <- 11; m <- 16
alpha <- 1 - 0.90
qf(c(alpha/2, 1- alpha/2), df1=n-1, df2=m-1)

## [1] 0.3515 2.5437
```

says that $P(0.3515 \leq (s_x^2/s_y^2)/(\sigma_x^2/\sigma_y^2) \leq 2.5437) = 0.9$ when $n = 11$ and $m = 16$. That is,

$$\frac{1}{2.5437}\frac{s_x^2}{s_y^2} < \frac{\sigma_x^2}{\sigma_y^2} < \frac{1}{0.3515}\frac{s_x^2}{s_y^2}$$

with 90% confidence.

Suppose $n = 10, m = 20, s_x = 2.3$, and $s_y = 2.8$. Find an 80% confidence interval for the ratio of σ_x/σ_y.

8.24 Assume our data, $x_1, x_2, \ldots, x_n$, is uniform on the interval $[0, \theta]$ (θ is an unknown parameter). Set max to be the maximum value in the data set. Then the quantity max/θ is pivotal with distribution

$$P\left(\frac{\text{max}}{\theta} < x\right) = x^n, \quad 0 \leq x \leq 1.$$

Thus $P(\text{max}/x < \theta) = x^n$. As θ is always bigger than $\text{max}(X)$, we can solve for $x^n = \alpha$ and get that θ is in the interval $[\text{max}, \text{max}/x]$ with probability $1 - \alpha$.

Use this fact to find a 90% confidence interval for the number of entries in the 2002 New York City Marathon. The place variable from the data set nyc.2002 contains the place of the runner in the sample and is randomly sampled from all the possible places.

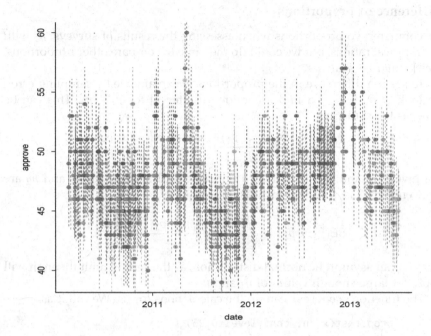

Figure 8.4: Fluctuating presidential approval ratings in United States from early 2010 through spring 2013.

8.4 Confidence intervals for differences

When we have two samples, we might ask whether the two came from the same population. For example, Figure 8.4 shows results for several polls on presidential approval rating from early 2010 to early 2013.[3] It is clear that the rating varies over time, but for any given time period the polls are all pretty much in agreement. This is to be expected, as the polls are tracking the same population proportion for a given time period. However, how can we tell if the differences between polls for different time periods are due to a change in the underlying population proportion or merely an artifact of sampling variation? One method is to estimate the difference between the population parameters computing confidence intervals to assess how confident we are about the estimate.

---

[3]A similar figure appeared in a February 9, 2004, edition of *Salon* (http://www.salon.com) for then President Bush. The data for this figure is in ObamaApproval, for the Bush numbers one has BushApproval.

Difference of proportions

We compare two proportions when assessing the results of surveys, as with the approval ratings, but we could do the same to compare other proportions, such as market shares.

To see if a difference in the proportions is explainable by sampling error, we look at $\hat{p}_1 - \hat{p}_2$ and find a confidence interval for $p_1 - p_2$. This can be done, as the statistic

$$Z = \frac{(\hat{p}_1 - \hat{p}_2) - (p_1 - p_2)}{SE(\hat{p}_1 - \hat{p}_2)} = \frac{observed - expected}{SE},$$

is a pivotal quantity with standard normal distribution when n_1 and n_2 are large enough. The standard error is known to be:

$$SE(\hat{p}_1 - \hat{p}_2) = \sqrt{\frac{\hat{p}_1(1 - \hat{p}_1)}{n_1} + \frac{\hat{p}_2(1 - \hat{p}_2)}{n_2}}.$$

Z has an asymptotic normal distribution, as the central limit theorem will apply for large enough values of n_1 and n_2.

The function prop.test can do the calculations for us. We call it as

```
prop.test(x, n, conf.level=0.95)
```

The data is specified in terms of counts, x, and sample sizes, n, using data vectors containing two entries. The results will differ slightly from the above description, as prop.test uses a continuity correction.

• Example 8.10: Comparing poll results
In a span of two weeks the same poll is taken. The first time, 1,000 people are interviewed, and 560 agree; the second time, 1,200 are interviewed, and 570 agree. Find a 95% confidence interval for the difference of proportions.

Rather than do the work by hand, we let prop.test find a confidence interval.

```
prop.test(x=c(560,570), n=c(1000,1200), conf.level=0.95)

##
##   2-sample test for equality of proportions with
##   continuity correction
##
## data:  c(560, 570) out of c(1000, 1200)
## X-squared = 15.44, df = 1, p-value = 8.53e-05
## alternative hypothesis: two.sided
## 95 percent confidence interval:
##   0.04231 0.12769
## sample estimates:
## prop 1 prop 2
##   0.560  0.475
```

We see that a 95% confidence interval is $(0.04231, 0.12769)$, which misses including 0. We conclude that there appears to be a real difference in the population parameters. ••

Difference of means

Many problems involve comparing independent samples to see whether they come from identical parent populations. A teacher could compare two sections of the same class to look for differences; a pharmaceutical company could compare the effects of two drugs; or a manufacturer could compare two samples taken at different times to monitor quality control.

Let $x_1, x_2, \dots, x_{n_x}$ and $y_1, y_2, \dots, y_{n_y}$ be the two samples with sample means $\bar{x}$ and $\bar{y}$ and sample variances s_x^2 and s_y^2. Assume the populations for each sample are normally distributed. The sampling distribution of $\bar{x} - \bar{y}$ is asymptotically normal, as each is asymptotically normal. As with the one-sample T-statistic, if we normalize by the standard error:

$$T = \frac{(\bar{x} - \bar{y}) - (\mu_x - \mu_y)}{SE(\bar{x} - \bar{y})} \tag{8.1}$$

the result will have an (approximate) t-distribution and for large values of n_x and n_y be approximately a standard normal.

The standard error of $\bar{x} - \bar{y}$ is computed differently depending on the assumptions. First, for independent random variables, the variance of a sum is the sum of a variance. This is used to show that $VAR(\bar{x} - \bar{y}) = \sigma_x^2/n_x + \sigma_y^2/n_y$. Taking a square root yields the standard deviation. It depends on two unknown population parameters—the two standard deviations. The standard error is found by replacing these by an appropriate estimate:

- When the two population variances are equal, the data can be pooled to give an estimate of the common variance σ^2. Pooling then uses more data to make an estimate, so is intuitively more accurate. Let s_p^2 be the pooled estimate. It is defined by

$$s_p^2 = \frac{(n_x - 1)s_x^2 + (n_y - 1)s_y^2}{n_x + n_y - 2}. \tag{8.2}$$

- When the population variances are not equal, the two sample standard deviations are used to estimate the respective population standard deviations.

The standard error is then

$$SE(\bar{x} - \bar{y}) = \begin{cases} \sqrt{s_p^2(1/n_x + 1/n_y)} & \text{if } \sigma_x = \sigma_y, \\ \sqrt{s_x^2/n_x + s_y^2/n_y} & \text{if } \sigma_x \neq \sigma_y. \end{cases} \tag{8.3}$$

The statistic T will have a sampling distribution given by the t-distribution. When the variances are equal, the sampling variation of s_p is smaller, as all the data is used to estimate σ. This is reflected in a larger value of the degrees of freedom. The values used are

$$\text{degrees of freedom} = \begin{cases} n_x + n_y - 2 & \text{if } \sigma_x = \sigma_y, \\[2ex] \left(\frac{s_x^2}{n_x} + \frac{s_y^2}{n_y}\right)^2 \cdot \left(\frac{(s_x^2/n_x)^2}{n_x-1} + \frac{(s_y^2/n_y)^2}{n_y-1}\right)^{-1} & \text{if } \sigma_x \neq \sigma_y. \end{cases}$$

$$(8.4)$$

(The latter value is the Welch approximation.)

Given this, the T-statistic is pivotal, allowing for the following confidence intervals.

> ## Confidence intervals for difference of means for two independent samples
>
> Let $x_1, x_2, \ldots, x_{n_x}$ and $y_1, y_2, \ldots, y_{n_y}$ be two independent samples with distribution $\text{Normal}(\mu_i, \sigma_i)$, $i = x$ or y. A $(1-\alpha) \cdot 100\%$ confidence interval of the form
>
> $$(\bar{x} - \bar{y}) \pm t^* \text{SE}(\bar{x} - \bar{y})$$
>
> can be found where t^* is given by the t-distribution. This is based on the sampling distribution of T given in Equation 8.1. This distribution is the t-distribution. The standard error and degrees of freedom differ depending on whether or not the variances are assumed equal in the model for the data. The standard error is given by Equation 8.3 and the degrees of freedom by Equation 8.4.

If the unsummarized data is available, the t.test function can be used to compute the confidence interval for the difference of means. It is used as

```
t.test(x, y, var.equal=FALSE, conf.level=0.95)
```

The data is contained in two data vectors, x and y. The assumption on the equality of variances is specified by the argument var.equal= with default of FALSE.

• **Example 8.11: Comparing independent samples**
In a clinical trial, a weight-loss drug is tested against a placebo to see whether the drug is effective. The amount of weight loss for each group is given by

```
x <- c(0, 0, 0, 2, 4, 5, 13, 14, 14, 14, 15, 17, 17)
y <- c(0, 6, 7, 8, 11, 13, 16, 16, 16, 17, 18)
boxplot(list(placebo=x, ephedra=y), col="gray") # compare spreads
```

Find a 90% confidence interval for the difference in mean weight loss.

As there is no expectation that the variability should be changed by the treatment, we inspect boxplots of the data (not shown). The assumption of equal variances appears reasonable prompting the use of t.test with the argument var.equal=TRUE.

```
confint(t.test(x,y, var.equal=TRUE))

## (-8.28, 2.70) with 95 percent confidence
```

By comparison, if we did not assume equal variances, then the computation yields:

```
confint(t.test(x,y))

## (-8.19, 2.61) with 95 percent confidence
```

With var.equal=TRUE, we have $13 + 11 - 2 = 22$ degrees of freedom. In this example, the approximate degrees of freedom in the unequal variance case is found to be 21.99—essentially identical. The default 95% confidence interval is $(-8.279, 2.699)$, so the difference of 0 is still in the confidence interval, even though the sample means differ quite a bit at first glance (8.846 versus 11.636). ••

Formula interface The t.test function has a formula interface. The specification is y ~ f, where f is a two-level factor. This can be convenient when the data is stored in the long format.

For example, the GaltonFamilies (HistData) data set lists the height and gender of 934 children. A 95% confidence interval for the difference in height by gender based on this sample (which is assumed to be a random sample from a population of interest) could be found with:

```
out <- t.test(childHeight ~ gender, GaltonFamilies, conf.level=0.95)
confint(out)

## (-5.45, -4.81) with 95 percent confidence
```

Matched samples

Sometimes we have two samples that are not independent. They may be paired or matched up in some way. A classic example in statistics involves the measurement of shoe wear. If we wish to measure shoe wear, we might give five kids one type of sneaker and five others another type and let them play for a while. Afterward, we could measure shoe wear and compare. The only problem is that variation in the way the kids play could mask small variations in the wear due to shoe differences. One way around this is to put mismatched pairs of shoes on all ten kids and let them play. Then, for each kid, the amount of wear on each shoe is related, but the difference should be solely attributable to the differences in the shoes.

If the two samples are $x_1, x_2, \ldots, x_n$ and $y_1, y_2, \ldots, y_n$, then the statistic

$$T = \frac{(\bar{x} - \bar{y}) - (\mu_x - \mu_y)}{\text{SE}(\bar{x} - \bar{y})}$$

is pivotal with a t-distribution. What is the standard error? As the samples are not independent, the standard error for the two-sample T-statistic is not applicable. Rather, it is just the standard error for the single sample $x_i - y_i$.

> ### Comparison of means for paired samples
>
> Let $x_1, x_2, \ldots, x_n$ and $y_1, y_2, \ldots, y_n$ be two samples. If the sequence of differences, $x_i - y_i$, is an *i.i.d.* sample from a Normal(μ, σ) distribution, then a $(1 - \alpha) \cdot 100\%$ confidence interval for the difference of means, $\mu_x - \mu_y$, is given by
>
> $$(\bar{x} - \bar{y}) \pm t^* s / \sqrt{n},$$
>
> where s is the sample standard deviation of the derived sample $x_i - y_i$ and t^* is found from the t-distribution with $n - 1$ degrees of freedom.

The t.test function can compute the confidence interval. If x and y store the data, the function may be called as either

```
t.test(x, y, paired=TRUE)      or      t.test(x - y)
```

We use the argument conf.level=... to specify the confidence level.

• Example 8.12: Comparing shoes

The shoes (MASS) data set contains shoe wear for ten children each wearing two different shoes. By comparing the differences, we can tell whether the two types of shoes have different mean wear amounts.

```
library(MASS)
names(shoes)

## [1] "A" "B"

### Alternately: with(shoes, t.test(A,B,conf.level=0.9,paired=TRUE))
with(shoes, {
  out <- t.test(A-B, conf.level=0.9)
  confint(out)
})

## (-0.63, -0.19) with 90 percent confidence
```

The 90% confidence interval does not include 0, indicating a certain confidence in the two population means differing. ••

Problems

8.25 Two different AIDS-treatment "cocktails" are compared. For each, the time it takes (in years) to fail is measured for seven randomly assigned patients. The data is:

Type	1	2	3	4	5	6	7	$\bar{x}$	s
Cocktail 1:	3.1	3.3	1.7	1.2	0.7	2.3	2.9	2.24	0.99
Cocktail 2:	1.8	2.3	2.2	3.5	1.7	1.6	1.4	2.13	0.69

Find an 80% confidence interval for the difference of means. What assumptions are you making on the data?

8.26 In determining the recommended dosage of AZT for AIDS patients, tests were done comparing efficacy for various dosages. If a low dosage is effective, then that would be recommended, as it would be less expensive and would have fewer potential side effects.

A test to decide whether a dosage of 1,200 mg is similar to one of 400 mg is performed on two random samples of AIDS patients. A numeric measurement of a patient's health is made, and the before-and-after differences are recorded after treatment:

400 mg group	7	0	8	1	10	12	2	9	5	2
1200 mg group	2	1	5	1	5	7	-1	8	7	3

Find a 90% confidence interval for the differences of the means. What do you assume about the data?

8.27 The following data is from IQ tests for pairs of twins that were separated at birth. One twin was raised by the biological parents, the other by adoptive parents.

Foster	80	88	75	113	95	82	97	94	132	108
Biological	90	91	79	97	97	82	87	94	131	115

Find a 90% confidence interval for the differences of mean. What do you assume about the data? In particular, are the two samples independent?

8.28 For the babies data set, the variable age contains the mother's age and the variable dage contains the father's age for several babies. Find a 95% confidence interval for the difference in mean age. Does it contain 0? What do you assume about the data?

8.5   Confidence intervals for the median

The confidence intervals for the mean are based on the fact that the distribution of the statistic

$$T = \frac{\bar{x} - \mu}{SE(\bar{x})}$$

is known. This is true when the sample is an *i.i.d.* sample from a normal population or one close to normal. However, many data sets, especially long-tailed skewed ones, are not like this. For these situations, *nonparametric* methods are preferred. That is, no parametric assumptions on the population distribution for the sample are made, although assumptions on its shape may apply.

Confidence intervals based on the binomial distribution

The binomial distribution can be used to find a confidence interval for the median for any continuous parent population. The key is to look at whether a data point is more or less than the median. As the median splits the area in half, the probability that a data point is more than the median is exactly $1/2$. (We need a continuous distribution with positive density over its range to say "exactly" here.) Let T count the number of data points more than the median in a sample of size n. Then T is a Binomial$(n, 1/2)$ random variable.

Let $x_{(1)}, x_{(2)}, \ldots, x_{(n)}$ be the sample after sorting from smallest to largest. A $(1 - \alpha) \cdot 100\%$ confidence interval is constructed by finding the largest $j \geq 1$ so that $P(x_{(j)} \leq M \leq x_{(n+1-j)}) \geq 1 - \alpha$. In terms of T, this becomes the largest

j so that $P(j \leq T \leq n - j) \geq 1 - \alpha$, which in turn becomes a search for the largest j with $P(T < j) < \alpha/2$. We can find this in the data after sorting.

- **Example 8.13: CEO compensation in 2013**

The data set ceo2013 contains compensation for CEOs at 200 companies. A random sample of size 10 is taken:

```
ceos <- ceo2013[sample(1:200, 10),
                c("company_name_ticker", "cash_compensation")]
```

Based on this, find a 90% confidence interval for the median of the full 200.

As compensation is far from normally distributed, we will use the sign test. First, let's cut down to a reasonable scale:

```
ceos <- round(ceos[,2]/1e6, 2)          # in millions
n <- length(ceos)
pbinom(0:n,n,1/2)
```

```
## [1] 0.0009766 0.0107422 0.0546875 0.1718750 0.3769531 0.6230469
## [7] 0.8281250 0.9453125 0.9892578 0.9990234 1.0000000
```

For a 90% confidence interval, $\alpha/2 = 0.05$. Thus, j is 2, as $P(T < 2) = 0.0107422$, but $P(T < 3) = 0.0546875$. Sorting the data we get:

```
sort(ceos)
```

```
## [1]  2.58  3.25  3.40  3.51  3.85  5.00  6.07  6.99  7.53 20.45
```

This gives a confidence interval of $(3.25, 7.53)$. (The actual answer is 5.31.)

This process can be parameterized by α with:

```
alpha <- 1 - 0.90
n <- length(ceos)
j <- qbinom(alpha/2, n, 1/2)
sort(ceos)[c(j, n+1-j)]
```

```
## [1] 3.25 7.53
```

••

Confidence intervals based on signed-rank statistic

The Wilcoxon signed-rank statistic allows for an improvement on the confidence interval given by counting the number of data points above the median. Its usage is valid when the X_i are assumed to be *symmetric about their median.* If this is so, then a data point is equally likely to be on the left or right of the median, and the distance from the median is independent of what side of the median the data point is on. If we know the median then we can rank the data by distance to the median. Add up the ranks for the values where the data is more than the median. The distribution of this statistic, under the assumption, can be computed and used to give confidence intervals. It is available in R under the family name signrank. In particular, qsignrank will return the quantiles.

This procedure is implemented in the wilcox.test function. Unlike with prop.test and t.test, to return a confidence interval when using wilcox.test we need to specify that a confidence interval is desired with the argument conf.int=TRUE.

- **Example 8.14: CEO confidence interval**
The data on CEOs is too skewed to apply this test, but after taking a log transform we will see a symmetric data set.

Thus we can apply the Wilcoxon method to the log-transformed data, and then transform back.

```
ans <- wilcox.test(log(ceos), conf.int=TRUE, conf.level=0.9)
confint(ans)

## (1.26, 1.98) with 90 percent confidence

confint(ans, transform=exp)              # inverse of log.

## (3.51, 7.26) with 90 percent confidence
```

This produces a smaller confidence interval at the same level, which should not be surprising—a stronger assumption about the shape of the data gives more confidence in the answer.   ••

Confidence intervals based on the rank-sum statistic

The *t*-test to compare samples is robust to nonnormality in the parent distribution but is still not appropriate when the underlying distributions are decidedly nonnormal. However, if the two distributions are the same up to a possible shift of center, then a confidence interval based on a nonparametric statistic can be given.

Let $f(x)$ be a density for a mean-zero distribution, and suppose we have two independent random samples: the first, $x_1, x_2, \ldots, x_{n_x}$, from a population with density $f(x - \mu_x)$, and the second, $y_1, y_2, \ldots, y_{n_y}$, from a population with density $f(x - \mu_y)$. The basic statistic, called the *rank-sum statistic*, looks at all possible pairs of the data and counts the number of times the x value is greater than or equal to the y value. If the population mean for the x values is larger than the population mean for the y values, this statistic will likely be large. If the mean is smaller, then the statistic will likely be small. The distribution of this statistic is given by R with the `wilcox` family and is used to give a confidence interval for the difference of the means.

The command `wilcox.test(x, y, conf.int=TRUE)` is used to find a confidence interval for the difference in medians of the two data sets.

- **Example 8.15: CEO pay in 2012**

In the last example we considered compensation for a random sample of CEOs in 2013. That data set also has past compensation. We take a sample from that and see if there is a change in the median.

```
ceos_past <- ceo2013[sample(1:200, 10), "cash_compensation_past"]
ceos_past <- round(ceos_past / 1e6, 2)  # in millions
ceos_past

## [1] 23.50  6.39 11.05  3.81  8.90  4.81 23.76  3.04  2.95  3.19

ceos                                     # 2013 data

## [1]  3.25  3.85  2.58  3.51  6.99 20.45  7.53  3.40  5.00  6.07
```

Figure 8.5 shows two data sets that are quite skewed, so confidence intervals based on the T-statistic would be inappropriate. Rather, as the two data sets have a similar shape, we find the confidence interval returned by `wilcox.test`. As before, we need to specify that a confidence interval is desired. To answer our question, we'll look at a 90% confidence interval and see if it contains 0.

```
wilcox.test(ceos, ceos_past, conf.int=TRUE, conf.level=0.9)

##
## 	Wilcoxon rank sum test
##
## data:  ceos and ceos_past
## W = 43, p-value = 0.6305
## alternative hypothesis: true location shift is not equal to 0
## 90 percent confidence interval:
##  -5.39  1.14
## sample estimates:
## difference in location
##                 -0.785
```

The 90% confidence interval contains a value of 0.

This example would be improved if we had matched or paired data—that is, the salaries for the same set of CEOs in the years 2012 and 2013—as

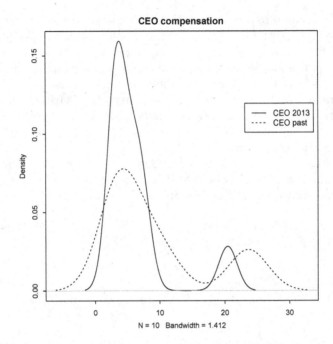

Figure 8.5: Densities of 2012 and 2013 CEO compensations indicating similarly shaped distributions with possible shift.

then differences in the sampling would be minimized. If that case is appropriate, then adding the argument paired=TRUE to wilcox.test computes a confidence interval based on the signed-rank statistic.   ••

Problems

8.29 The commuter revisited: the commuter records 20 commute times to gauge her median travel time. The data has a sample median of 24 and is summarized in this stem-and-leaf diagram:

```
> stem(commutes)
2 | 1112233444444
2 | 5569
3 | 113
```

If the data is appropriate for t.test, use that to give a 90% confidence interval for the mean. Otherwise use wilcox.test (perhaps after a transform) to give a confidence interval for the median.

8.30 The data set u2 contains song lengths for several albums by the band U2. How might you interpret the lengths as a sample from a larger population? Use wilcox.test to construct a 95% confidence interval for the difference of population means between the album *October* and the album *The Joshua Tree*.

8.31 The data set cfb contains a sampling of the data contained in the Survey of Consumer Finances. For the variables AGE and INCOME find 95% confidence intervals for the median.

8.32 We can simulate the signed-rank distribution and see that it applies for any symmetric distribution regardless of tail length. The following will simulate the distribution for $n = 20$ using normal data.

```
n <- 20
M <- 250
res <- replicate(M, {
  x <- rnorm(n)
  sum(rank(abs(x))[x>0])                    # only add positive values
})
```

This can be plotted with

```
hist(res, probability=TRUE)
x <- 40:140
lines(x, dsignrank(x,n))                    # density-like, but discrete
```

If you change the line x <- rnorm(x) to x <- rt(n,df=2), the underlying distribution will be long tailed, and short tailed if you change it to x <- runif(n,-1,1). Do both, and then compare all three samples. Are they different or the same? What happens if you use skewed data, such as x <- rexp(n)-1?

9

Significance tests

Finding a confidence interval for a parameter is one form of statistical inference. A significance test, or test of hypothesis, is another. Rather than specify a range of values for a population parameter, a significance test assumes a value for the population parameter and then computes a probability based on a sample given that assumption.

In Section 7.2 of Chapter 7 we discussed a hypothetical experiment to test if consuming honey during exercise could increase performance. To look at that question, two cohorts were formed by randomizing the available subjects and for one of the cohorts a treatment was applied and results measured. The question of whether the treatment had an effect was asked. The investigation assumed the role of the skeptic—all differences were due to the random assignment and not the treatment. From this assumption and the randomization in the experiment, a distribution of possible values was computed that allowed us to quantify the question in terms of a probability, in particular there was about an 8.57% chance of seeing the observed difference or something more extreme.

That example showed a plausible means of turning a simple question into one that lends itself to analysis. Namely, we assume some model for how the data is produced and weigh how likely it is to produce the observed data. The approach in that example is a prototype for a significance test. Before formalizing a vocabulary surrounding significance tests, we next discuss a familiar framework that highlights the approach.

• Example 9.1: A criminal trial
The ideas behind a significance test can be illustrated by analogy to a criminal trial in the United States—as seen on TV. Imagine the following simplified scenario: a defendant is charged with a crime and must stand trial. During the trial, a prosecutor and defense lawyer respectively try to convince the jury that the defendant is either guilty or innocent. The jury is supposed to be unbiased. When deciding the defendant's fate, the jurors are instructed to *assume* that the defendant is innocent unless *proven* guilty beyond *a shadow of a doubt*. At the end of the trial the jurors decide the guilt or innocence of the defendant based on the strength of their belief in the assumption of his innocence given the evidence. If the jurors believe it very unlikely that

an innocent person could have evidence to the contrary, they will find the defendant "guilty." If it is not so unlikely, they will rule "not guilty."

The system is not foolproof. A guilty man can go free if he is found not guilty, and an innocent man can be erroneously convicted. The frequency with which these errors occur depends on the threshold used to find guilt. In a criminal trial, to decrease the chance of an erroneous guilty verdict, the stringent *shadow of a doubt* criterion is used. In a civil trial, this phrasing is relaxed to *a preponderance of the evidence*. The latter makes it easier to err with a truly innocent person but harder to err with a truly guilty one.   ••

Let's rephrase the above example in terms of significance tests. The assumption of innocence is replaced with the *null hypothesis*, H_0. This stands in contrast to the *alternative hypothesis*, H_A. This would be an assumption of guilt in the trial analogy. In a trial, this alternative is not used as an assumption; it only gives a direction to the interpretation of the evidence. The determination of guilt by a jury is not proof of the alternative, only a failure of the assumption of innocence to explain the evidence well enough. A guilty verdict is more accurately called a verdict of "not plausibly innocent."

The performer of a significance test seeks to determine whether the null hypothesis is reasonable given the available data. The evidence is replaced by an experiment that produces a *test statistic*. The probability that the test statistic is the observed value *or is more extreme* as implied by the alternative hypothesis is calculated *using the assumptions of the null hypothesis*. This probability is called the *p-svalue*. This is like the weighing of the evidence—the jury calculating the likelihood that the evidence agrees with the assumption of innocence.

The calculation of the p-value is called a *significance test*. The p-value is based on both the sampling distribution of the test statistic under H_0 and the single observed value of it during the trial. In words, we have

$$p\text{-value} = P(\text{test statistic is the observed value or is more extreme} \mid H_0).$$

The p-value helps us decide whether differences in the test statistic from the null hypothesis are attributable to chance or sampling variation, or to a failure of the null hypothesis. If a p-value is small, the test is called *statistically significant*, as it indicates that the null hypothesis is unlikely to produce more extreme values than the observed one. Small p-values cast doubt on the null hypothesis; large ones do not.

What is "large" or "small" depends on the area of application, but there are some standard levels that are used. Some R functions will mark p-values with significance stars, as described in Table 9.1. Although these are useful for quickly identifying significance, the cutoffs are arbitrary, settled on more for ease of calculation than actual relevance.

p-value range	significance stars	common description
$[0, 0.001]$	***	extremely significant
$(0.001, 0.01]$	**	highly significant
$(0.01, 0.05]$	*	statistically significant
$(0.05, 0.1]$	.	could be significant
$(0.1, 1]$		not significant

Table 9.1: Level of significance for range of p-values.

In some instances, as with a criminal trial, a decision is made based on the p-value. A juror is instructed that a defendant, to be found guilty, must be thought guilty beyond a shadow of a doubt. A significance test is less vague, as a *significance level* is specified that the p-value is measured against. A typical significance level is 0.05 or one in twenty. If the p-value is less than the significance level, then the null hypothesis is sometimes said to be *rejected*, or viewed as false. If the p-value is larger than the significance level, then the null hypothesis is *accepted*.

The words "reject" and "accept" are perhaps more absolute than they should be. When rejecting the null, we don't actually prove the null to be false or the alternative to be true. All that is shown is that the null hypothesis is unlikely to produce values more extreme than the observed value. When accepting the null we don't prove it is true, we just find that the evidence is not too unlikely if the null hypothesis is true.

By specifying a significance level, we indirectly find values of the test statistic that will lead to rejection. This allows us to specify a *rejection region* consisting of all values for the observed test statistic that produce p-values smaller than the significance level. The boundaries between the acceptance and rejection regions are called *critical values*. The use of a rejection region avoids the computation of a p-value: reject if the observed value is in the rejection region and accept otherwise. We prefer, though, to find and report the p-value rather than issue a simple verdict of "accept" or "reject."

This decision framework has been used historically in scientific endeavors. Researchers may be investigating whether a specific treatment has an effect. They might construct a significance test with a null hypothesis of the treatment having no effect, against the alternative hypothesis of some effect. (In this case, the alternative hypothesis is commonly known as the *research hypothesis*.) The significance test then determines the reasonableness of the assumption of no effect. If this is rejected, there has been no proof of the research hypothesis, only that the null hypothesis is not supported by the data.

In the example of honey intake and exercise performance, the null hypothesis, or the skeptic's approach, is there is no effect. The alternative is that there is a positive effect. The p-value is computed using the randomization distribution, which is found assuming the null hypothesis is true. The p-value of 0.0857 is small, especially considering the size of the data set, but not below the common 0.05 significance level. If you were asked if the null hypothesis is accepted or rejected, the answer in this case would be "accepted."

As with a juried trial, the system is not foolproof. When a decision is made based on the p-value, mistakes can happen. If the null hypothesis is falsely rejected, it is a *Type-I error* (innocent man is found guilty). If the null hypothesis is false, it may be falsely accepted (guilty man found not guilty). This is a *Type-II error*.

Here is another example to illustrate the process.

• Example 9.2: Which mean?

Imagine we have a widget-producing machine that sometimes goes out of calibration. The calibration is measured in terms of a mean for the widgets. How can we tell if the machine is out of calibration by looking at the output of a single widget?

Assume, for simplicity, that the widgets produced are random numbers that usually come from a normal distribution with mean 0 and variance 1. When the machine slips out of calibration, the random numbers come from normal distribution with mean 1 and variance 1. Based on the value of a single one of these random numbers, how can we decide whether the machine is in calibration or not?

This question can be approached as a significance test. We might assume that the machine is in calibration (has mean 0), unless we see otherwise based on the value of the observed number.

Let x be the random number. The model for the data is that x comes from a normal distribution with unknown mean. The null hypothesis is that this is 0, the alternative is that this is 1.

We usually write this as

$$H_0 : \mu = 0, \quad H_A : \mu = 1,$$

where the assumption that the observed data is drawn from a normal distribution with variance 1 is implicit.

Now, suppose we observe a value 0.7 from the machine. Is this evidence that the machine is out of calibration?

The p-value in this case is the probability that a Normal$(0,1)$ random variable produces a value of 0.7 or *more*. This is 1 - pnorm(0.7,0,1), or 0.2420. Why this probability? The calculation is done under the null hypothesis, so a normal distribution with mean 0 and variance 1 is used. The observed value of the test statistic is 0.7. Larger values than this are more extreme, given the alternative hypothesis. This p-value is not very small, and there is no evidence that the null hypothesis is false. It may be, if the alternative were true, that a value of 0.7 or less is pnorm(.7,1,1), or 0.3821, so it, too, is not unlikely. (See Figure 9.1.)

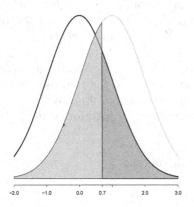 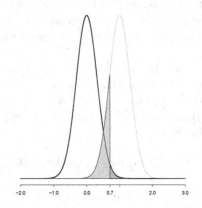

Figure 9.1: The two figures show the area to the right of 0.7 for a mean-0 normal with standard deviation 1 and $1/\sqrt{10}$ and the area to the left for a mean-1 normal with similar spread. The left graphic points out that it is hard to determine which of the two distributions a value of 0.7 would have come from, as the spreads overlap quite a bit. This is not the case, when the standard deviations are smaller, as illustrated on the right.

Even though 0.7 is closer to the mean of 1 than the mean of 0, it is really not conclusive evidence that the null hypothesis (the assumption of calibration) is incorrect. The problem is that the two distributions are so "close" together. It would be much easier to decide between the two if the means were 10 units apart instead of just 1 (with the same variance). Alternatively, if the standard deviations were smaller, the same would be true. This can be achieved by taking averages, as we know that the standard deviation of an average is $\sigma/\sqrt{n}$, or smaller than the population standard deviation by the factor $1/\sqrt{n}$.

With this in mind, suppose our test statistic is now the sample mean of a random sample of ten widgets. How does our thinking change if the sample mean is now 0.7?

The p-value looks similar, $P(\bar{x} \geq 0.7 \mid H_0)$, but when we compute, we use the sampling distribution of $\bar{x}$ under H_0, or Normal$(0, 1/\sqrt{10})$. The p-value is now 0.0134, as found by 1 - pnorm(0.7,0,1/sqrt(10)). This is illustrated in the right graphic of Figure 9.1. Now the evidence is convincing that the machine is out of calibration.   ●●

The above example illustrates the steps involved in finding a p-value:

1. Specify some model for the underlying data. Our typical assumptions will involve a parameterized family for the population, such as the normal distribution.

2. Identify H_0 and H_A, the null and alternative hypotheses.

3. Specify a test statistic that discriminates between the two hypotheses. The sampling distribution of this statistic needs to be knowable under the assumptions of the null hypothesis.

4. Collect the data, then find the observed value of the test statistic.

5. Using H_A, specify values of the test statistic that are "extreme" under H_0 in the direction of H_A. The p-value will be the probability of the event that the test statistic is the observed value or more extreme.

6. Calculate the p-value under the null hypothesis. The smaller the p-value, the stronger the evidence is against the null hypothesis.

9.1 Significance test for a population proportion

A researcher may wish to know whether a politician's approval ratings are falling, or whether unemployment rate is rising, or whether the poverty rate is changing. In many cases, a known proportion exists. What is asked for is a comparison against this known proportion. A test of proportion can be used to help answer these questions.

Assume p_0 reflects the historically true proportion of some variable of interest. A researcher may wish to test whether the current unknown proportion, p, is different from p_0. The hypothesis are

$$H_0 : p = p_0,$$

versus an alternative hypothesis on p. Possible alternatives are

$$H_A : p > p_0, \; H_A : p < p_0, \; \text{or} \; H_A : p \neq p_0.$$

If the survey is a random sample from the target population, the number of successes, x, is binomially distributed, $\hat{p} = x/n$, and $\hat{p}$ will be approximately normal for large enough values of n. We might use $\hat{p}$ directly as a test statistic, but it is more common to standardize $\hat{p}$, yielding the following:

$$Z = \frac{\hat{p} - \mathsf{E}(\hat{p} \mid H_0)}{\mathsf{SD}(\hat{p} \mid H_0)} = \frac{\hat{p} - p_0}{\sqrt{p_0(1 - p_0)/n}}.$$

We use the notation $\mathsf{E}(\hat{p} \mid H_0)$ to remind ourselves that we use the null hypothesis when calculating this expected value. In this case, if we assume H_0 is true, then p_0 is the expected value of $\hat{p}$ and is assumed to be known. Thus, we can use $\mathsf{SD}(\hat{p} \mid H_0) = \sqrt{(p_0(1 - p_0))/n}$ in our test statistic (as compared to the use of $\mathsf{SE}(\hat{p})$ when we found confidence intervals for p).

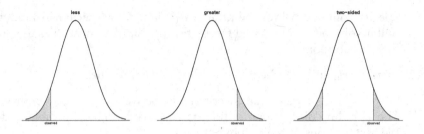

Figure 9.2: The computation of the p-value depends on the alternative hypothesis, which determines what is meant by "more extreme." The three common alternatives are illustrated in the figure.

The p-value varies based on the alternative hypothesis. This is because what is meant by "more extreme" for the value of the test statistic depends on H_A. In this instance there are three cases:

$$p\text{-value} = \begin{cases} P(\widehat{p} \leq \text{observed value} \mid H_0) & \text{if } H_A : p < p_0, \\ P(\widehat{p} \geq \text{observed value} \mid H_0) & \text{if } H_A : p > p_0, \quad (9.1) \\ P(|\widehat{p} - p_0| \geq |\text{observed value} - p_0| \mid H_0) & \text{if } H_A : p \neq p_0. \end{cases}$$

The first two cases are "one-sided" or "one-tailed," the last "two-sided" or "two-tailed." The absolute values in the third case can be confusing but are there to say that large differences in either direction of p_0 are "more extreme." Figure 9.2 illustrates the areas.

Significance test for a population proportion

A significance test for an unknown proportion between

$$H_0 : p = p_0, \quad H_A : p < p_0, \; p > p_0, \text{ or } p \neq p_0$$

can be performed with test statistic

$$Z = \frac{\widehat{p} - p_0}{\sqrt{p_0(1 - p_0)/n}}.$$

If $\widehat{p}$ is based on a simple random sample and n is large enough, Z has a standard normal distribution under the null hypothesis. The p-values can be computed from (9.1) or with `prop.test`.

- **Example 9.3: Poverty-rate increase**
In the year 2000, the poverty rate in the United States was around 11.7 percent. By 2010, this rate had risen to 15.00 percent. Does it continue to rise? In 2011 the rate was measured at 15.13 percent, as reported by the United States Census Bureau. A national census takes place every decade. Suppose the different rates for 2000 and 2010 came from a census, so would essentially be population parameters. For the Census Bureau to decide the 2011 rate, a random sampling scheme is employed. Assume that the numbers come from a simple random sample (they don't, the Census Bureau needs more sophisticated strategies), so that we can use the binomial model for our analysis, and that the sample size for 2011 was 150,000.[1]

We investigate whether the 15.13% figure for the year 2011 shows an increase from the year-2010 figure. The null hypothesis is that it is the same as the 15.00% amount of 2010; the alternative is that the new figure is greater than the old:

$$H_0 : p = 0.1500, \quad H_A : p > 0.1500.$$

A test statistic is based on the proportion living in poverty of a random sample of size 150,000. In the sample, 22,695, or a proportion of .1513, were found to be in poverty. Is this difference significant?

We are interested in rising rates, so the direction of the alternative would be that the rate is .1500 *or higher*, as larger values support this one-sided alternative. The p-value is $P(\hat{p} \geq .1500 \mid H_0)$, which is found in R with

```
phat <- 22695 / 150000                    # 0.1513
p0 <- 0.1500; n <- 150000
SD <- sqrt(p0 * (1-p0)/n)
pnorm(phat, mean=p0, sd=SD, lower.tail=FALSE) # p-value

## [1] 0.07926
```

Thus the p-value is 0.08. Some papers reported this as "not statistically significant," presumably using the $\alpha = 0.05$ standard for this assessment. ••

Using prop.test to compute p-values

The calculation above is done by hand. The pre-loaded stats package in R has many built-in functions to do significance tests more directly. The R function for the above test of proportions is prop.test. This same function was also used to find confidence intervals under similar assumptions.

The prop.test function needs a few arguments. A template for usage to perform a significance test is

```
prop.test(x, n, p=..., alternative="two.sided")
```

---

[1]The Census Bureau reported a 90% C.I. with a margin of error 0.002.

The value for x is the sample frequency; in our case, 22,695. The value of n is the sample size 150,000. These are the same as when we used this function to find confidence intervals.

To perform a significance test, the null and alternative hypotheses must be specified. The null is done with the p argument; for our example p = 0.1500. The alternative hypothesis is specified with the argument alternative, which we abbreviate to alt. This argument has one of these values: "less", "greater", or "two.sided". The default is two.sided. As $H_A : p > 0.1500$, we will use "greater". This argument is common to many of the functions in R that perform significance tests.

To illustrate, the above calculation is done with

```
prop.test(x=22695, n=150000, p=.1500, alternative="greater")

##
##  1-sample proportions test with continuity correction
##
## data:  22695 out of 150000, null probability 0.15
## X-squared = 1.978, df = 1, p-value = 0.0798
## alternative hypothesis: true p is greater than 0.15
## 95 percent confidence interval:
##  0.1498 1.0000
## sample estimates:
##      p
## 0.1513
```

The p-value, and the null and alternative hypotheses are repeated in the output. In addition, a confidence interval is given, as is a sample estimate that we labeled phat. The p-value is slightly different from our hand calculation, as a continuity correction is used by R.

It isn't any more difficult to test the alternative hypothesis, that the rate has changed, or $H_A : p \neq p_0$. This is done by specifying the alternative as two.sided:

```
out <- prop.test(x=22695, n=150000, p=0.1500, alternative="two.sided")
out$p.value                               # just the p-value

## [1] 0.1596
```

The p-value is different—it is twice as big—as we would guess by looking at the symmetry in Figure 9.2.

Problems

9.1 United States federal law on dietary supplements requires that the Food and Drug Administration (FDA) prove a supplement harmful in order to ban its sale. In contrast, for a new prescription drug, a pharmaceutical company must prove the product is safe.

Write null and alternative hypotheses for a hypothetical significance test by the FDA when testing a dietary supplement. Do you think the same standard should be used for both dietary supplement and new prescription drugs?

9.2 The samhda (UsingR) data set contains information on marijuana usage among children as collected at the Substance Abuse and Mental Health Data Archive. The variable marijuana indicates whether the individual has ever tried marijuana. A 1 means yes, a 2 no. If it used to be that 50% of the target population had tried marijuana, does this data indicate an increase in marijuana usage? Do a significance test of proportions to decide.

9.3 A new drug therapy is tested. Of 50 patients in the study, 40 had no recurrence in their illness after 18 months. With no drug therapy, the expected percentage of no recurrence would have been 75%. Does the data support the hypothesis that this percentage has increased? What is the p-value?

9.4 In the United States in 2007, the proportion of adults age 21–24 who had no medical insurance was 28.1 percent. A survey of 75 recent college graduates in this age range finds that 40 percent are without insurance. Does this support a difference from the nationwide proportion? Perform a test of significance and report the p-value. Is it significant?

9.5 On a number of highways a toll is collected for permission to travel on the roadway. To lessen the burden on drivers, electronic toll-collection systems are often used. An engineer wishes to check the validity of one such system. She arranges to survey a collection unit for single day, finding that of 5,760 transactions, the system accurately read 5,731. Perform a one-sided significance test to see if this is consistent with a 99.9% accuracy rating at the 0.05 significance level. (Do you have any doubts that the normal approximation to the binomial distribution should apply here?)

9.6 In an example on the poverty rate a count of 22,695 in a survey of 150,000 produced a p-value of 0.079. For the same size sample, what range of counts would have produced a p-value less than 0.05? (Start by asking what observed proportions in the survey would have such a p-value.)

9.7 Historically, a car from a given company has a 10% chance of having a significant mechanical problem during its warranty period. A new model of the car is being sold. Of the first 25,000 sold, 2,700 have had an issue. Perform a test of significance to see whether the proportion of all of these new cars that will have a problem is more than 10%. What is the p-value?

9.2   Significance test for the mean (t-tests)

Significance tests can also be constructed for the unknown mean of a parent population. The hypotheses take the form

$$H_0 : \mu = \mu_0, \quad H_A : \mu < \mu_0, \ \mu > \mu_0, \ \text{or } \mu \neq \mu_0.$$

For certain populations, a useful test statistic is

$$T = \frac{\bar{x} - \mathrm{E}(\bar{x}\,|\,H_0)}{\mathrm{SE}(\bar{x}\,|\,H_0)} = \frac{\bar{x} - \mu_0}{s/\sqrt{n}} = \frac{\text{observed} - \text{expected}}{\mathrm{SE}}.$$

T takes a common form where the expected value and the standard error are found under the null hypothesis.

In the case of normally distributed initial data, the sampling distribution of T under the null hypothesis is known to be the t-distribution with $n-1$ degrees of freedom. If n is large enough, the sampling distribution of T is a standard normal by the central limit theorem. As both the t-distribution and normal distribution are similar for large n, the following applies to both assumptions.

Test of significance for the population mean

If the data $x_1, x_2, \ldots, x_n$ are a random sample from a Normal(μ, σ) distribution, or n is large enough for the central limit theorem to apply, a test of significance for

$$H_0 : \mu = \mu_0, \quad H_A : \mu < \mu_0, \ \mu > \mu_0, \ \text{or } \mu \neq \mu_0$$

can be performed with test statistic

$$T = \frac{\bar{x} - \mu_0}{s/\sqrt{n}}.$$

Let $t = (\bar{x} - \mu_0)/(s/\sqrt{n})$ be the observed value of the test statistic. The p-value is computed by

$$p\text{-value} = \begin{cases} \mathrm{P}(T \leq t \,|\, H_0) & H_A : \mu < \mu_0, \\ \mathrm{P}(T \geq t \,|\, H_0) & H_A : \mu > \mu_0, \\ \mathrm{P}(|T - \mu_0| \geq |t - \mu_0| \,|\, H_0) & H_A : \mu \neq \mu_0. \end{cases}$$

In R, the function t.test can be used to compute the p-value with unsummarized data, as in

```
t.test(x, mu=..., alternative="two.sided")
```

The null hypothesis is specified by a value for the argument mu. The alternative is specified as appropriate by alt="less", alt="greater", or alt="two.sided" (the default).

• Example 9.4: SUV gas mileage

A consumer group wishes to see whether the actual mileage of a new SUV matches the advertised 17 miles per gallon. The group suspects it is lower. To test the claim, the group fills the SUV's tank and records the mileage. This is repeated ten times. The results are presented in a stem-and-leaf diagram:

```
##
##    The decimal point is at the |
##
##    11 | 4
##    12 |
##    13 | 1
##    14 | 77
##    15 | 0569
##    16 | 08
```

Does this data support the null hypothesis that the miles per gallon rating is 17 or the alternative, that it is less?

The data is assumed to be normal, and the stem-and-leaf plot shows no reason to doubt this. The null and alternative hypotheses are

$$H_0 : \mu = 17, \quad H_A : \mu < 17.$$

This is a one-sided test. The p-value will be computed from those values of the test statistic less than the observed value, as these are more extreme given the alternative hypothesis.

```
mpg <- c(11.4, 13.1, 14.7, 14.7, 15.0, 15.5, 15.6, 15.9, 16.0, 16.8)
xbar <- mean(mpg); s <- sd(mpg); n <- length(mpg)
c(xbar=xbar, s=s, n-n)                          # summaries

##    xbar      s
## 14.870  1.572  0.000

SE <- s/sqrt(n)
obs <- (xbar - 17)/SE
pt(obs, df = 9, lower.tail = T)

## [1] 0.001018
```

The p-value is very small and discredits the claim of 17 miles per gallon (assuming our observations are a representative sample), as the difference of $\bar{x}$ from 17 is not well explained simply by sampling variation.

The above calculations could be done using t.test as follows:

```
t.test(mpg, mu = 17, alternative="less")

##
##  One Sample t-test
##
## data:  mpg
## t = -4.285, df = 9, p-value = 0.001018
## alternative hypothesis: true mean is less than 17
## 95 percent confidence interval:
##    -Inf 15.78
## sample estimates:
## mean of x
##      14.87
```

The output contains the same p-value (up to rounding), plus a bit more information, including the observed value of the test statistic, a one-sided confidence interval, and $\bar{x}$ (the estimate for μ). ••

It is easy to overlook the entire null hypothesis. We explicitly write the assumption that $\mu = \mu_0$. However, we have also assumed a model for our data (a normal distribution) and that our data comes from a random sample. With these assumptions, the test statistic has a known sampling distribution. The t-statistic is robust to small differences in the assumed normality of the population, but a really skewed population distribution (e.g., non-normal) would still be a poor candidate for this significance test unless n is large. To identify such situations, it is recommended that you plot the data prior to doing any analysis, to ensure that it is appropriate.

• **Example 9.5: Rising textbook costs?**
A college bookstore claims that, on average, a college student will pay $500 per semester for textbooks and supplies. A student group investigates this claim by randomly selecting and interviewing ten students and finding the amount spent on textbooks and supplies for each. (A job made easier in the U.S. since the passing of the Higher Education Opportunity Act in 2008.)
    The data collected is

Costs	304, 431, 385, 987, 303, 480, 455, 724, 642, 506

Do a test of significance of $H_0 : \mu = 500$ against the alternative hypothesis $H_A : \mu > 500$.
    We enter the data with:

```
costs <- c(304, 431, 385, 987, 303, 480, 455, 724, 642, 506)
```

We will assume a random sample was used, though in practice that may not have been the case. Since n is small, we assess the shape of the sample with a stem-and-leaf plot:

```
stem(costs)

##
##    The decimal point is 2 digit(s) to the right of the |
##
##    2 | 009
##    4 | 3681
##    6 | 42
##   .8 | 9
```

We see no compelling reason to doubt a normality assumption, though expenses often carry a skew, as it is easier to spend too much, but hard to spend too little (and still buy the books).

We then use t.test, with "greater" for the alternative giving:

```
t.test(costs, mu=500, alternative="greater")

##
##   One Sample t-test
##
## data:  costs
## t = 0.3253, df = 9, p-value = 0.3762
## alternative hypothesis: true mean is greater than 500
## 95 percent confidence interval:
##   399.4   Inf
## sample estimates:
## mean of x
##     521.7
```

The p-value is not small. The data gives us little reason to doubt that the null hypothesis applies to this data.   ••

Power

The language of "accept" or "reject" after a significance test is performed is based on a comparison of the p-value with α, a predetermined significance level. The value of α is chosen to give a desired probability of "rejecting" the null hypothesis when it is a true description of the data. Rejecting in this

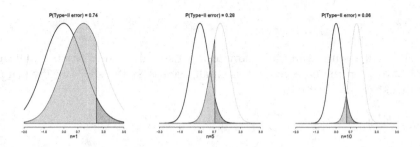

Figure 9.3: The probability of a Type-II error depends on the sample size. In the figure, this probability is represented by the area to the left of z^* for the probability distribution centered at 1. For $n = 1$, this error is likely, for $n = 10$ it is unlikely.

case is known as a Type-I error. There is a Type-II error which is made when the null hypothesis is "accepted," even though it is not the case. Of course, one would like both Type-I and Type-II errors to not be made, or at least be uncommon. By choosing α, the likelihood of Type-I errors is known. What about Type-II errors?

We return to the comparison of Figure 9.1. In that figure the null hypothesis ($\mu = 0$) is represented by a probability distribution centered at 0. The alternative is a simple alternative ($\mu = 1$—not $\mu > 0$), and so can also be represented by a probability distribution, in this case centered at 1. The scale of this distribution is determined by n, the sample size—larger samples make the distributions narrower.

In Figure 9.3 for different n we let z^* represent the critical value under the null for which if the observed value is more than z^*, then one-sided test will be rejected when $\alpha = 0.05$ (found with qnorm(1-alpha, mu=0, sd=1/sqrt(n))). The area to the right under the null, α, is the probability of a Type-I error. The shaded area to the left of z^* under the alternative is the probability of accepting the null, when the alternative is true, that is the probability of a Type-II error.

A natural approach is to specify an acceptable probability for a Type-II error (often called β). A typical value is $\beta = 0.20$. The power of a test, for a given specific value of the alternative, is $1 - \beta$. With both α and β specified, n is chosen large enough that the probability of the errors are so bounded. Power computations are generally made prior to data collection, so that a minimal sample size is taken to ensure the desired characteristics.

The computational details are performed by the power.t.test function. There are several arguments that need specifying: the single value of the alternative is specified through delta which is the difference from the mean specified in the null (1 in our figure), the number of tails used (alternative="one.sided"), the population standard deviation (1 in our figure), the value for α (sig.level), the value for $1 - \beta$ (power), the type of test (type="one.sample"):

```
alpha <- 0.05; beta <- 0.20
power.t.test(delta=1, sd=1,
             sig.level=alpha, power=1-beta,
             type="one.sample", alternative="one.sided")

##
##      One-sample t test power calculation
##
##              n = 7.728
##          delta = 1
##             sd = 1
##      sig.level = 0.05
##          power = 0.8
##    alternative = one.sided
```

The value of n reported is 7.7. As such, it would take a sample of size 8 to have a test where delta=1 with a power of 0.80. Smaller deltas require larger sample sizes.

The above required a value for the population standard deviation. In general this isn't known, especially before any data is collected. One needs to replace this with a reasonable estimate.

Problems

9.8 A study of the average salaries of New York City residents was conducted for 770 different jobs. It was found that massage therapists average $58,260 in yearly income. Suppose the study surveyed 25 massage therapists and had a standard deviation of $3,250. Perform a significance test of the null hypothesis that the average massage therapist makes $55,000 per year against the one-sided alternative that it is more. Assume the data is normally distributed.

9.9 The United States Department of Energy conducts weekly phone surveys on the price of gasoline sold in the United States. Suppose one week the sample average was $4.03, the sample standard deviation was $0.42, and the sample size was 800. Perform a one-sided significance test of $H_0 : \mu = 4.00$ against the alternative $H_A : \mu > 4.00$.

9.10 The variable sat.m in the data set stud.recs (UsingR) contains math SAT scores for a group of students sampled from a larger population. Test the null hypothesis that the population mean score is 500 against a two-sided alternative. Would you "accept" or "reject" at a 0.05 significance level?

9.11 In the babies (UsingR) data set, the variable dht records the father's height for the sampled cases. Do a significance test of the null hypothesis that the population mean height is 68 inches against an alternative that it is taller. Remove the values of 99 from the data, as these indicate missing data.

9.12 A consumer-reports group is testing whether a gasoline additive changes a car's gas mileage. A test of seven cars finds an average improvement of 0.5 miles per gallon with a standard deviation of 3.77. Is this difference significantly greater than 0? Assume the values are normally distributed.

9.13 The data set OBP (UsingR) contains on-base percentages for the 2002 Major League Baseball season. Do a significance test to see whether the mean on-base percentage is 0.330 against a two-sided alternative.

9.14 The data set normtemp (UsingR) contains measurements of 130 healthy, randomly selected individuals. The variable temperature contains normal body temperature. Does the data appear to come from a normal distribution? If so, perform a t-test to see if the commonly assumed value of 98.6°F is correct. (Studies have suggested that 98.2°F is actually more accurate.)

9.15 A PBS report was done on a study which looked if doctors who received free samples were more likely to prescribe more expensive drug treatments. It was reported that the average cost of a first visit for acne or rosacea was over $450 for medication for the physicians that use samples compared to a clinic where we there are no samples and the average cost was only 200-hundred dollars. Suppose the number of physicians surveyed was 8 and the variance was $100 dollars. Is the difference statistically significant?

9.16 A one-sided, one-sample t-test will be performed. What sample size is needed to have a power of 0.80 for a significance level of 0.05 if delta=0.05 and the population standard deviation is assumed to be 5?

9.3   Significance tests and confidence intervals

You may have noticed that the R functions for performing a significance test for a population proportion or mean are the same functions used to compute confidence intervals. This is no coincidence, as performing a significance test and constructing a confidence interval both make use of the same test statistic, although in different ways.

Suppose we have a random sample from a normally distributed population with mean μ and variance σ^2. We can use the sample to find a confidence interval for μ, or we could use the sample to do a significance test of

$$H_0 : \mu = \mu_0, \quad H_A : \mu \neq \mu_0.$$

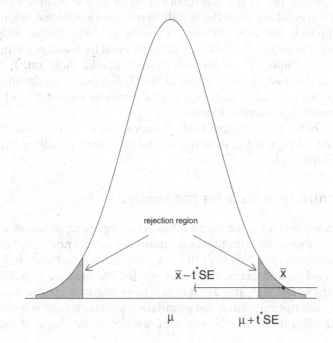

Figure 9.4: If $\bar{x}$ is in the two-sided rejection region, then a confidence interval around $\bar{x}$ does not contain μ

In either case, the T statistic

$$T = \frac{\bar{x} - \mu}{\text{SE}(\bar{x})}$$

is used to make the statistical inference. The two approaches are related by the following:

> a two-sided significance test with significance level α will be rejected if and only if the $(1 - \alpha) \cdot 100\%$ confidence interval around $\bar{x}$ does not contain μ_0.

To see why, suppose α is given. The confidence interval uses t^* found from

$$P(-t^* \leq T \leq t^*) = 1 - \alpha.$$

From this, the confidence interval will not contain μ_0 if the value of T is more than t^* or less than $-t^*$. This same relationship is used to find the critical values defining the boundaries of the rejection region. If the observed value of T is more than t^* or less than $-t^*$, then the observed value is in the rejection region, and the null hypothesis is rejected. This is illustrated in Figure 9.4.

Many people prefer the confidence interval to the p-value of the significance test for good reasons. If the null hypothesis is that the mean is 16, but the true mean is just a bit different, then the probability that a significance test will fail can be made arbitrarily close to 1 just by making n large enough. The confidence interval, on the other hand, would show much better that the value of the mean is likely close to 16. The language of significance tests, however, is more flexible and allows us to consider more types of problems. Both approaches are useful to have.

R is agnostic: it can return both the confidence interval and the p-value when asked, although the defaults for the functions usually return just the confidence interval.

9.4   Significance tests for the median

The significance test for the mean relies on a large sample, or on an assumption that the parent distribution is normally (or nearly normally) distributed. In the situation where this isn't the case, we can use test statistics similar to the ones used to find confidence intervals for the median. Significance tests based on these test statistics are referred to as *nonparametric* tests, as they do not make assumptions about the population parameters to calculate the test statistic (though there may be assumptions about the shape of the distribution).

The sign test

The sign test is a simple test for the median of a distribution that has no assumptions on the parent distribution except that it is continuous with positive density. Let H_0 suppose that the median is m. If we count the number of data points higher than the median, we get a number that will have a Binomial$(n, 1/2)$ distribution, as under H_0, a data point is equally likely to be more or less than the median. This leads to the following test.

Sign test for the median

Assume $x_1, x_2, \cdots x_n$ are from a continuous distribution with positive density. A significance test of the hypotheses

$$H_0: \text{median} = m, \quad H_A: \text{median} < m, > m, \text{or} \neq m,$$

can be performed with test statistic

$$T = \text{the number of } X_i \text{ with } X_i > m.$$

If the data has values equal to m, then delete those values from the data set. Under H_0, T has a Binomial$(n, 1/2)$ distribution. Large values of T support the alternative that the median is greater than M; small values of T support the alternative that the median is smaller than M. For two-sided alternatives, large or small values of T support H_A.

The p-value is calculated by

$$p\text{-value} = \begin{cases} P(T \geq \text{obs} \,|\, H_0) & H_A : \text{median} > m \\ P(T \geq n - \text{obs} \,|\, H_0) & H_A : \text{median} < m \\ P(T \geq \max(\text{obs}, n - \text{obs}) \,|\, H_0) & H_A : \text{median} \neq m. \end{cases}$$

In R, the test statistic can be computed using sum. The p-values are found using pbinom(k). However, as $P(T \geq k) = 1 - P(T \leq k - 1)$, the p-value is found with 1 - pbinom(k - 1, n, 1/2).

• **Example 9.6: Length of cell-phone calls**
Suppose a cell-phone bill contains data for the number of minutes per call:

Length (mins)	2, 1, 3, 3, 3, 3, 1, 3, 16, 2, 2, 12, 20, 3, 1

Is this data consistent with an assumption that the population median call length is 5 minutes with an alternative that the median length is less than 54?

The hypothesis test is

$$H_0 : \text{the median} = 5, \quad H_A : \text{the median} < 5.$$

The data is clearly nonnormal, so a t-test is inappropriate. A sign test can be used. Here, small values of T support the alternative.

```
calls <- c(2, 1, 3, 3, 3, 3, 1, 3, 16, 2, 2, 12, 20, 3, 1)
obs <- sum(calls > 5)          # find observed value of T
n <- length(calls)
1 - pbinom(n-obs - 1, n, 1/2)  # we want P(T >= 12) = 1 - P(T <= 11)

## [1] 0.01758
```

We get a p-value of 0.0176, which indicates the data is not consistent with the null hypothesis. ••

The signed-rank test

The signed-rank test is an improvement to the sign test when the population is symmetric, but not close enough to normal to use a t-test. Assume H_0 : median $= m$. If X_i are from a continuous distribution with density $f()$ that is symmetric about m, then not only is X_i equally likely to be on either side of m, but the distance from m is independent of the side. Thus, if we rank all the data by its distance to m, the sum corresponding to the values larger than m may be viewed as a random sample of a certain size from the numbers 1 through n. The distribution of this sum can be characterized, so the sum of the ranks can be an effective test statistic.

The Wilcoxon signed-rank test for the median

If the data, $x_1, x_2, \ldots, x_n$, form an *i.i.d.* sample from a continuous, symmetric distribution, then a significance test of the hypotheses

$$H_0 : \text{the median} = m, \quad H_A : \text{median} < m, > m, \text{ or } \neq m$$

can be performed with test statistic

$$T = \sum_{i:x_i > m} \text{rank}(|x_i - m|).$$

Under H_0, the distribution of T can be calculated. Large values of T support the alternative hypothesis H_A : median $> m$.

In R, the function `wilcox.test` performs the test as

```
wilcox.test(x, mu=..., alternative="two.sided")
```

The data is contained in x, the null hypothesis is specified by the argument mu, and the alternative is specified with the argument `alternative`. This argument takes a value of `"less"`, `"greater"`, or `"two.sided"` (the default value). If desired, the distribution of T is given by the function `psignrank`.

A typical application of the signed-rank test is to use it after transforming the data to make it look like the parent distribution is symmetric. Transformations which don't affect the order of the data, such as a logarithm, have the median transform in a direct way.

The assumption of a density f, implies that the data should not have any ties, though this is common in practice. The `wilcox.test` will make a complaint about inexact p-values when the data has ties. The `wilcox_test` function from the `coin` package (which implements a general framework for such tests) adjusts for tied data.

● **Example 9.7: Number of recruits**
In salmon populations, there is a relationship between the number of spawners and the subsequent number of "recruits" that they produce. A common model involves two parameters, which describe how many recruits there are when there are few spawners and how many there are when there are many spawners. The data set `salmon.rate` (UsingR) contains simulated data on one

of the parameters. A plot of the data shows that a normal population assumption is not correct; rather, the population appears to be lognormal.

Perform a significance test of

$$H_0 : \text{median} = 0.005, \quad H_A : \text{median} > 0.005.$$

After taking logs, we can see that the data is symmetric, so the signed-rank test can apply to the log-transformed data. The significance test of this data is

$$H_0 : \text{median} = \log(0.005), \quad H_A : \text{median} > \log(0.005).$$

```
wilcox.test(log(salmon.rate), mu=log(0.005), alt="greater")

##
##  Wilcoxon signed rank test with continuity correction
##
## data:  log(salmon.rate)
## V = 2077, p-value = 0.065
## alternative hypothesis: true location is greater than -5.298
```

A somewhat small p-value is found.

To contrast, the p-value for the sign test is found with these commands:

```
T <- sum(salmon.rate > .005); n <- length(salmon.rate)
1 - pbinom(T - 1, n, 1/2)

## [1] 0.1361
```

This p-value is larger. This may seem surprising, but the Wilcox test has an additional assumption, which generally yields less conservative p-values. Conservative here implies larger p-values, so harder to "reject" (make a Type-I error). ••

Problems

9.17 The exec.pay (UsingR) data set contains data on the salaries of CEOs at 199 top companies in the United States. The amounts are in \$10,000s. The data is not symmetric. Do a sign test to determine whether the median pay is more than \$220,000.

9.18 Repeat the previous exercise, using the signed-rank test on the log-transformed data. Do you reach the same conclusion?

9.19 The babies (UsingR) data set contains data covering many births. In-formation included is the gestation period, and a factor indicating whether the mother was a smoker. Extracting the gestation times for mothers who smoked during pregnancy can be done with this command:

```
smokers <- subset(babies, smoke == 1 & gestation != 999)
```

Perform a significance test of the null hypothesis that the average gesta-tion period is 40 weeks against a two-sided alternative. Explain what test you used, and why you chose that one.

9.20 If the sign test has fewer assumptions on the population, why wouldn't we always use that instead of a *t*-test? The answer lies in the power of the sign test to detect when the null hypothesis is false. The sign test will not reject a false null as often as the *t*-test. The following commands will per-form a simulation comparing the two tests on data that has a Normal(1,2) distribution. The significance tests performed are both

$$H_0 : \mu = 0, \quad H_A : \mu > 0.$$

Run the simulation. Is there a big difference between the two tests?

```
m <- 200; n <- 10

out <- replicate(m, {
  x <- rnorm(n, mean=1, sd=2)
  ttest <- t.test(x, mu=0, alt = "greater")$p.value
  sgntest <- 1 - pbinom(sum(x > 0) - 1, n, 1/2)
  c(t.test   = ifelse(ttest   < 0.05, 1, 0),
     sign.test= ifelse(sgntest < 0.05, 1, 0))
})

res.t <- out["t.test",]
res.sign <- out["sign.test",]
results <- c(t = sum(res.t)/m, sign=sum(res.sign)/m)
```

9.5   Two-sample tests of proportion

In the previous sections our tests of significance compared a sample to some assumed value of the population and determined whether the sample sup-ported the null hypothesis. This assumes some specific knowledge about the

population parameters. In many situations, we'd like to compare two parameters. In this section we consider how. This can be useful in many different contexts: comparing polling results taken over different periods of time,
surveying results after an intervention such as an advertising campaign, or
comparing attitudes among different ethnic groups.

In a previous example, we compared the 2011 poverty rate, which was
found by a sample, with the 2010 poverty rate which was assumed known
from a census. In reality, the 2010 data was not derived from a national census, but from a sample of similar size, say 160,000. To compare the 2011 rate
to the 2010 rate, we would compare two samples. How do we handle this
with a significance test?

Let $\widehat{p}_1$ be the estimated 2010 poverty rate and $\widehat{p}_2$ be the estimated 2011
poverty rate. We wish to perform a significance test of

$$H_0 : p_1 = p_2, \quad H_A : p_1 < p_2$$

using the values of $\widehat{p}_1$ and $\widehat{p}_2$ in the test statistic. If we think of the test as one
of differences, we can rephrase it as

$$H_0 : p_1 - p_2 = 0, \quad H_A : p_1 - p_2 < 0.$$

A natural test statistic would follow the form observed minus expected
over the SE:

$$Z = \frac{(\widehat{p}_1 - \widehat{p}_2) - E(\widehat{p}_1 - \widehat{p}_2 \mid H_0)}{SE(\widehat{p}_1 - \widehat{p}_2 \mid H_0)}.$$

We assume that the surveys were a simple random sample from the population, so that the number responding favorably, x_i, has a binomial distribution with $n = n_i$ and $p = p_i$ for $i = 1,2$. (So $\widehat{p}_i = x_i/n_i$.) Thus, the expectation
in Z is simply $p_1 - p_2 = 0$ under the null hypothesis. The standard error is
found from the standard deviation under the null hypothesis

$$SD(\widehat{p}_1 - \widehat{p}_2 \mid H_0) = \sqrt{\frac{p_1(1 - p_1)}{n_1} + \frac{p_2(1 - p_2)}{n_2}} = \sqrt{p(1 - p)(\frac{1}{n_1} + \frac{1}{n_2})},$$

where $p = p_1 = p_2$ under the null hypothesis. The value of p is not assumed
in H_0, so we estimate it and use the *standard error* instead. To estimate p—
under the null hypotheses—it makes sense to use the entire sample as we are
assuming both come from a population with parameter p:

$$\widehat{p} = \frac{\text{total who are favorable}}{\text{total size of both samples}} = \frac{x_1 + x_2}{n_1 + n_2} = \frac{n_1\widehat{p}_1 + n_2\widehat{p}_2}{n_1 + n_2}.$$

This leaves

$$Z = \frac{\widehat{p}_1 - \widehat{p}_2}{\sqrt{\widehat{p}(1 - \widehat{p})(\frac{1}{n_1} + \frac{1}{n_2})}}. \tag{9.2}$$

Two-sample test of proportions

If we have sample proportions for two random samples, a significance test of

$$H_0 : p_1 = p_2, \quad H_A : p_1 < p_2, \, p_1 > p_2, \text{ or } p_1 \neq p_2$$

can be carried out with test statistic Z given by (9.2). Under H_0, Z has a standard normal distribution if n_1 and n_2 are sufficiently large. Large values of Z support the alternative $p_1 > p_2$; small values support $p_1 < p_2$.

In R, the function prop.test will perform a two-sample test of proportions:

```
prop.test(x, n, alternative="two.sided")
```

The data is specified by a vector of values with x storing the counts and n the sample size. There is no need to specify a null hypothesis, as it is always the same. The alternative hypothesis is specified by one of alt="less", alt="greater", or alt="two.sided" (the default).

• **Example 9.8: Poverty rate, continued**
Assume the 2010 poverty rate of 15.00% was derived from a random sample of 160,000 people, and the 2011 poverty rate of 15.13% was derived from a random sample of 150,000. Is the difference between the proportions statistically significant?

Let $\widehat{p}_1 = 0.1500$ and $\widehat{p}_2 = 0.1513$ be the given sample proportions. Our null and alternative hypotheses are

$$H_0 : p_1 = p_2, \quad H_A : p_1 < p_2.$$

We can use prop.test using $n\widehat{p}$ to give the frequencies of those in the sample

```
phat <- c(0.1500, 0.1513)        # the sample proportions
n <- c(160000, 150000)           # the sample sizes
n * phat                         # the counts

## [1] 24000 22695

prop.test(n * phat, n, alternative="less")

##
##   2-sample test for equality of proportions with
##   continuity correction
```

```
##
## data:  n * phat out of n
## X-squared = 1.012, df = 1, p-value = 0.1571
## alternative hypothesis: less
## 95 percent confidence interval:
##  -1.0000000  0.0008212
## sample estimates:
## prop 1 prop 2
## 0.1500 0.1513
```

The p-value agrees with the statement made earlier that the difference is not statistically significant.

If we were to do this computation "by hand," rather than by using prop.test, we would find:

```
p <- sum(n * phat)/sum(n)        # (n_1p_1 + n_2p_2)/(n_1 + n_2)
obs <- (phat[1]-phat[2]) / sqrt(p*(1-p)*sum(1/n))
obs

## [1] -1.011

pnorm(obs)

## [1] 0.1559
```

This also gives a small p-value, though not the same. The difference is due to a continuity correction used by prop.test.   ••

Problems

9.21 A cell-phone store has sold 150 phones of Brand A and had 14 returned as defective. Additionally, it has sold 125 phones of Brand B and had 15 phones returned as defective. Is there statistical evidence that Brand A has a smaller chance of being returned than Brand B?

9.22 In the year 2001, a poll of 600 people found that 250 supported gay marriage. A 2013 poll of 1,050 found 52% in support. Do a test of significance to see whether the difference in proportions is statistically significant.

9.23 There were two advance screenings of a new movie. The first audience was composed of a classical music radio station's listeners, the second a rock-and-roll music station's listeners. Suppose the audience size was 350 for each screening. If 82% of the audience at the first screening rated the movie favorably, but only 70% of second audience did, is this difference statistically

Group	n	AMS sufferers
placebo	22	18
ginkgo biloba	22	3

Table 9.2: Data on acute mountain sickness. Source: *Aviation, Space, and Environmental Medicine* 67, 445–452, 1996.

significant? Can you assume that each audience is a random sample from the population of the respective radio station listeners?

9.24 The HIP mammography study was one of the first and largest studies of the value of mammograms. The study began in New York in the 1960s and involved 60,000 women randomly assigned to two groups—one that received mammograms, and one that did not. The women were then observed for the next 18 years. Of the 30,000 who had mammograms, 153 died of breast cancer; of the 30,000 who did not, 196 died of breast cancer. Compare the two sample proportions to see whether there is a statistically significant difference between the death rates of the two groups. (There is debate about the validity of the experimental protocol.)

9.25 Ginkgo biloba extract is widely touted as a miracle cure for several ailments, including acute mountain sickness (AMS), which is common in mountaineering. A randomized study took 44 healthy subjects to the Himalayas; half received the extract (80 mg twice/day) and half received placebos. Each group was measured for AMS. The results of the study are given in Table 9.2. Compute a p-value for a two-sided alternative.

9.26 Immediately after a ban on using of hand-held cell phones while driving was implemented, compliance with the law was measured. A random sample of 1,250 found that 98.9% were in compliance. A year after the implementation, compliance was again measured. A sample of 1,100 drivers found 96.9% in compliance. Is the difference in proportions statistically significant?

9.27 In 2011, the Centers for Disease Control and Prevention studied distracted driving habits, as an additional, 387,000 people were injured in motor vehicle crashes involving a distracted driver in 2011, compared to 416,000 people injured in 2010 (http://www.cdc.gov/motorvehiclesafety/distracted_driving/). They found that 31% of those surveyed between 18 and 64 had read or sent a text message while driving in the past 30 days.

Suppose the 2011 survey involved 1,000 subjects. A student in 2013 does a follow-up study of 100 students at her college and finds the 41% read or sent messages. Is there statistical evidence that the rate for the college is higher than the national rate in 2011?

9.28 The start of a May 5, 2004, *New York Times* article reads

> In the wake of huge tobacco tax increases and a ban on smoking in bars, the number of adult smokers in New York City fell 11 percent from 2002 to 2003, one of the steepest short-term declines ever measured, according to surveys commissioned by the city.

The article continues, saying that the surveys were conducted using methods—the questions and a random dialing approach—identical to those done annually by the federal Centers for Disease Control and Prevention. Each survey used a large sample of 10,000 people, giving a stated margin of error of 1 percentage point.

The estimated portion of the population that smoked in 2002 was 21.6%; the estimated proportion in 2003 was 19.3%. Are these differences significant at the 0.01 level? (By 2010, the rate had fallen to 14%.)

9.6 Two-sample tests of center

A physician may be interested in knowing whether the time to recover for one surgery is shorter than that for another surgery. A taxicab driver might wish to know whether the time to travel one route is faster than the time to travel another. A consumer group might wish to know whether gasoline prices are similar in two different cities. Or a government agency might want to know whether consumer spending is similar in two different states. All of these questions could be approached by taking random samples from the respective populations and comparing. We consider the situation when the question of issue can be boiled down to a comparison of the centers of the two populations. We can use a significance test to compare centers in the same manner as we compare two population proportions.

However, as there are more possibilities for types of populations considered, there are more test statistics to consider.

Suppose $x_i, i = 1,\ldots,n_x$ and $y_j, j = 1,\ldots,n_y$ are random samples from the two populations of interest. A significance test to compare the centers of their parent distributions would use the hypotheses

$$H_0 : \mu_x = \mu_y, \quad H_A : \mu_x < \mu_y, \ \mu_x > \mu_y, \ \text{or} \ \mu_x \neq \mu_y. \quad (9.3)$$

A reasonable test statistic depends on the assumptions placed on the parent populations. If the populations are normally distributed or nearly so, and the samples are independent of each other, then a t-test can be used. If the populations are not normally distributed, then a nonparametric Wilcoxon test may be appropriate. If the samples are not independent but paired off in some way, then a paired test might be called for.

Two-sample tests of center with normal populations

Suppose the two samples are independent with normally distributed populations. As $\bar{x}$ and $\bar{y}$ estimate μ_x and μ_y respectively, the observed value of $\bar{x} - \bar{y}$ should be a good estimate for the expected difference $\mu_x - \mu_y$. We use this to form a test statistic. As both sample means have normally distributed sampling distributions, a natural test statistic is then

$$T = \frac{(\bar{x} - \bar{y}) - E(\bar{x} - \bar{y} \mid H_0)}{\mathrm{SE}(\bar{x} - \bar{y} \mid H_0)} = \frac{\text{observed} - \text{expected}}{\mathrm{SE}}.$$

Under H_0, the expected value of the difference is 0. The standard error is found from the formula for the standard deviation, which is based on the independence of the samples:

$$\mathrm{SD}(\bar{x} - \bar{y} \mid H_0) = \sqrt{\frac{\sigma_x^2}{n_x} + \frac{\sigma_y^2}{n_y}}.$$

As with confidence intervals, the estimate used for the population variances depends on an assumption of equal variances.

If the two variances are assumed equal, the all the data is pooled to estimate $\sigma = \sigma_x = \sigma_y$ using

$$s_p = \sqrt{\frac{(n_x - 1)s_x^2 + (n_y - 1)s_y^2}{n_x + n_y - 2}}. \tag{9.4}$$

The standard error used is

$$\mathrm{SE}(\bar{x} - \bar{y}) = s_p \sqrt{\frac{1}{n_x} + \frac{1}{n_y}}. \tag{9.5}$$

With this, T has a t-distribution with $n - 2$ degrees of freedom.

If the population variances are not assumed to be equal, then we estimate σ_x with s_x and σ_y with s_y to get

$$\mathrm{SE}(\mu_x - \mu_y) = \sqrt{\frac{s_x^2}{n_x} + \frac{s_y^2}{n_y}}. \tag{9.6}$$

Additionally, we use the Welch approximation for the degrees of freedom as described in Chapter 8. This again yields a test statistic that is described by the t-distribution under the null hypothesis, though with a degrees of freedom between $n - 2$ and the smaller of $n_1 - 1$ and $n_2 - 1$, the latter value being useful to give a conservative estimate for the degrees of freedom.

t-tests for comparison of means of independent samples

Assume $x_1, x_2, \ldots, x_{n_x}$ and $y_1, y_2, \ldots, y_{n_y}$ are independent random samples from $\mathrm{Normal}(\mu_i, \sigma_i)$ distributions, where $i = x$ or

y. A significance test of

$$H_0 : \mu_x = \mu_y, \quad H_A : \mu_x < \mu_y, \; \mu_x > \mu_y, \; \text{or} \; \mu_x \neq \mu_y$$

can be done with test statistic T. T will have the t-distribution with a specified number of degrees of freedom under H_0. Larger values of T support $H_A : \mu_x > \mu_y$.

The function t.test will perform the significance test. It is used with the arguments

```
t.test(x, y, alternative="two.sided", var.equal=FALSE)
```

The data is specified in two data vectors, x and y. There is no need to specify the null hypothesis, as it is always the same. The alternative is specified by "less", "greater", or "two.sided" (the default). The argument var.equal=TRUE is given to specify the equal-variance case. The default is to assume unequal variances.

Formula interface For the two-sample t-test, the t.test function also has a model formula interface for cases where the data is stored in long format with the values in a variable, say x, and a two-level factor, say f. In this case, the call is t.test(x ~ f, data=dataset, alternative="...").

• Example 9.9: Differing dosages of AZT
AZT was the first FDA-approved antiretroviral drug used in the care of HIV-infected individuals. The common dosage is 300 mg twice daily. Higher dosages cause more side effects. But are they more effective? A study done in 1990 compared dosages of 300 mg, 600 mg, and 1,500 mg (source http://www.aids.org). The study found higher toxicity with greater dosages, and, more importantly, that the lower dosage may be equally effective.

The p24 antigen can stimulate immune responses. The measurement of p24 levels for the 300 mg and 600 mg groups is given in Table 9.3. Perform a t-test to determine whether there is a difference in means, assuming the dosage does not change the variance.

Let μ_x be the mean of the 300 mg group, and μ_y the mean of the 600 mg group. We can test the hypotheses

$$H_0 : \mu_x = \mu_y, \quad H_A : \mu_x \neq \mu_y$$

with a t-test. First, we check to see whether the assumption of a common variance and normality seems appropriate by looking at two density plots:

```
m300 <- c(284, 279, 289, 292, 287, 295, 285, 279, 306, 298)
m600 <- c(298, 307, 297, 279, 291, 335, 299, 300, 306, 291)
```

Amount					p24 level				
300 mg	284	279	289	292	287	295	285	279	306	298
600 mg	298	307	297	279	291	335	299	300	306	291

Table 9.3: Levels of p24 in mg for two treatment groups.

```
plot(density(m300))
lines(density(m600), lty=2)
```

The graph (Figure 9.5) shows two density estimates that indicate normally distributed populations with similar spreads. As such, the *t*-test looks appropriate.

```
t.test(m300, m600, var.equal=TRUE)

##
##   Two Sample t-test
##
## data:  m300 and m600
## t = -2.034, df = 18, p-value = 0.05696
## alternative hypothesis: true difference in means is not equal to 0
## 95 percent confidence interval:
##   -22.1584    0.3584
## sample estimates:
## mean of x mean of y
##      289.4      300.3
```

The *p*-value is 0.05696 for the two-sided test. This suggests a difference in the mean values, but it is not statistically significant at the 0.05 significance level. A look at the reported confidence interval for the difference of the means shows a wide range of possible value for $\mu_x - \mu_y$. We conclude that this data is consistent with the assumption of no mean difference.

How would this have been different if we did not assume equal variances?

```
t.test(m300, m600)

##
##   Welch Two Sample t-test
##
## data:  m300 and m600
## t = -2.034, df = 14.51, p-value = 0.06065
## alternative hypothesis: true difference in means is not equal to 0
## 95 percent confidence interval:
```

```
## -22.3557   0.5557
## sample estimates:
## mean of x mean of y
##     289.4      300.3
```

In this example, the same observed value of the test statistic (marked t) is found as in the equal-variance case, as (9.5) and (9.6) yield identical standard errors when the two sample sizes are the same. We get a larger p-value, though, as the degrees of freedom differ.

The t.test function has no assumption of equal variance as the default. Though making this assumption can produce smaller p-values, it should only be done if applicable and before considering the data.   ••

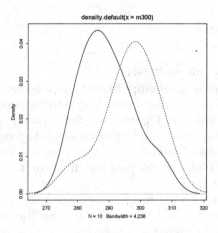

Figure 9.5: Density plots to compare variances and shapes of the 300 mg dosage (solid) and the 600 mg dosage (dashed).

Matched samples

There are times when two samples depend on each other in some way, for example, samples from twin studies, where identical or fraternal twins are used as pairs, so that genetic or environmental factors can be controlled. For this, the usual two-sample t-test is not applicable. We mention two examples.

• **Example 9.10: Twin studies**
Suppose, an industry-sponsored clinical trial "demonstrated" that Finasteride inhibits male-pattern hair loss. Their approach used involved two treatment groups: one received a Finasteride treatment, the other a placebo. A

randomized, double-blind study was performed. Hair loss was measured by photographs, hair counts, and questionnaires.

What was different about this study was the use of identical twins for the treatment groups. For each pair of twins, one was randomly assigned to the treatment group and the other to the control group. This allowed the researchers to "control" for genetic differences—differences that might be so great that the true effect of the Finasteride treatment could be hidden. The researchers stated

> As identical twins share the same genetic makeup, compari-
> son between the responses of each subject in a twin pair, when
> one receives drug and the other receives placebo, allows for rigor-
> ous examination of the effects due to drug treatment in a limited
> number of subjects.

••

• **Example 9.11: Pre- and post-tests**
Outcomes assessment is an attempt to measure whether a certain action ac-tually does what it is intended to do. For example, does a statistics book actually work for teaching R? Or, does a statistics class make you understand the difference between mere sampling variation and a true effect? One way to assess the effectiveness of something is with a pre-test and a post-test. If the scores are markedly better on the post-test, then we may be able to attribute the change to the teaching.

Imagine a class takes a pre-test and a post-test. Each student has two test scores, x_i for the first test and the matching y_i for the second. How can we test whether there is a difference in the means? We might be tempted to use the t-test, but we should be careful, as the two samples are not independent. This assumption of independence was used implicitly when computing the standard error in the test statistic. Besides, what is really important is the change in the scores $x_i - y_i$. ••

For paired data, even if there are large variations within the samples, we can still test a difference in the means by using a one-sample test applied to the data, $x_i - y_i$.

Significance tests for paired samples   If the two samples $x_1, x_2, \ldots, x_n$ and $y_1, y_2, \ldots, y_n$ are matched so that the differences $x_i - y_i$ are an *i.i.d.* sample, then the significance test of hypotheses

$$H_0 : \mu_x = \mu_y, \quad H_A : \mu_x < \mu_y, \ \mu_x \neq \mu_y, \ \text{or} \ \mu_x > \mu_y$$

Group	Score								
Finasteride treatment	5	3	5	7	4	4	7	4	3
Placebo	2	3	2	4	2	2	3	4	2

Table 9.4: Assessment for hair loss on 1–7 scale for twin study.

becomes a significance test of

$$H_0 : \mu = 0, \quad H_A : \mu < 0, \mu > 0, \quad \text{or} \quad \mu \neq 0.$$

If the differences have a normally distributed population, a t-test can be used. If the differences are from a symmetric distribution, the Wilcoxon signed-rank test can be used. Otherwise, the sign test can be used, where μ is interpreted as the difference of medians.

In R, both the t.test and wilcox.test functions have an argument to indicate paired tests should be performed.

• **Example 9.12: Twin studies continued**
For the Finasteride study, photographs are taken of each head. They are assessed using a standard methodology. This results in a score between 1 and 7: 1 indicating greatly decreased hair growth and 7 greatly increased. Simulated data, presented as pairs, is in Table 9.4.

We can assess the differences with a paired t-test as follows:

```
Finasteride <- c(5, 3, 5, 6, 4, 4, 7, 4, 3)
placebo     <- c(2, 3, 2, 4, 2, 2, 3, 4, 2)
t.test(Finasteride, placebo, paired=TRUE, alternative="two.sided")

##
##  Paired t-test
##
## data:  Finasteride and placebo
## t = 4.154, df = 8, p-value = 0.003192
## alternative hypothesis: true difference in means is not equal to 0
## 95 percent confidence interval:
##   0.8403 2.9375
## sample estimates:
## mean of the differences
##                   1.889
```

We see a very small p-value, indicating that the result is significant. The null hypothesis of no effect is in doubt. ••

Test	Score									
Pre-test	77	56	64	60	57	53	72	62	65	66
Post-test	88	74	83	68	58	50	67	64	74	60

Table 9.5: Pre- and post-test scores.

● **Example 9.13: Pre- and post-tests, continued**
To test whether a college course is working, a pre- and post-test is arranged
for the students. The results are given in Table 9.5. Compare the scores with a
t-test. First, assume that the scores are randomly selected from the two tests.
Next, assume that they are pairs of scores for ten students.

For each, we test the hypotheses that

$$H_0 : \mu_1 = \mu_2, \quad H_A : \mu_1 < \mu_2,$$

and we assume that the data is normally distributed.

If we assume that the scores are random samples from the two test populations, then the usual t-test is used. We first make a boxplot, to decide
whether the variances are equal (not shown), and then we apply the test.

```
pre  <- c(77, 56, 64, 60, 57, 53, 72, 62, 65, 66)
post <- c(88, 74, 83, 68, 58, 50, 67, 64, 74, 60)
boxplot(pre, post)
out <- t.test(pre, post, var.equal=TRUE, alternative="less")
out$p.value

## [1] 0.1139
```

The p-value is small but not significant.

If we assume these scores are paired off, then we focus on the differences.
This gives a much smaller p-value:

```
out <- t.test(pre, post, paired=TRUE, alternative="less")
out$p.value

## [1] 0.04564
```

This time, the difference is significant at the 0.05 level.

If small samples are to be used, it can often be advantageous to use paired
samples, rather than independent samples.   ●●

The Wilcoxon rank-sum test for equality of center

The two-sample t-test tests whether two independent samples have the same
center when both samples are drawn from a normal distribution. However,

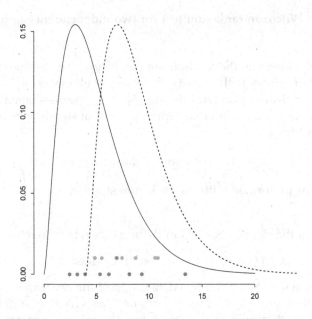

Figure 9.6: Two random samples from similar distributions indicated by points with different shades. The lower ranked ones come primarily from the distribution shifted to the left.

there are many situations in which the parent populations may be heavily skewed or have heavy tails. Then the t-test is not appropriate. However, if it is assumed that our two samples come from two distributions that are identical up to a shift of center, then the Wilcoxon rank-sum test can be used to perform a significance test to test whether the centers are identical.

Suppose $f(x)$ is a density of a continuous distribution with mean 0. Further, assume that the x_i are a random sample from a population with density $f(x - \mu_1)$ (so it has mean μ_1), and that the y_j are a random sample from a population with density $f(x - \mu_2)$. Figure 9.6 shows two samples where the μ's are different. The darker distribution is shifted to the left, and its sample, indicated with darker dots, has most of its values to the left of the other sample. This would not be expected if the two populations were identical.

The rank-sum statistic quantifies this, allowing for a significance test. First, rank all the data from smallest to largest. Then, let W be the sum of the ranks corresponding to the x_i. Large values of W support the alternative hypothesis that $\mu_x > \mu_y$, small values the alternative that $\mu_x < \mu_y$. What large and small mean is given by the distribution of W under the null hypothesis. This is called the rank-sum distribution and is returned by the R family wilcox.

Wilcoxon rank-sum test for two independent samples

Assume that the random sample, $y_1, y_2, \ldots, y_{n_x}$, comes from a distribution with density $f(\cdot - \mu_x)$, and that $y_1, y_2, \ldots, y_{n_y}$ from a distribution with density $f(\cdot - \mu_y)$ (same-shaped density, but perhaps different centers). A test of significance of the hypotheses

$$H_0 : \mu_1 = \mu_2, \quad H_A : \mu_1 < \mu_2, \ \mu_1 > \mu_2, \ \text{or} \ \mu_1 \neq \mu_2$$

can be performed with the rank-sum statistic.

To perform the significance test in R, the `wilcox.test` function is used as

```
wilcox.test(x, y, alternative="two.sided")
```

The variables x and y store the two data sets, and the alternative is specified, as usual, with one of "less", "greater", or "two.sided" (the default). The `wilcox.test` function will also work in the case when there are ties in the data.

• **Example 9.14: Comparing grocery checkers**
A grocery store's management wishes to compare checkout personnel to see if there is a difference in their checkout times. A small sample of data comparing two checkers' times (in minutes) is given by:

Checker	Checkout times								
Checker A	5.8	1.0	1.1	2.1	2.5	1.1	1.0	1.2	3.2	2.7
Checker B	1.5	2.7	6.6	4.6	1.1	1.2	5.7	3.2	1.2	1.3

Compare the mean checkout times.
We use the `wilcox.test` function after verifying that the model assumptions are met.

```
A <- c(5.8, 1.0, 1.1, 2.1, 2.5, 1.1, 1.0, 1.2, 3.2, 2.7)
B <- c(1.5, 2.7, 6.6, 4.6, 1.1, 1.2, 5.7, 3.2, 1.2, 1.3)
plot(density(A))
lines(density(B))
```

The graph (not shown) suggests that the populations are skewed with long tails. As such, the t-test assumptions are not met. However, we also see that the samples appear to have densities with the same shape, so the rank-sum test is available. A two-sided test can be performed with

```
wilcox.test(A,B)

## Warning: cannot compute exact p-value with ties

##
##  Wilcoxon rank sum test with continuity correction
##
## data:  A and B
## W = 34, p-value = 0.2394
## alternative hypothesis: true location shift is not equal to 0
```

The *p*-value is not significant.   ••

Problems

9.29 A 2003 study at the Cleveland Clinic compared the effects of two choles-
terol drugs, atorvastatin and pravastatin, on middle-aged heart-disease patients. It was found that the atorvastatin treatment group had an average
LDL level of 79 after treatment, whereas the pravastatin group had an average LDL level of 110. Suppose the two groups contained 250 patients each,
and the sample standard deviations were 25 for the atorvastatin group and 20
for the pravastatin. If the populations are assumed to be normally distributed,
perform a two-sample test to compare whether the mean LDL levels for atorvastatin are lower than those for pravastatin, or whether the differences are
explainable by chance variation.

9.30 A test to determine whether echinacea is beneficial in treating the common cold was set up as follows. If a child reported cold symptoms, then he
was randomly assigned to be given either echinacea or a placebo. Recovery
time was measured and is summarized with:

group	n	$\bar{x}$	s
echinacea	200	5.3	2.5
placebo	207	5.4	2.5

Is this statistical evidence that children in the echinacea group had a
quicker recovery?

9.31 For the babies (UsingR) data set, the variable age contains the recorded
mom's age and dage contains the dad's age for several different cases in the
sample. Do a significance test of the null hypothesis of equal ages against a
one-sided alternative that the dads are older in the sampled population.

9.32 The data set normtemp (UsingR) contains body measurements for 130 healthy, randomly selected individuals from some parent population. The variable temperature contains normal body temperature data and the variable gender contains gender information, with male coded as 1 and female as 2. Is the sample difference across the two groups statistically significant?

9.33 Students wishing to graduate must achieve a specific score on a standardized test. Those failing must take a course and then attempt the test again. Suppose 12 students are enrolled in the extra course and their two test scores are given by

Student						scores						
Pre-test	17	12	20	12	20	21	23	10	15	17	18	18
Post-test	19	25	18	18	26	19	27	14	20	22	16	18

Assuming these students represent a random sample from some larger population, perform a t-test to see if there would be any improvement in the population mean scores following such a class. If you assume equal variances or a paired test, explain why.

9.34 Water-quality researchers wish to measure biomass/chlorophyll ratio for phytoplankton (in milligrams per liter of water). There are two possible tests, one less expensive than the other. To see whether the two tests give the same results, ten water samples were taken and each was measured both ways, providing the data below.

Method					measurement					
method 1	45.9	57.6	54.9	38.7	35.7	39.2	45.9	43.2	45.4	54.8
method 2	48.2	64.2	56.8	47.2	43.7	45.7	53.0	52.0	45.1	57.5

Do a t-test to see if there is a difference in the means of the measured amounts. If you assume equal variances or a paired test, explain why.

9.35 The shoes (MASS) data set contains a famous data set on shoe wear. Ten boys wore two different shoes each, then measurements were taken on shoe wear. The wear amounts are stored in variables A and B. First make a scatterplot of the data, then compare the mean wear for the two types of shoes using the appropriate t-test.

9.36 The Galton (HistData) data set contains data used by Francis Galton in 1885. Each data point contains a child's height and an average of his or her parents' heights. Assuming the data is a random sample for a population of interest, perform a t-test to see if there is a difference in the population mean height. Assume the paired t-test is appropriate. What problems are there with this assumption?

9.37 The question of equal variances comes up when we perform a two-sample t-test. We've answered this based on a graphical exploration. The F-test for equal variances of two normal populations can be used to test formally for equality. The test statistic is the ratio of the sample variances, which under the null hypothesis of equal variances has an F-distribution. This test is carried out by the function var.test. A two-sided test ($H_A : \sigma_1^2 \neq \sigma_2^2$) is performed with the command var.test(x,y).

   Do a two-sided test for equality of variance on the data in Table 9.3.

Goodness of fit

In this chapter we return to problems involving categorical data. We previously summarized such data using tables. Here we discuss a significance test for the distribution of the values in a table. The test statistic will be based on how well the actual counts for each category fit the expected counts.

Such tests are called goodness-of-fit tests, as they measure how well the distribution of the data fits a probability model. In this chapter we will also discuss goodness-of-fit tests for continuous data. For example, we will learn a significance test for investigating whether a data set is normally distributed.

10.1 The chi-squared goodness-of-fit test

In a public-opinion poll, there are often more than two possible opinions. For example, suppose a survey of registered voters is taken to see which candidate is likely to be elected in an upcoming election. For simplicity, we assume there are two candidates, a Republican and a Democrat. A prospective voter may choose one of these or may be undecided. If 100 people are surveyed, and the results are 35 for the Republican, 40 for the Democrat, and 25 undecided, is the difference between the Republican and Democratic candidate significant?

The multinomial distribution

Before answering a question about significance, we need a probability model, so that p-value calculations can be made. The above example is a bit different from the familiar polling model. When there are just two categories to choose from we use the binomial model as our probability model; in this case, with more categories, we generalize and use the *multinomial model*.

Assume we have k categories to choose from, labeled 1 through k. We pick one of the categories at random, with probabilities specified by $p_1, p_2, \ldots, p_k$; p_i gives the probability of selecting category i. We must have $p_1 + p_2 + \cdots + p_k = 1$. If all the p_i equal $1/k$, then each category is equally likely (like rolling a die). Picking a category with these probabilities produces a single random value; repeat this selection n times, with each pick being independent, to get n values. A table of values will report the frequencies. Call these table entries $y_1, y_2, \ldots, y_k$. These k numbers sum to n. The joint distribution of these random variables is called the *multinomial distribution*.

• Example 10.1: Using `sample` to simulate multinomial data

We can create multinomial data in R with the `sample` function. For example, an M&M's bag is filled using colors drawn from a fixed ratio. A bag of 30 can be filled as follows:[1]

```
cols <- c("blue", "brown", "green", "orange", "red", "yellow",
          "purple")
prob <- c(1, 1, 1, 1, 2, 2, 2)         # ratio of colors
prob <- prob / sum(prob)
n <- 30
bagful <- sample(cols, n, replace=TRUE, prob=prob)
table(bagful)

## bagful
##   blue  brown  green orange purple    red yellow
##      3      3      4      4      3      8      5
```

••

A formula for the multinomial distribution is similar to that for the binomial distribution except that more factors are involved, as there are more categories to choose from. The distribution can be specified as follows:

$$P(y_1 = y_1, \ldots, y_k = y_k) = \binom{n}{y_1} \binom{n - y_1}{y_2} \cdots \binom{n - y_1 - y_2 - \cdots - y_{k-1}}{y_k} p_1^{y_1} \cdots p_k^{y_k}.$$

As an example, consider the voter survey. Suppose we expected the percentages to be 35% Republican, 35% Democrat, and 30% undecided. What is the probability in a survey of 100 likely voters that we see 35, 40, and 25, respectively? It is

$$P(y_1 = 35, y_2 = 40, y_3 = 25) = \binom{100}{35} \binom{65}{40} \binom{25}{25} (0.35)^{35} (0.35)^{40} (0.3)^{25}.$$

This value can be found directly with

```
choose(100,35)*choose(65,40)*choose(25,25) * .35^35 * .35^40 * .30^25

## [1] 0.00386
```

---

[1]The `sample` function needs to be called with `replace=TRUE` to sample with replacement. The `mosaic` package provides a convenience wrapper `resample` which uses this default.

(We could skip the last choose factor, as $\binom{j}{j} = 1$ for any j.) The dmultinom function can also be used for the above computation. This small value is the probability of the observed value, but it is not a p-value. A p-value also includes the probability of seeing more extreme values than the observed one. We would still need to specify what that means to compute a p-value.

Pearson's χ^2-statistic

Trying to use the multinomial distribution directly to answer a problem about the p-value is difficult, as the variables y_i are correlated—they add to n. If one value is large the others are more likely to be small, so the meaning of "extreme" in calculating a p-value is not immediately clear. As an alternative, the problem of testing whether a given set of probabilities could have produced the data is done as before: by comparing the observed value with the expected value and then normalizing to get something with a known distribution.

Each y_i is a random variable telling us how many of the n choices were in category i. If we focus on a single i, then y_i is seen to be Binomial(n, p_i) with an expected value of np_i. Based on this, a good statistic might be

$$\sum_{i=1}^{k} (y_i - np_i)^2.$$

This gives the total discrepancy between the observed and the expected. We use the square as $\sum(y_i - np_i) = 0$. This sum gets larger when a category is larger or smaller than expected. So a larger-than-expected value contributes, and any correlated smaller-than-expected values do, too. As usual, we scale this by the right amount to yield a test statistic with a known distribution. In this case, each term is divided by the expected amount, producing *Pearson's chi-squared statistic* (written using the Greek letter "*chi*"):

$$\chi^2 = \sum_{i=1}^{k} \frac{(y_i - np_i)^2}{np_i} = \sum \frac{(\text{observed} - \text{expected})^2}{\text{expected}}. \tag{10.1}$$

If the multinomial model is correct, then the *asymptotic* distribution of y_i is known to be the chi-squared distribution with $k - 1$ degrees of freedom. The number of degrees of freedom coincides with the number of free ways we can specify the values for p_i in the null hypothesis. Here we are free to choose $k - 1$ of the values but not k, as the values must sum to 1.

The chi-squared distribution is a good fit to the sampling distribution of the statistic if the expected cell counts are all five or more. Figure 10.1 shows a simulation and a histogram of the corresponding χ^2-statistic, along with a theoretical density, when $n = 20$ and np is 5 or more for all the p's.

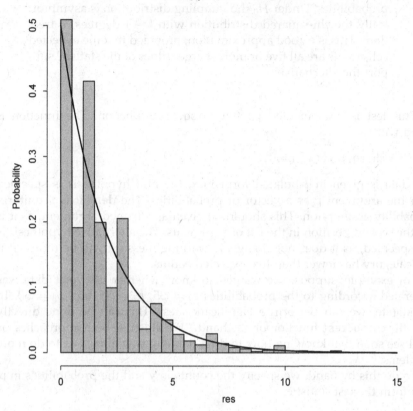

Figure 10.1: Simulation of χ^2-statistic with $n = 20$ and probabilities $3/12$, $4/12$, and $5/12$. The chi-squared density with $2 = k - 1$ degrees of freedom is added.

Using this statistic as a test statistic allows us to construct a significance test. Larger values are now considered more extreme, as they imply more discrepancy from the predicted amount.

The chi-squared significance test for goodness of fit

Let $y_1, y_2, \ldots, y_k$ be the observed cell counts in a table that arise from random sampling. Suppose their joint distribution is described by the multinomial model with probabilities $p_1, p_2, \ldots, p_k$. A significance test of

$$H_0 : p_1 = \pi_1, \cdots, p_k = \pi_k, \quad H_A : p_i \neq \pi_i \text{ for at least one } i$$

can be performed with the χ^2-statistic. The π_i are specified probabilities. Under H_0 the sampling distribution is asymptotically the chi-squared distribution with $k-1$ degrees of freedom. This is a good approximation, provided that the expected cell counts are all five or more. Large values of the statistic support the alternative.

This test is implemented by the chisq.test function. The function is called with

```
. chisq.test(x, p=...)
```

The data is given in tabulated form in x; the null hypothesis is specified with the argument p as a vector of probabilities. The default is a uniform probability assumption. This should be given as a named argument, as it is not the second position in the list of arguments. The alternative hypothesis is not specified, as it does not change. A warning message will be returned if any category has fewer than five expected counts.

For example, suppose we wanted to know whether the voter data was generated according to the probabilities $p_1 = .35$, $p_2 = .35$, and $p_3 = .30$. To investigate, we can perform a significance test. This can be done directly with the chisq.test function or "by hand." We illustrate both approaches, as we'll see soon that knowing how to do it the long way allows us to do more problems.

To do this by hand, we specify the counts in y and the probabilities in p, then form the test statistic:

```
y <- c(35, 40, 25)
p <- c(35, 35, 30)            # ratios
p <- p/sum(p)                 # proportions
n <- sum(y)
chi2 <- sum( (y - n*p)^2 / (n*p) )
chi2

## [1] 1.548

pchisq(chi2, df=3 - 1, lower.tail=FALSE)

## [1] 0.4613
```

In contrast, the above could have been done with

```
chisq.test(y, p=p)

##
##  Chi-squared test for given probabilities
```

```
##
## data:  y
## X-squared = 1.548, df = 2, p-value = 0.4613
```

The function returns the value of the test statistic (after X-squared), the degrees of freedom, and the p-value.

• **Example 10.2: Teen smoking**
The samhda (UsingR) data set contains information about health behavior for school-age children. For example, the variable amt.smoke measures how often a child smoked in the previous month. There are seven levels: a 1 means he smoked every day and a 7 means not at all. Values 98 and 99 indicate missing data. See ?samhda for a description. We investigate whether the sample proportions are statistically different from the probabilities:

$$p_1 = 0.15, p_2 = 0.05, p_3 = 0.05, p_4 = 0.05, p_5 = 0.10, p_6 = 0.20, p_7 = 0.40.$$

A test of significance can be constructed as follows:

```
library(UsingR)
amt <- with(samhda, amt.smoke[amt.smoke < 98])
y <- table(amt)
y

## amt
##   1    2    3    4    5    6    7
##  32    7   13   10   14   43  105

ps <- c(0.15, 0.05, 0.05, 0.05, 0.10, 0.20, 0.40)
chisq.test(y, p=ps)

##
##   Chi-squared test for given probabilities
##
## data:  y
## X-squared = 7.938, df = 6, p-value = 0.2427
```

The p-value of 0.2427 is not significant. There is no evidence that the population proportions differ from those specified by the null hypothesis.   ••

Partially specified null hypotheses

In the example with voting data, we might be interested in assessing whether the Republican candidate differences from the Democrat can be attributed to sampling variation. That is, we would want to test the hypotheses

$$H_0 : p_1 = p_2, \quad H_A : p_1 \neq p_2.$$

These, too, can be tested with the χ^2-statistic, but we need to specify what we mean by "expected," as under H_0 this is not fully specified. (The value of p_1 and p_2 depends on p_3, which isn't specified.)

To do so, we will use any values completely specified by the null hypothesis; for those values that aren't (e.g., p_3 above), we estimate using the null hypothesis to pool our data as appropriate. For this problem, none of the p_i values are fully specified. To estimate $\hat{p}_1 = \hat{p}_2$, we use both of the cell counts through $(y_1 + y_2)/(2n)$. This leaves $\hat{p}_3 = y_3/n = 1 - \hat{p}_1 - \hat{p}_2$. Then the χ^2-statistic in this case becomes

$$\chi^2 = \sum_{i=1}^{k} \frac{(y_i - n\hat{p}_i)^2}{n\hat{p}_i}.$$

Again, if all the expected counts are large enough, this will have an approximately chi-squared distribution. There is only one degree of freedom in this problem, as only one thing is left to estimate, namely the value $p = p_1 = p_2$. Once we specify a value of p, then, by the assumptions in the null hypothesis, all the p_i are decided.

We get the p-value in our example as follows:

```
y <- c(35, 40, 25)
n <- sum(y)
phat1 <- phat2 <- sum(y[1:2])/(2*n)
phat3 <- 1 - phat1 - phat2
phat <- c(phat1, phat2, phat3)
#
obs <- sum((y - n*phat)^2/(n*phat))
obs
```

```
## [1] 0.3333
```

```
pchisq(obs, df =1 , lower.tail=FALSE)
```

```
## [1] 0.5637
```

The difference is not statistically significant.

In general, the χ^2-statistic can be used in significance tests where the null specifies some relationship among the multinomial probabilities. The asymptotic distribution of the test statistic under the null hypothesis will be chi-squared. The degrees of freedom will depend on the number of values that we are free to specify.

Candidate	party	poll amount	actual
Schwarzenegger	Republican	315	48.6
Bustamante	Democrat	197	31.5
McClintock	Republican	141	12.5
Camejo	Green	39	2.8
Huffington	Independent	16	0.6
Other	–	79	4.0

Table 10.1: California gubernatorial recall election.

Problems

10.1 A die is rolled 100 times and yields these frequencies

	1	2	3	4	5	6
count	13	17	9	17	18	26

Is this a fair die? Answer using a significance test with $H_0 : p_i = 1/6$ for each i, and $H_A : p_i \neq 1/6$ for at least one i.

10.2 Table 10.1 contains the results of a poll of 787 registered voters and the actual race results (in percentages of total votes) in the 2003 gubernatorial recall election in California.

Is the sample data consistent with the actual results? Answer this using a test of significance.

10.3 A package of M&M's candies is filled from batches that contain a specified percentage of each of six colors. These percentages are given in the mandms (UsingR) data frame. Assume a package of candies contains the following color distribution: 15 blue, 34 brown, 7 green, 19 orange, 29 red, and 24 yellow. Perform a chi-squared test with the null hypothesis that the candies are from a milk chocolate package. Repeat assuming the candies are from a Peanut package. Based on the p-values, which would you suspect is the true source of the candies?

10.4 The pi2000 (UsingR) data set contains the first 2,000 digits of π. Perform a chi-squared significance test to see if the digits appear with equal probability.

10.5 A simple trick for determining what language a document is written in is to compare the letter distributions (e.g., the number of z's) to the known proportions for a language. For these proportions, we use the familiar letter frequencies given in the frequencies variable of the scrabble (UsingR) data set. These are an okay approximation to those in the English language.

	a	e	i	o	u
Count	28	39	23	22	11
Scrabble frequency	9	12	9	8	4

Table 10.2: Vowel distribution and Scrabble frequency.

For simplicity (see ?scrabble for more details), we focus on the vowel distribution of a paragraph from R's webpage appearing below. The counts and Scrabble frequencies are given in Table 10.2.

> R is a language and environment for statistical computing and graphics. It is a GNU project which is similar to the S language and environment which was developed at Bell Laboratories (formerly AT&T, now Lucent Technologies) by John Chambers and colleagues. R can be considered as a different implementation of S. There are some important differences, but much code written for S runs unaltered under R.

Perform a chi-squared goodness-of-fit test to see whether the distribution of vowels appears to be from English.

10.6 The names of common stars are typically Greek or Arab in derivation. The bright.stars (UsingR) data set contains 96 names of common stars. Perform a significance test on the letter distribution to see whether they could be mistaken for English words.

The letter distribution can be found with:

```
all.names <- paste(bright.stars$name, sep="", collapse="")
x <- unlist(strsplit(tolower(all.names), ""))
letter.dist <- sapply(letters, function(i) sum(x == i))
```

The English-letter frequency is found using the scrabble (UsingR) data set with:

```
ps <- scrabble$frequency[1:26]
ps <- ps/sum(ps)
```

10.7 The number of murders by day of week in New Jersey during 2011 and 2003 is shown in Table 10.3.

1. For the 2011 data, perform a significance test to test the null hypothesis that a murder is equally likely to occur on any given day.

Year	Sun	Mon	Tues	Wed	Thurs	Fri	Sat
2003	53	42	51	45	36	37	65
2011	63	53	50	51	55	52	56

Table 10.3: Number of murders by day of week in New Jersey in 2003 and 2011. (Source: http://www.njsp.org.)

2. Again, for the 2011 data perform a significance test of the null hypothesis that murders happen on each weekday with equal probability; similarly on the weekends, but not necessarily with the same probability.

For each test, write down explicitly the null and alternative hypotheses.

10.8 A large bag of M&M's is opened and some of the colors are counted: 41 brown, 48 orange, 105 yellow, and 58 green. Test the partially specified null hypothesis that the probability of brown is equal to the probability of orange. What do you conclude?

10.9 The data for Figure 10.1 was simulated using the following commands:

```
M <- 2000; n <- 20
p <- c(3,4,5)/12

res <- replicate(M, {
  x <- sample(1:3, n, replace=TRUE, prob=p)
  y <- sapply(1:3, function(i) sum(x==i))
  expected <- n * p
  chi <- sum( (y - expected)^2/expected )
  chi
})

col <- rgb(.7,.7,.7,.75)
hist(res, prob=TRUE, breaks=50,
    col=col, ylab="Probability", xlab="res",
    main="Chi-squared simulation")
curve(dchisq(x, df=length(p)-1), add=TRUE, lwd=2)
```

The sampling distribution of χ^2 is well approximated by the chi-squared distribution, with $k-1$ degrees if the expected cell counts are all five or more. Do a simulation like the above, only with $n = 5$. Does the fit seem right? Repeat with $n = 20$ using the different probabilities p=c(1,19,20)/40. Does the fit seem right?

10.10 When $k = 2$ you can algebraically simplify the χ^2-statistic. Show that it simplifies to

$$\chi^2 = \left(\frac{\hat{p}_1 - p_1}{\sqrt{p_1(1 - p_1)/n}} \right)^2 .$$

This is the square of the statistic used in the one-sample test of proportion and is asymptotically a single-squared normal or a chi-squared random variable with 1 degree of freedom. Thus, in this case, the chi-squared test is equivalent to the test of proportions.

10.2   The chi-squared test of independence

In a two-way contingency table we are interested in the relationship between the variables. In particular, we ask whether the levels of one variable effect the distribution of the other variable. That is, are they independent random variables in the same sense that we defined an independent sequence of random variables?

For example, in the seat-belt-usage data from Table 3.3, does the fact that a parent has her seat belt buckled effect the chance that the child's seat belt will be buckled?

The differences appear so dramatic that the answer seems to be obvious. We can set up a significance test to help make the decision formal, using a method that can be used when the data does not tell such a clear story.

To approach this question with a significance test, we need to state the null and alternative hypotheses, a test statistic, and a probability model.

First, our model for the sampling is that each observed car follows some specified probability that is recorded in any given cell. These probabilities don't change from observation to observation, and the outcome of one does not effect the distribution of another. That is, we have an *i.i.d.* sequence. Then a multinomial model applies. Fix some notation. Let n_r be the number of rows in the table (the number of levels of the row variable), n_c be the number of columns, and y_{ij} record the frequency of the (i,j) cell. Let p_{ij} be the cell probability for the ith row and jth column from a model for the data. The marginal probabilities are denoted p_i^r and p_j^c where, for example, $p_i^r = p_{i1} + p_{i2} + \cdots + p_{in_j}$.

Our null hypothesis is that the column variable should be *independent* of the row variable. When stated in terms of the cell probabilities, this says that $p_{ij} = p_i^r p_j^c$. This is consistent with the notion that independence means multiply.

Our hypotheses can be stated informally as:

H_0 : the variables are independent, H_A : the variables are not independent.

In terms of our notation, we can rewrite the null hypothesis as $H_0 : p_{ij} = p_i^r p_j^c$.

	Child		
Parent	buckled	unbuckled	marginal
buckled	56	8	64
unbuckled	2	16	18
marginal	58	24	82

Table 10.4: Seat-belt usage in California with marginal distributions

The χ^2-statistic,

$$\chi^2 = \sum \frac{(\text{observed} - \text{expected})^2}{\text{expected}},$$

can still be used as a test statistic after we have estimated each p_{ij} in order to compute the "expected" counts. Again we use the data and the assumptions to estimate the p_{ij}. Basically, the data is used to estimate the marginal probabilities, and the assumption of independence allows us to estimate the p_{ij} from there.

The marginal probabilities are estimated by the marginal distributions of the data. For our example these are added to Table 3.3 to give Table 10.4.

The estimate for $p_1^r = P(\text{parent is buckled})$ is $\widehat{p}_1^r = 64/82$, and for $p_2^r = P(\text{parent is unbuckled})$ it is $\widehat{p}_2^r = 18/82$. Similarly, for p_j^c we have $\widehat{p}_1^c = 58/82$ and $\widehat{p}_2^c = 24/82$. As usual, we've used a "hat" for estimated values.

With these estimates, we can use the relationship $p_{ij} = p_i^r p_j^c$ to find the estimate $\widehat{p}_{ij} = \widehat{p}_i^r \widehat{p}_j^c$. For our seat-belt data we have the estimates in Table 10.5. In order to show where the values come from, the values have not been simplified.

With this table we can compute the expected amounts in the ijth cell with $n\widehat{p}_{ij}$. This is often written $R_i C_j / n$, where R_i is the row sum and C_j the column sum, as this simplifies computations by hand.

With the expected amounts now known, we form the χ^2-statistic as:

$$\chi^2 = \sum_{i=1}^{n_r} \sum_{j=1}^{n_c} \frac{(y_{ij} - n\widehat{p}_{ij})^2}{n\widehat{p}_{ij}}. \tag{10.2}$$

Under the hypothesis of multinomial data and the independence of the variables, the sampling distribution of χ^2 will be the chi-squared distribution with $(n_r - 1) \cdot (n_c - 1)$ degrees of freedom. Why this many? In general, we subtract one degree of freedom from $n_r \cdot n_c - 1$ for each estimated parameter. As there are $n_r - 1 + n_c + 1$ estimated parameters, the value for the degrees of freedom is $n_r \cdot n_c - 1 - (n_r - 1 + n_c + 1) = n_r \cdot n_c - n_r - n_c + 1 = (n_r - 1) \cdot (n_c - 1)$.

	Child		
Parent	buckled	unbuckled	marginal
buckled	$\frac{64}{82} \cdot \frac{58}{82}$	$\frac{64}{82} \cdot \frac{24}{82}$	$\frac{64}{82}$
unbuckled	$\frac{18}{82} \cdot \frac{58}{82}$	$\frac{18}{82} \cdot \frac{24}{82}$	$\frac{18}{82}$
marginal	$\frac{58}{82}$	$\frac{24}{82}$	1

Table 10.5: Seat-belt usage in California with estimates $\hat{p}_{ij}$ for the corresponding p_{ij}.

We now have all the pieces to formulate the problem in the language of a significance test.

The chi-squared test of independence

Let $y_{ij}, i = 1, \ldots, n_r, j = 1, \ldots, n_c$ be the cell frequencies in a two-way contingency table for which the multinomial model applies. A significance test of

H_0 : the two variables are independent

H_A : the two variables are not independent

can be performed using the chi-squared test statistic (10.2). Under the null hypothesis, this statistic has a sampling distribution that is approximated by the chi-squared distribution with $(n_r - 1)(n_c - 1)$ degrees of freedom. The p-value is computed using $P(\chi^2 \geq$ observed value $| H_0)$.

In R this test is performed by the chisq.test function. If the data is summarized in a table or a matrix in the variable x the usage is

chisq.test(x).

If the data is unsummarized and is stored in two variables x and y where the ith entries match up, then the function can be used as

chisq.test(x, y).

Alternatively, the data could be summarized first using table. (For table and xtabs objects, the summary method for these objects will also report this test.)
For each usage, the null and alternative hypotheses are not specified, as they are the same each time the test is used.

The argument simulate.p.value=TRUE will return a p-value estimated using a Monte Carlo simulation. This is used if the expected counts in some cells are too small to use the chi-squared distribution to approximate the sampling distribution of χ^2.

To illustrate, the following will do the chi-squared test on the seat-belt data. This data is summarized, so we first need to make a table. We use rbind to combine rows.

```
seatbelt <- rbind(c(56,8), c(2,16))
seatbelt

##      [,1] [,2]
## [1,]   56    8
## [2,]    2   16

chisq.test(seatbelt)

##
## 	Pearson's Chi-squared test with Yates' continuity
## 	correction
##
## data:  seatbelt
## X-squared = 36, df = 1, p-value = 1.978e-09
```

The minuscule p-value is consistent with our observation that the two variables are not independent.

• Example 10.3: Teen smoking and gender
The samhda (UsingR) data set contains survey data on 590 children. The variables gender and amt.smoke contain information about the gender of the participant and how often the participant smoked in the last month. Are the two variables independent?
We compute a p-value for the hypotheses

$$H_0 : \text{the two variables are independent,}$$
$$H_A : \text{the two variables are not independent}$$

using the χ^2-statistic.
In this example we use xtabs to make a table, then apply chisq.test. The xtabs function allows us to use the convenient subset argument to eliminate the data for which the values are not applicable.

```
tbl <- xtabs( ~ gender + amt.smoke,     # no left side in formula
             subset = amt.smoke < 98 & gender !=7,
             data=samhda)
tbl
```

```
##          amt.smoke
## gender  1  2  3  4  5  6  7
##      1 16  3  5  6  7 24 64
##      2 16  4  8  4  7 19 40

chisq.test(tbl)

## Warning: Chi-squared approximation may be incorrect

##
##  Pearson's Chi-squared test
##
## data:  tbl
## X-squared = 4.147, df = 6, p-value = 0.6568
```

The significance test shows no reason to doubt the hypothesis that the two variables are independent.

The warning message is due to some expected counts being small. Could this significantly change the p-value reported? A p-value based on a simulation may be computed.

```
chisq.test(tbl,simulate.p.value=TRUE)

##
##  Pearson's Chi-squared test with simulated p-value (based
##  on 2000 replicates)
##
## data:  tbl
## X-squared = 4.147, df = NA, p-value = 0.6707
```

The p-value is not changed significantly.   ••

The chi-squared test of homogeneity

How can we assess the effectiveness of a drug treatment? Typically, there is a clinical trial, with each participant randomly allocated to either a treatment group or a placebo group. If the results are measured numerically, a t-test may be appropriate to investigate whether any differences in means are significant. When the results are categorical, we see next how to use the χ^2-statistic to test whether the distributions of the results are the same.

Stanford University Medical Center conducted a study to determine whether the antidepressant Celexa can help stop compulsive shopping. Twenty-four compulsive shoppers participated in the study: twelve were given a placebo and twelve a dosage of Celexa daily for seven weeks. After

	Much worse	Worse	Same	Much improved	Very much so
Celexa	0	2	3	5	2
placebo	0	2	8	2	0

Table 10.6: Does Celexa treatment cut down on compulsive shopping?

this time the individuals were surveyed to determine whether their desires to shop had been curtailed. Data simulated from a preliminary report is given in Table 10.6.

Does this indicate that the two samples have different distributions?

We formulate this as a significance test using hypotheses:

$$H_0 : \text{the two distributions are the same}$$
$$H_A : \text{the two distributions are different.}$$

We use the χ^2-statistic. Again we need to determine the expected amounts, as they are not fully specified by H_0.

Let the random variable be the column variable, and the category that breaks up the data be the row variable in our table of data. For row i of the table, let p_{ij} be the probability that the random variable (the survey result) will be in the jth level of the random variable. We can rephrase the hypotheses as

$$H_0 : p_{ij} = p_j \text{ for all rows } i, \quad H_A : p_{ij} \neq p_j \text{ for some } i, j.$$

If we let n_i be the number of counts in each row (R_i before), then the expected amount in the (i,j) cell under H_0 should be $n_i p_j$. We don't specify the value of p_j in the null hypothesis, so it is estimated. Under H_0 all the data in the jth column of our table is binomial with n and p_j, so an estimator for p_j would be the column sum divided by n: C_j/n. Based on this, the expected number in the (i,j)-cell would be

$$e_{ij} = n_i \widehat{p}_j = \frac{R_i C_j}{n}.$$

This is the same formula as the chi-squared test of independence.

As the test statistic and its sampling distribution under H_0 are the same as with the test of independence, the chi-squared significance tests of homogeneity and independence are identical in implementation despite the differences in the hypotheses.

Before proceeding, let's combine the data so that there are three outcomes: "worse," "same," and "better."

```
celexa <- c(2, 3, 7)
placebo <- c(2, 8, 2)
x <- rbind(celexa, placebo)
colnames(x) <- c("worse", "same", "better")
x

##            worse same better
## celexa        2    3      7
## placebo       2    8      2

chisq.test(x)

## Warning: Chi-squared approximation may be incorrect

##
##   Pearson's Chi-squared test
##
## data:  x
## X-squared = 5.051, df = 2, p-value = 0.08004
```

The warning notes that one or more of the expected cell counts is less than five, indicating a possible discrepancy with the asymptotic distribution used to find the p-value. We can use a simulation to find the p-value, instead of using the chi-squared distribution approximation, as follows:

```
chisq.test(x, simulate.p.value=TRUE)

##
##   Pearson's Chi-squared test with simulated p-value (based
##   on 2000 replicates)
##
## data:  x
## X-squared = 5.051, df = NA, p-value = 0.1014
```

In both cases, the p-value is small but not tiny.

Problems

10.11 A number of drivers were surveyed to see whether they had been in an accident during the previous year, and, if so, whether it was a minor or major accident. The results are tabulated by age group in Table 10.7. Do a chi-squared hypothesis test of independence for the two variables.

10.12 The airquality data set contains measurements of air quality in New York City. We wish to see if ozone levels are independent of temperature. First we gather the data, using complete.cases to remove missing data from our data set.

| | \multicolumn{3}{c}{Accident type} | | |
Age	none	minor	major
under 18	67	10	5
18–25	42	6	5
26–40	75	8	4
41–65	56	4	6
over 65	57	15	1

Table 10.7: Type of accident by age.

```
aq <- airquality[complete.cases(airquality),]
aq <- transform(aq,
  te = cut(Temp, quantile(Temp)),
  oz = cut(Ozone,quantile(Ozone))
)
xtabs(~ te + oz, data=aq)

##          oz
## te        (1,18] (18,31] (31,62] (62,168]
##   (57,71]     15       9       3        0
##   (71,79]     10      10       7        1
##   (79,84.5]    4       6      11        5
##   (84.5,97]    0       0       6       22
```

Perform a chi-squared test of independence on the two variables te and oz. Does the data support an assumption of independence?

10.13 The following table contains data on the severity of injuries sustained during car crashes.

| | | \multicolumn{4}{c}{Injury level} | | | |
		none	minimal	minor	major
Seat belt	yes	12,813	647	359	42
	no	65,963	4,000	2,642	303

The data is tabulated by whether or not the passenger wore a seat belt. Are the two variables independent?

10.14 The data set oral.lesion (UsingR) contains data on location of an oral lesion for three geographic locations. This data set appears in an article by Mehta and Patel about differences in p-values in tests for independence when the exact or asymptotic distributions are used. Compare the p-values found by `chisq.test` when the asymptotic distribution of the sampling distribution is used to find the p-value and when a simulated value is used. Are the p-values similar? If not, which do you think is more accurate? Why?

10.15 In an effort to increase student retention, many colleges have tried "block programs." Assume that 100 students are broken into two groups of "50" at random. Fifty are in a block program; the others are not. The number of years each student attends the college is then measured. The following records the data:

Program	1 year	2 year	3 year	4 year	5+ years
nonblock	18	15	5	8	4
block	10	5	7	18	10

We wish to test whether a block program makes a difference in retention, assuming this sample is representative of a future population. Perform a chi-squared test of significance to investigate whether the distributions are homogeneous.

10.16 Table 10.3 lists the number of murders in New Jersey by day of week for the years 2003 and 2011. Is there a statistically significant difference in the distributions?

10.3 Goodness-of-fit tests for continuous distributions

When finding confidence intervals for a sample we were concerned about whether or not the data was sampled from a normal distribution. To investigate, we made a quantile plot or histogram and eyeballed the result. In this section, we see how to compare a continuous distribution with a theoretical one using a significance test.

The chi-squared test is used for categorical data. We can try to make it work for continuous data by "binning." That is, as in a construction of a histogram, we can choose some bins and count the number of data points in each. Now the data can be thought of as categorical, and the test can be used for goodness of fit.

This is fine in theory but works poorly in practice. The Kolmogorov-Smirnov test will be a better alternative in the continuous distribution case.

Kolmogorov-Smirnov test

Suppose we have a random sample $x_1, x_2, \ldots, x_n$ from some continuous distribution. (There should be no ties in the data.) Let $f(x)$ be the density and X some other random variable with this density. The *cumulative distribution*

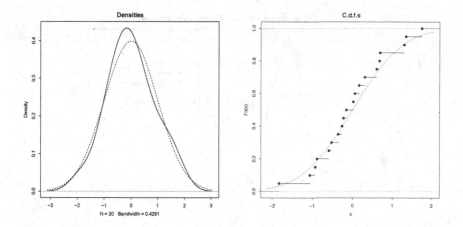

Figure 10.2: For a sample of size 20 from a normally distributed population, both sample and theoretical densities and cumulative distribution functions are drawn.

function for X is $F(x) = P(X \leq x)$, or the area to the left of x under the density of X.

The c.d.f. can be defined the same way when X is discrete. In that case it is computed from the p.d.f. by summing: $P(X \leq x) = \sum_{y \leq x} f(y)$.

For a sample, $x_1, x_2, \ldots, x_n$, the empirical distribution is the distribution generated by sampling from the data points. This becomes:

$$F_n(x) = \frac{\#\{i : x_i \leq x\}}{n}.$$

The function $F_n(x)$ can easily be plotted in R (e.g, Figure 10.2) using the ecdf function in the stats package. This function is used in a manner similar to the density function: the return value is plotted in a new figure using plot or may be added to the existing plot using lines.

If the data is from the population with c.d.f. F, then we would expect that F_n is close to F is some way. But what does "close" mean? In this context, we have two different functions of x. Define the distance between them as the largest difference they have:

$$D = \text{maximum in } x \text{ of } |F_n(x) - F(x)|.$$

The surprising thing is that with only the assumption that F is continuous, D has a known sampling distribution called the Kolmogorov-Smirnov distribution. This is illustrated in Figure 10.3, where the sampling distribution of the statistic for $n = 25$ is simulated for several families of random data. In each case, we see the same distribution. This fact allows us to construct a significance test using the test statistic D. In addition, a similar test can be done to compare two independent samples.

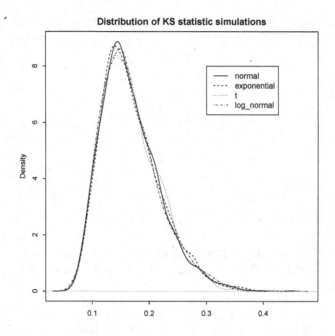

Figure 10.3: Density estimates for sampling distribution of the Kolmogorov-Smirnov statistic with $n = 25$ for normal, exponential, t, and log-normal data.

The Kolmogorov-Smirnov goodness-of-fit test

Assume $x_1, x_2, \ldots, x_n$ is an *i.i.d.* sample from a continuous distribution with c.d.f. $F(x)$. Let $F_n(x)$ be the empirical c.d.f. A significance test of

$$H_0 : F(x) = F_0(x), \qquad H_A : F(x) \neq F_0(x)$$

can be constructed with test statistic D. Large values of D support the alternative hypothesis.

In R, this test is implemented in the function ks.test. Its usage follows this pattern:

```
ks.test(x, y="name", ...)
```

The variable x stores the data. The argument y is used to set the family name of the distribution in H_0. It has a character value of "name" containing the "p" function that returns the c.d.f. for the family (e.g., "pnorm" or "pt"). The ...

argument allows the specification of the assumed parameter values. These depend on the family name and are specified as named arguments, as in mean=1, sd=1. The parameter values should not be estimated from the data, as this effects the sampling distribution of D.

If we have two *i.i.d.* independent samples, $x_1, \ldots, x_n$ and $y_1, \ldots, y_m$, from two continuous distributions F^X and F^Y, then a significance test of

$$H_0 : F^X = F^Y, \qquad H_A : F^X \neq F^Y$$

can be constructed with a similar test statistic:

$$D = \text{maximum in } x \text{ of } |F_n^X(x) - F_m^Y(x)|.$$

In this case, the ks.test can be used as

```
ks.test(x, y)
```

where x and y store the data.

We illustrate with some simulated data.

```
x <- rnorm(100, mean=5, sd=2)
ks.test(x, "pnorm", mean=0, sd=2)              # "wrong" parameters

##
## One-sample Kolmogorov-Smirnov test
##
## data:  x
## D = 0.7991, p-value < 2.2e-16
## alternative hypothesis: two-sided

ks.test(x, "pnorm", mean=5, sd=2)$p.value  # correct parameters

## [1] 0.02613

x = runif(100, min=0, max=5)
ks.test(x, "punif", min=0, max=6)$p.value  # "wrong" parameters

## [1] 0.0014

ks.test(x, "punif", min=0, max=5)$p.value  # correct parameters

## [1] 0.4299
```

The *p*-values are significant only when the parameters do not match the known population ones.

• **Example 10.4: Difference in SAT scores**
The data set stud.recs (UsingR) contains math and verbal SAT scores for

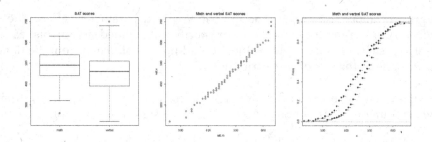

Figure 10.4: Three plots comparing the distribution of math and verbal SAT scores in the stud.recs (UsingR) data set.

some students (sat.m and sat.v). Assuming naively that the two samples are independent, are the samples from the same population of scores?

First, we make a q-q plot, a side-by-side boxplot, and a plot of the e.c.d.f.'s for the data, to see whether there is any merit to the question.

```
library(UsingR)
sat.m <- stud.recs$sat.m; sat.v <- stud.recs$sat.v
```

```
boxplot(list(math=sat.m, verbal=sat.v), main="SAT scores")
qqplot(sat.m, sat.v, main="Math and verbal SAT scores")
plot(ecdf(sat.m), main="Math and verbal SAT scores")
lines(ecdf(sat.v), lty=2)
```

The graphics are in Figure 10.4. The q-q plot shows similarly shaped distributions, but boxplots show that the centers appear to be different. Consequently, the cumulative distribution functions do not look that similar. The Kolmogorov-Smirnov test detects this and returns a small p-value.

```
ks.test(sat.m, sat.v)

## Warning: p-value will be approximate in the presence of ties

##
##   Two-sample Kolmogorov-Smirnov test
##
## data:  sat.m and sat.v
## D = 0.2125, p-value = 0.001456
## alternative hypothesis: two-sided
```

●●

The Shapiro-Wilk test for normality

The Kolmogorov-Smirnov test for a univariate data set works when the distribution in the null hypothesis is fully specified prior to our looking at the data. In particular, any assumptions on the values for the parameters should not depend on the data, as this can change the sampling distribution. Figure 10.5 shows the sampling distribution of the Kolmogorov-Smirnov statistic for Normal$(0,1)$ data and the sampling distribution of the Kolmogorov-Smirnov statistic for the same data when the sample values of $\bar{x}$ and s are used for the parameters of the normal distribution (instead of 0 and 1). The figure was generated with this simulation:

```
res <- replicate(2000, {
  x <- rnorm(25, mean=0, sd=1)
  c(ks.test(x, pnorm, mean=mean(x), sd=sd(x))$statistic,
    ks.test(x, pnorm, mean=0,       sd=1)$statistic)
})
plot(density(res[1,]), main="K-S sampling distribution", ylab="")
lines(density(res[2,]), lty=2)
legend(0.2, 12, legend=c("estimated", "exact"), lty=1:2)
```

(To retrieve just the value of the test statistic from the output of ks.test we take advantage of the fact that its return value is a list with one component named statistic containing the desired value. This is why the syntax ks.test(...)$statistic is used.)

A consequence is that we can't use the Kolmogorov-Smirnov test to test for normality of a data set unless we know the parameters of the underlying distribution.[2] The Shapiro-Wilk test allows us to perform that analysis. This test statistic is based on the ideas behind the quantile-quantile plot, which we've used to gauge normality. Its definition is a bit involved, but its usage in R is not.

The Shapiro-Wilk test for normality

If $x_1, x_2, \ldots, x_n$ is an *i.i.d.* sample from a continuous distribution, a significance test of

$$H_0 : \text{parent distribution is normal,}$$
$$H_A : \text{the parent distribution is not normal}$$

can be carried out with the Shapiro-Wilk test statistic.

---

[2]The Lilliefors test, implemented by lillie.test in the contributed package nortest, will make the necessary adjustments to use this test statistic. As well, the nortest package implements other tests of normality.

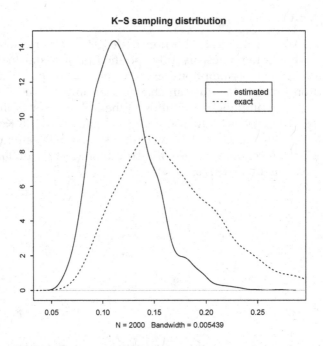

Figure 10.5: The sampling distribution for the Kolmogorov-Smirnov statistic when the parameters are estimated (solid line) and when not.

In R, the function shapiro.test will perform the test. The usage is simply

        shapiro.test(x)

where the data vector x contains the sample data.

• Example 10.5: Normality of SAT scores
For the SAT data in the stud.recs (UsingR) data set, we saw in Example 10.3 that the two distributions are different. Are they normally distributed? We can answer with the Shapiro-Wilk test:

```
shapiro.test(stud.recs$sat.m)

##
##  Shapiro-Wilk normality test
##
## data:  stud.recs$sat.m
## W = 0.9898, p-value = 0.3055

shapiro.test(stud.recs$sat.v)
```

```
##
##   Shapiro-Wilk normality test
##
## data:   stud.recs$sat.v
## W = 0.994, p-value = 0.752
```

In each case, the *p*-value is not statistically significant. There is no evidence in the data that the assumption of it being a random sample from a normal population should be doubted. ••

• **Example 10.6: Is on-base percentage normally distributed?**
The data set OBP (UsingR) records the on-base percentage for all players in the 2002 Major League Baseball season. It appears bell shaped except for one outlier. Does the data come from a normally distributed population?
   Using the Shapiro-Wilk test gives us

```
shapiro.test(OBP)$p.value
```

```
## [1] 1.206e-07
```

The difference from normality is statistically significant. Perhaps this is due to the one outlier. We investigate:

```
shapiro.test(OBP[OBP<.5])$p.value
```

```
## [1] 0.006404
```

The conclusion is the same. However, note the dramatic difference in the *p*-value that just one outlier makes. The statistic is not very resistant. ••

In defining the *t*-test, it was assumed that the data is sampled from a normal population. This is because the sampling distribution of the *t*-statistic is known under this assumption. However, this would not preclude us from using the *t*-test to perform statistical inference on data that has failed a formal test for normality. For small samples the *t*-test may apply, as the distribution of the *t*-statistic is robust to small changes in the assumptions on the parent distribution. If the parent distribution is not normal but also not too skewed, then a *t*-test can be appropriate. For large samples, the central limit theorem may apply, making a *t*-test valid.

Finding parameter values using fitdistr

If we know a data set comes from a known distribution and would like to estimate the parameter values, we can use the convenient fitdistr function from the MASS library. This function estimates the parameters for a wide family of distributions. The function is called with these arguments:

```
fitdistr(x, densfun=family.name, start=list(...))
```

We specify the data as a data vector, x; the family is specified by its full name (unlike that used in ks.test); and, for many of the distributions, reasonable starting values are specified using a named list. The fitdistr function fits the parameters by a method called maximum-likelihood. Often this coincides with using the sample mean or standard deviation to estimate the parameters, but in general it allows for a uniform approach to this estimation problem and associated inferential problems.

● **Example 10.7: Exploring** fitdistr
The data set babyboom (UsingR) contains data on the births of 44 children in a one-day period at a hospital in Brisbane, Australia. The variable wt records the weights of each newborn. A histogram suggests that the data comes from a normally distributed population. We can use fitdistr to find estimates for the parameters μ and σ, which for the normal distribution are the population mean and standard deviation.

```
library(MASS)
fitdistr(babyboom$wt,"normal")

##      mean       sd
##    3275.95    522.00
##   ( 78.69) (  55.65)
```

These estimates include standard errors in parentheses computed using a normal approximation. These can be employed to give confidence intervals for the estimates.

This estimate for the mean and standard deviation could also be done directly, as it coincides with the sample mean and sample standard deviation. However, the standard errors are new. To give a different usage, we look at the variable running.time, which records the time of day of each birth. The time differences between successive births are called the inter-arrival times. We first find the inter-arrival times using diff:

```
inter = diff(babyboom$running.time)
```

We fit the gamma distribution to the data. The gamma distribution generalizes the exponential distribution. It has two parameters, a shape and a rate. A value of 1 for the shape coincides with the exponential distribution. The fitdistr function does not need starting values for the gamma distribution.

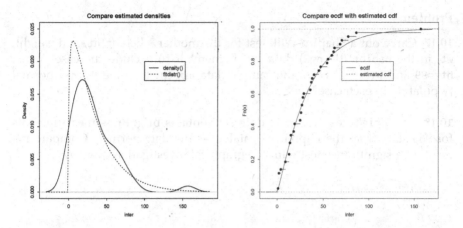

Figure 10.6: Empirical and theoretical densities and cumulative distribution functions for the inter-arrival times in the babyboom data set.

```
out <- fitdistr(inter, "gamma")
out

##      shape        rate
##   1.208846    0.036350
##  (0.233207)  (0.008632)
```

Finally, we look at density estimates and cumulative distribution functions with the following commands (Figure 10.6):

```
plot(density(inter), ylim=c(0, 0.025),
     main="Compare estimated densities", xlab="inter")
curve(dgamma(x, shape=out$estimate["shape"],
          rate=out$estimate["rate"]), add=TRUE, lty=2)
legend(100, 0.020, legend=c("density()", "fitdistr()"),lty=1:2)
#
plot(ecdf(inter),
 main="Compare ecdf with estimated cdf", xlab="inter")
curve(pgamma(x, shape=1.208593, rate=0.036350), add=TRUE)
legend(70, 0.8, legend=c("ecdf", "estimated cdf"), lty=1:2)
```

●●

Problems

10.17 Carry out a Shapiro-Wilk test for the mother's height, ht, and weight, wt, in the babies (UsingR) data set. Remember to exclude the cases when ht==99 and wt==999. Does the sample data appear to come from a normal population in each case?

10.18 The brightness (UsingR) data set contains brightness measurements for 966 stars from the Hipparcos catalog. Is the data normal? Compare the result of a significance test with the graphical investigation done by

```
hist(brightness, prob=TRUE)
lines(density(brightness))
curve(dnorm(x, mean(brightness), sd(brightness)), add=TRUE)
```

10.19 The variable temperature in the data set normtemp (UsingR) contains normal body temperature measurements for 130 healthy, randomly selected individuals. Test the assumption that the sample data of normal body temperature comes from a normal distribution?

10.20 The rivers data set contains the length of 141 major rivers in North America. Fit this distribution using the gamma distribution and fitdistr. How well does the gamma distribution fit the data? Answer by graphing the empirical and estimated densities.

10.21 Find parameter estimates for μ and σ for the variables sat.m and sat.v in the stud.recs (UsingR) data set. Assume the respective populations are normally distributed.

10.22 How good is the Kolmogorov-Smirnov test at rejecting the null when it is false? The following command will do 1000 simulations of the test when the data is not normal, but long-tailed and symmetric.

```
res <- replicate(1000, ks.test(rt(25, df=3),"pnorm")$p.value)
```

(The syntax above is using the fact that ks.test returns a list of values with one component named p.value.) What percentage of the trials have a *p*-value less than 0.05?

Try this with the exponential distribution (that is, replace rt(25,df=3) with rexp(25)-1). Is it better when the data is skewed?

10.23 A key to understanding why the Kolmogorov-Smirnov statistic has a sampling distribution that does not depend on the underlying parent population (as long as it is continuous) is the fact that if $F(x)$ is the c.d.f. for a random variable X, then $F(X)$ is uniformly distributed on $[0,1]$.

This can be proved algebraically using inverse functions, but instead we see how to simulate the problem to gain insight. The following line will illustrate this for the normal distribution:

```
qqplot(pnorm(rnorm(100)), runif(100))
```

The qqplot should be nearly straight if the distribution is uniform. Change the distribution to some others and see that you get a nearly straight line in each case. For example, the *t*-distribution with 5 degrees of freedom would be investigated with

```
qqplot(pt(rt(100,df=5),df=5),runif(100))
```

Try the uniform distribution, the exponential distribution, and the lognormal distribution (lnorm).

10.24 Is the Shapiro-Wilk test resistant to outliers? Run the following commands and decide whether the presence of a single large outlier (the 5) changes the ability of the test to detect normality.

```
shapiro.test(c(rnorm( 100), 5))
shapiro.test(c(rnorm(1000), 5))
shapiro.test(c(rnorm(4000), 5))
```

Linear regression

In Chapter 3 we looked at the simple linear regression model,

$$y_i = \beta_0 + \beta_1 x_i + \epsilon_i,$$

as a way to summarize a linear relationship between pairs of data (x_i, y_i). In this chapter we return to this model. We begin with a review and then further the discussion using the tools of statistical inference. Additionally, we will see that the methods developed for this model extend readily to the multiple linear regression model where there is more than one predictor.[1]

11.1 The simple linear regression model

Many times we assume that an increase in a predictor variable will correspond to an increase (or decrease) in the response variable. A basic model for this is the simple linear regression model:

$$y_i = \beta_0 + \beta_1 x_i + \epsilon_i.$$

The y variable is called the response variable and the x variable the predictor variable, covariate, or regressor.

As a statistical model, this says that the value of y_i depends on three things: that of x_i, the function $\beta_0 + \beta_1 x$, and the value of the random variable ϵ_i. The model says that for a given value of x, the corresponding value of y can be found by first applying the function to x and then adding the random error term ϵ_i.

To be able to make statistical inference, we assume that the error terms, ϵ_i, are *i.i.d.* and have a Normal$(0, \sigma)$ distribution. This assumption can be rephrased as an assumption on the randomness of the response variable. If the x values are fixed, then the distribution of y_i is normal with mean $\mu_{y|x} = \beta_0 + \beta_1 x_i$ (depending of the values of x) and variance σ^2 (not depending on the values of x). This can be expressed as y_i has a Normal$(\beta_0 + \beta_1 x_i, \sigma)$

---

[1]There is a large literature on using R for modeling such as described here and related extensions. For example, all of these books are quite informative: [57], [29], [22], [25], [20], [21], and [48]. The text [36] introduces R through a modeling approach.

distribution. If the x values are random, the model assumes that, conditionally on knowing these random values, the same is true about the distribution of the y_i.

Estimating the parameters in simple linear regression

One goal when modeling is to "fit" the model by estimating the parameters based on the sample. For the regression model the method of least squares is used. With an eye toward a more general usage, suppose we have several predictors, $x_1, x_2, \ldots, x_k$; several parameters, $\beta_0, \beta_1, \ldots, \beta_p$; and some function, f, which gives the mean for the variables y_i. That is, the statistical model

$$y_i = f(x_{1i}, x_{2i}, \ldots, x_{ki} \mid \beta_1, \beta_2, \ldots, \beta_p) + \epsilon_i.$$

The method of least squares finds values for the β's that minimize the squared difference between the actual values, y_i, and those predicted by the function f. That is, the following sum is minimized:

$$\sum_i [y_i - f(x_{1i}, x_{2i}, \ldots, x_{ki} \mid \beta_0, \beta_1, \ldots, \beta_p)]^2.$$

For the simple linear regression model, the formulas are not difficult to write (they are given below). For the more general model, even if explicit formulas are known, we don't present them.

The simple linear regression model for y_i has three parameters, β_0, β_1, and σ^2. The least-squares estimators for these are

$$\widehat{\beta}_1 = \frac{\sum (x_i - \bar{x})(y_i - \bar{y})}{\sum (x_i - \bar{x})^2}, \tag{11.1}$$

$$\widehat{\beta}_0 = \bar{y} - \widehat{\beta}_1 \bar{x}, \qquad \text{and} \tag{11.2}$$

$$\widehat{\sigma}^2 = \frac{1}{n-2} \sum [y_i - (\widehat{\beta}_0 + \widehat{\beta}_1 x_i)]^2. \tag{11.3}$$

We call $\widehat{y} = \widehat{\beta}_0 + \widehat{\beta}_1 x$ the prediction line; a value $\widehat{y}_i = \widehat{\beta}_0 + \widehat{\beta}_1 x_i$ the predicted value for x_i; and the difference between the actual and predicted values, $e_i = y_i - \widehat{y}_i$, the *residual*. The *residual sum of squares* is denoted RSS and is equal to $\sum_i e_i^2$.

Quickly put, the regression line is chosen to minimize the residual sum of squares, RSS; it has slope $\widehat{\beta}_1$, intercept $\widehat{\beta}_0$, and goes through the point $(\bar{x}, \bar{y})$. Furthermore, the estimate for σ^2 is $\widehat{\sigma}^2 = \text{RSS}/(n-2)$.

Figure 11.1 shows a data set simulated from the equation $y_i = 1 + 2x_i + \epsilon_i$, where $\beta_0 = 1, \beta_1 = 2$, and $\sigma^2 = 3$. Both the line $y = 1 + 2x$ and the regression line $\widehat{y} = 0.329 + 2.158 \cdot x$, predicted by the data, are drawn. They are different, of course, as one of them depends on the random sample. Keep in mind

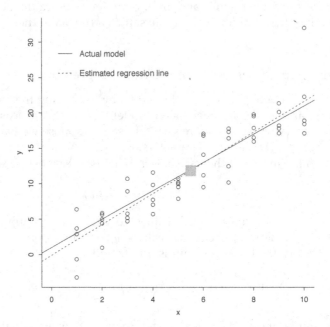

Figure 11.1: Simulation of model $y_i = 1 + 2x_i + \epsilon_i$. The regression line based on the data is drawn with dashes. The big square marks the value $(\bar{x}, \bar{y})$.

that the data is related by the true model, but if all we have is the data, the estimated model is given by the regression line. Our task of inference is to decide how much the regression line can tell us about the underlying true model.

Using lm to find the estimates

In Chapter 3 we learned how to fit the simple linear regression model using lm. The basic usage is of the form

```
lm(formula, data=..., subset=...)
```

Linear models are fit using R's model formulas, of which we have already seen a few examples.

The basic format for a formula is

```
response ~ predictor
```

The ~ (tilde) is read "is modeled by" and is used to separate the response from the predictor(s). The response variable can have regular mathematical expressions applied to it, but for the predictor variables the regular notations

+, -, *, /, and ^ have different meanings. A + means to add another term to the model, - means to drop a term, more or less coinciding with the symbols' common usage. But *, /, and ^ are used differently. If we want to use regular mathematical notation for the predictor we must insulate the symbols' usage with the I function, as in I(x^2).

As is usual with functions using model formulas, the data argument allows the variable names to reference those in the specified data frame, and the subset argument can be used to restrict the indices of the variables used by the modeling function.

By default, the lm function will print out the estimates for the coefficients. Much more is returned, but needs to be explicitly asked for. Usually, we store the results of the model in a variable, so that it can subsequently be queried for more information.

In Chapter 3 we fit a regression model to maximum heart rate by age with:

```
res.mhr <- lm(maxrate ~ age, data=heartrate)
res.mhr

##
## Call:
## lm(formula = maxrate ~ age, data = heartrate)
##
## Coefficients:
## (Intercept)          age
##      210.048       -0.798
```

These coefficients can be used directly for predictions. For example, a 50-year-old male would have a predicted maximum heart rate of:

```
208.36 - 0.76 * 50

## [1] 170.4
```

Extractor functions for lm

The lm function is reticent, but we can coax out more information as needed. This is done using extractor functions. Useful ones are summarized in Table 11.1.

These functions are passed an object returned by a modeling function, such as lm. These are "generic functions" which may have slightly different implementations depending on what type of model object is passed as the first object.

To illustrate, the estimate for σ^2 can be found using the resid function to retrieve the residuals from the model fitting:

Function	Description
summary	returns summary information about the regression
plot	makes diagnostic plots
coef	returns the coefficients
confint	returns confidence intervals for the coefficients
vcov	estimated covariance between parameter estimates
residuals	returns the residuals (can be abbreviated resid)
fitted	returns fitted values, $\hat{y}_i$
·deviance	returns RSS
predict	performs predictions
anova	finds various sums of squares
AIC	is used for model selection
model.matrix	matrix used to fit model mathematically

Table 11.1: Generic extractor functions for many of R's modeling functions, including lm.

```
n <- length(heartrate$age)
sum( resid(res)^2 ) / (n-2)

## [1] 31.05
```

Or, the RSS part can be found directly through deviance:

```
deviance(res)/ (n - 2)

## [1] 31.05
```

Problems

11.1 For the Cars93 (MASS) data set, answer the following:

1. For MPG.highway modeled by Horsepower, find the simple regression co-efficients. What is the predicted mileage for a car with 225 horsepower?

2. Fit the linear model with MPG.highway modeled by Weight. Find the predicted highway mileage of a 6,400 pound HUMMER H2 and a 2,524 pound MINI Cooper.

3. Fit the linear model Max.Price modeled by Min.Price. Why might you expect the slope to be around 1?

Can you think of any other linear relationships among the variables?

Age 2 (in.)	39	30	32	34	35	36	36	30
Adult (in.)	71	63	63	67	68	68	70	64

Table 11.2: Height as two-year-old and as an adult.

11.2 For the data set MLBattend (UsingR) concerning Major League Baseball attendance, fit a linear model of attendance modeled by wins. What is the predicted increase in attendance if a team that won 80 games last year wins 90 this year?

11.3 People often predict children's future height by using their 2-year-old height. A common rule is to double the height. Table 11.2 contains data for eight people's heights as 2-year-olds and as adults. Using the data, what is the predicted adult height for a 2-year-old who is 33 inches tall?

11.4 The galton (UsingR) data set contains data collected by Francis Galton in 1885 concerning the influence a parent's height has on a child's height. Fit a linear model for a child's height modeled by his parent's height. Make a scatterplot with a regression line. (Is this data set a good candidate for using jitter?) What is the value of $\widehat{\beta}_1$, and why is this of interest?

11.5 Formulas (11.1), (11.2), and the prediction line equation can be rewritten in terms of the correlation coefficient, r, as

$$\frac{\widehat{y}_i - \bar{y}}{s_y} = r \frac{x_i - \bar{x}}{s_x}.$$

Thus the five summary numbers: the two means, the standard deviations, and the correlation coefficient are fundamental for regression analysis.

This is interpreted as follows. Scaled differences of $\widehat{y}_i$ from the mean $\bar{y}$ are less than the scaled differences of x_i from $\bar{x}$, as $|r| \leq 1$. That is, "regression" toward the mean, as unusually large differences from the mean are lessened in their prediction for y.

For the data set galton (UsingR) use scale on the variables parent and child, and then model the height of the child by the height of the parent. What are the estimates for r and β_1?

11.2   Statistical inference for simple linear regression

If the simple regression model is appropriate for our data, then statistical inferences can be made about the unknown parameters.

Statistical inferences

If the linear model seems appropriate for the data, statistical inference is possible. What is needed is an understanding of the sampling distribution of the estimators.

To investigate these sampling distributions, we performed simulations of the model $y_i = x_i + \epsilon_i$, using x <- rep(1:10,10) and y <- rnorm(100, x, 5). Figure 11.2 shows the resulting regression lines for the different simulations. For reference, a single result of the simulation is plotted using a scatterplot. There is wide variation among the regression lines. In addition, histograms of the simulated values of $\widehat{\beta}_0$ and $\widehat{\beta}_1$ are shown.

We see from the figure that the estimators are random but not arbitrary. Both $\widehat{\beta}_0$ and $\widehat{\beta}_1$ are normally distributed, with respective means β_0 and β_1. Furthermore, $(n-2)\widehat{\sigma}^2/\sigma^2$ has a χ^2-distribution with $n-2$ degrees of freedom.

We will use the fact that the following statistics have a t-distribution with $n-2$ degrees of freedom:

$$\frac{\widehat{\beta}_0 - \beta_0}{SE(\widehat{\beta}_0)}, \qquad \frac{\widehat{\beta}_1 - \beta_1}{SE(\widehat{\beta}_1)}. \tag{11.4}$$

The standard errors are found from the known formulas for the variances of the $\widehat{\beta}_i$:

$$SE(\widehat{\beta}_0) = \widehat{\sigma} \left(\sum \frac{x_i^2}{\sum (x_i - \bar{x})^2} \right)^{1/2}, \quad SE(\widehat{\beta}_1) = \frac{\widehat{\sigma}}{\sqrt{\sum (x_i - \bar{x})^2}}. \tag{11.5}$$

(Recall that, $\widehat{\sigma}^2 = RSS/(n-2)$.)

Marginal t-tests

We can find confidence intervals and construct significance tests from the statistics in (11.4) and (11.5). For example, a significance test for

$$H_0 : \beta_1 = b, \quad H_A : \beta_1 \neq b$$

is carried out with the test statistic

$$T = \frac{\widehat{\beta}_1 - \beta_1}{SE(\widehat{\beta}_1)}.$$

Under H_0, T has the t-distribution with $n-2$ degrees of freedom.

A similar test for β_0 would use the test statistic $(\widehat{\beta}_0 - \beta_0)/SE(\widehat{\beta}_0)$.

When the null hypothesis is $\beta_1 = 0$ or $\beta_0 = 0$ we call these tests *marginal t-tests*, as they test whether the parameter is necessary for the model without consideration of the other parameters involved.

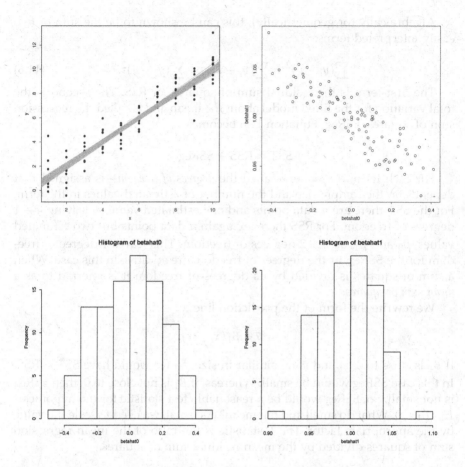

Figure 11.2: Four plots produced from a simulation finding the least squares regression coefficients from a known model. The upper left plot regression lines for 100 simulations from the model $y_i = x_i + \epsilon_i$. The plotted points show a single realization of the paired data during the simulation. The upper right plot shows a scatterplot of the points $(\widehat{\beta}_0, \widehat{\beta}_1)$. The lower left and right plots are histograms of $\widehat{\beta}_0$ and $\widehat{\beta}_1$.

The F-test

An alternate test for the null hypothesis $\beta_1 = 0$ can be done using a different but related approach that generalizes to the multiple-regression problem.

One goal of modeling is the attempt to explain the variation in the response variable using one or more predictor variables. The total variation in the y values about the mean is

$$\text{SST} = \text{total sum of squares} = \sum (y_i - \bar{y})^2.$$

Algebraically (or geometrically), this can be shown to be the sum of two easily interpreted terms:

$$\sum(y_i - \bar{y})^2 = \sum(y_i - \hat{y}_i)^2 + \sum(\hat{y}_i - \bar{y})^2. \tag{11.6}$$

The first term is the residual sum of squares, or RSS. The second is the total variation for the fitted model about the mean and is called the regression sum of squares, SSReg. Equation 11.6 becomes

$$SST = RSS + SSReg.$$

For each term, a number—called the *degrees of freedom*—is assigned that depends on the sample size and the number of estimated values in the term. For the SST there are n data points and one estimated value, $\bar{y}$, leaving $n - 1$ degrees of freedom. For RSS there are again n data points but two estimated values, $\hat{\beta}_0$ and $\hat{\beta}_1$, so $n - 2$ degrees of freedom. This leaves 1 degree of freedom for the SSReg, as the degrees of freedom are additive in this case. When a sum of squares is divided by its degrees of freedom it is referred to as a *mean sum of squares*.

We rewrite the form of the prediction line to:

$$\hat{y}_i = \bar{y} + \hat{\beta}_1(x_i - \bar{x}).$$

If $\hat{\beta}_1$ is close to 0, $\hat{y}_i$ and $\bar{y}$ are similar in size, so we would have SST $\approx$ RSS. In this case SSReg would be small. Whereas, if $\hat{\beta}_1$ is not close to 0, then SSReg is not small. So, SSReg would be a reasonable test statistic for the hypothesis $H_0 : \beta_1 = 0$. What do small and big mean? As usual, we need to scale the value by the appropriate factor. The F-statistic is the ratio of the mean regression sum of squares divided by the mean residual sum of squares.

$$F = \frac{SSReg/1}{RSS/(n-2)} = \frac{SSReg}{\hat{\sigma}^2}. \tag{11.7}$$

Under the null hypothesis $H_0 : \beta_1 = 0$, the sampling distribution of F is known to be the F-distribution with 1 and $n - 2$ degrees of freedom.

This allows us to make the following significance test.

F-test for $\beta_1 = 0$

A significance test for the hypotheses

$$H_0 : \beta_1 = 0, \qquad H_A : \beta_1 \neq 0$$

can be made with the the test statistic

$$F = \frac{SSReg}{\hat{\sigma}^2}.$$

> Under the null hypothesis, the statistic F has the F-distribution with 1 and $n - 2$ degrees of freedom. Larger values of F are more extreme, so the p-value for this test is computed from $P(F \geq \text{observed value} \mid H_0)$.

The F-statistic can be rewritten as

$$F = \left(\frac{\widehat{\beta_1}}{\text{SE}(\widehat{\beta_1})} \right)^2 .$$

Under the assumption $\beta_1 = 0$, this is the square of one of the t-distributed random variables of Equation 11.4. For simple linear regression the two tests of $H_0 : \beta_1 = 0$, the marginal t-test and the F-test, are equivalent. However, we will see that with more predictors, the two tests are different.

R^2—the coefficient of determination

The decomposition of the total sum of squares into the residual sum of squares and the regression sum of squares in Equation 11.6 allows us to interpret how well the regression line fits the data. If the regression line fits the data well, then the residual sum of squares, $\sum(y_i - \widehat{y}_i)^2$, will be small. If there is a lot of scatter about the regression line, then RSS will be big. To quantify this, we can divide by the total sum of squares, leading to the definition of the *coefficient of determination*:

$$R^2 = 1 - \frac{\text{RSS}}{\text{SST}} = 1 - \frac{\sum(y_i - \widehat{y}_i)^2}{\sum(y_i - \bar{y}_i)^2} = \frac{\sum(\widehat{y}_i - \bar{y})^2}{\sum(y_i - \bar{y})^2}. \tag{11.8}$$

This is close to 1 when the linear regression fit is good and close to 0 when it is not.

When the simple linear regression model is appropriate this value is interpreted as the proportion of the total response variation explained by the regression. That is, $R^2 \cdot 100\%$ of the variation is explained by the regression line. When R^2 is close to 1, most of the variation is explained by the regression line, and when R^2 is close to 0, not much is.

This interpretation is similar to that given for the Pearson correlation coefficient, r, in Chapter 3. This is no coincidence: for the simple linear regression model $r^2 = R^2$.

The *adjusted R^2* divides the sums of squares by their degrees of freedom. For the simple regression model, these are $n - 2$ for RSS and $n - 1$ for SST. This is done to penalize models that get better values of R^2 by using more predictors. This is of interest when multiple predictors are used.

Using `lm` to find values for a regression model

Here we illustrate how R can be used to directly compute these values and, alternatively, how these values are returned by the `lm` object and its extractor methods.

Confidence intervals

For example, based on the distribution of $\hat{\beta}_0$, a 95% confidence interval for β_0 can be found with:

$$\hat{\beta}_0 \pm t^* SE(\hat{\beta}_0).$$

Using the values in our example, this could be found with

```
res.mhr <- lm(maxrate ~ age, data=heartrate)

betahat0 <- coef(res.mhr)[1]        # first coefficient
n <- nrow(heartrate)
sigmahat <- sqrt( sum(resid(res.mhr)^2) / (n - 2))
SE <- with(heartrate,
           sigmahat*sqrt(sum(age^2) / (n*sum((age - mean(age))^2)))
          )
SE

## [1] 2.867

tstar <- qt(1 - 0.05/2, df=n - 2)

betahat0 + c(-1, 1) * tstar * SE

## [1] 203.9 216.2
```

Standard error

The summary method for `lm` objects provides most of the values related to the model, including, for example, the standard error just computed. Find SE in the `Coefficients:` part of the output under the column labeled `Std. Error`.

```
summary(res.mhr)

##
## Call:
## lm(formula = maxrate ~ age, data = heartrate)
##
## Residuals:
##      Min      1Q Median      3Q     Max
```

```
## -8.926 -2.538  0.388  3.187  6.624
##
## Coefficients:
##              Estimate Std. Error t value Pr(>|t|)
## (Intercept)  210.048       2.867    73.3  < 2e-16 ***
## age           -0.798       0.070   -11.4  3.8e-08 ***
## ---
## Signif. codes:  0 '***' 0.001 '**' 0.01 '*' 0.05 '.' 0.1 ' ' 1
##
## Residual standard error: 4.58 on 13 degrees of freedom
## Multiple R-squared:  0.909,Adjusted R-squared:  0.902
## F-statistic:  130 on 1 and 13 DF,  p-value: 3.85e-08
```

By reading the standard error from this output, a 95% confidence interval for β_1 may be more easily found than the one for β_0 above:

```
betahat1 <- coef(res.mhr)[2]              # second coefficient
SE <- 0.06996281                          # read from summary
tstar <- qt(1 - 0.05/2, df=n - 2)
betahat1 + c(-1, 1) * tstar * SE
```

```
## [1] -0.9489 -0.6466
```

The two coefficients in this model are returned by the coef method:

```
coef(res.mhr)
```

```
## (Intercept)          age
##    210.0485      -0.7977
```

The coef method called on the *summary* of the model returns a matrix with the standard errors included:

```
coef(summary(res.mhr))
```

```
##              Estimate Std. Error t value  Pr(>|t|)
## (Intercept) 210.0485    2.86694   73.27 2.124e-18
## age          -0.7977    0.06996  -11.40 3.848e-08
```

which can be used to programmatically extract the standard errors, as with:

```
coef(summary(res.mhr))["age", "Std. Error"]
```

```
## [1] 0.06996
```

The above shows how to do the work piece-by-piece. If that isn't of interest, the confint method can do both of these computations directly:

```
confint(res.mhr)

##                 2.5 %    97.5 %
## (Intercept) 203.8548  216.2421
## age           -0.9489   -0.6466
```

Significance tests

The summary function for lm objects displays more than the standard errors. For each coefficient a marginal t-test is performed. This is a two-sided hypothesis test of the null hypothesis that $\beta_i = 0$ against the alternative that $\beta_i \neq 0$. We see in this case that both are rejected with very low p-values (as to be expected as we expect an intercept around 220 and slope around -1). These small p-values are flagged in the output of summary with significance stars.

Other t-tests are possible. For example, we can test the null hypothesis that the slope is -1 with the commands

```
mu0 <- -1
T.obs <- (betahat1 - mu0)/SE
T.obs

##    age
## 2.891

2*pt(abs(T.obs), df=n-2, lower.tail=FALSE)

##     age
## 0.01262
```

This is a small p-value, indicating that the model with slope -1 is unlikely to have produced this data or anything more extreme than it.

Finding $\hat{\sigma}^2$, R^2

The estimate for $\hat{\sigma}$ is marked Residual standard error and is labeled with $13 = 15 - 2$ degrees of freedom. The degrees of freedom are contained in the df.residual component of the model object. The estimate for $\hat{\sigma}$ can be computed directly with:

```
sigma2 <- sum(resid(res.mhr)^2) / res.mhr$df.residual
sqrt(sigma2)                              # sigma hat

## [1] 4.578
```

The value of $R^2 =$ cor(age,mhr)^2 is given in the output along with an adjusted value.

F-test for $\beta_1 = 0$

Finally, the *F*-statistic is calculated. As this is given by $(\hat{\beta}_1/SE(\hat{\beta}_1))^2$, it can be found directly with:

```
(-0.7595 / 0.0561)^2

## [1] 183.3
```

The significance test $H_0 : \beta_1 = 0$ with two-sided alternative is performed and again returns a tiny *p*-value.

The sum of squares to compute *F* are also given as the output of the anova extractor function.

```
anova(res.mhr)

## Analysis of Variance Table
##
## Response: maxrate
##
##            Df Sum Sq Mean Sq F value  Pr(>F)
## age         1   2725    2725     130 3.8e-08 ***
## Residuals  13    272      21
## ---
## Signif. codes:  0 '***' 0.001 '**' 0.01 '*' 0.05 '.' 0.1 ' ' 1
```

These values in the column headed Sum Sq are SSReg and RSS. The total sum of squares, SST, would be the sum of the two.

A short summary The summary function can feel a bit verbose at times. The following function will be used in the sequel to tighten the display up to show just the coefficient information:

```
short_summary <- function(x) {
  x <- summary(x)
  cmat <- coef(x)
  printCoefmat(cmat)
}
```

Predicting the response with predict

The function predict is used to make different types of predictions. A template for its usage with lm objects is

```
predict(res, newdata=..., interval=..., level = ...)
```

The value of res is the output of an lm model. We call this res below, but we can use any valid name. Any changes to the values of the predictor are given to the argument newdata in the form of a data frame with names that match those used in the model formula. The arguments interval and level are set when prediction or confidence intervals are desired.

The simplest usage, predict(res), returns the predicted values (the $\hat{y}_i$'s) for the data. Predictions for other values of the predictor are specified using a data frame whose variable names match the variables used in the predictor side of the model, as this example illustrates:

```
res.mhr <- lm(maxrate ~ age, data=heartrate)
predict(res.mhr, newdata=data.frame(age=42))

##     1
## 176.5
```

This finds the predicted maximum heart rate for a 42-year-old. The age part of the data frame call is important. Variable names in the data frame supplied to the newdata argument must exactly match the variable names used when the model object was produced.

To assess whether the simple regression model is appropriate for the data we use a graphical approach.

Testing the model assumptions

The simple linear regression model, $y_i = \hat{\beta}_0 + \hat{\beta}_1 x_i + \epsilon_i = \mu_{y|x} + \epsilon_i$, places assumptions on the data set that we should verify before proceeding with any statistical inference. In particular, the linear model should be appropriate for the mean value of the y_i, and the error distribution should be normally distributed and independent.

Just as we looked at graphical evidence when investigating assumptions about normally distributed populations when performing a *t*-test, we will consider graphical evidence to assess the appropriateness of a regression model for the data. The plot method for lm (?plot.lm) objects can be used to plot 6 different diagnostic plots. We consider the four that are produced by default.[2]

The biggest key to assessing the aptness of the model is found in the residuals. The residuals are not an *i.i.d.* sample, as they sum to 0 and they do not have the same variance. The *standardized residuals* rescale the residuals to have unit variance.

---

[2]In using plot to produce the diagnostic plots it is convenient to first issue the command par(mfrow=c(2,2)). This sets up the plot device to have four panes for graphics added row by row.

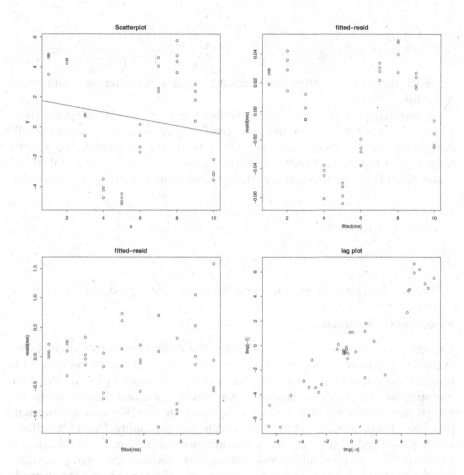

Figure 11.3: Four graphs showing problematic linear models. Scatterplot in upper left shows linear model is incorrect. Fitted versus residual plot in upper right shows a nonlinear trend. Fitted versus residual plot in lower left shows non-constant variance. Lag plot in lower right shows correlations in error terms.

Assessing the linear model for the mean

A scatterplot of the data with the regression line can show quickly whether the linear model seems appropriate for the data. If the general trend is not linear, either a transformation or a different model is called for. An example of a cyclical trend (which calls for a transformation of the data) is the upper-left plot in Figure 11.3 and is made with these commands:

```
x <- rep(1:10,4)
y <- rnorm(40, mean=5*sin(x), sd=1)
```

```
plot(y ~ x)
abline(lm(y ~ x))
```

When there is more than one predictor variable, a scatterplot will not be as useful.

A residual plot can also show whether the linear model is appropriate and can be made with more than one predictor. As well, it can detect small deviations from the model that may not show up in a scatterplot. The upper-right plot in Figure 11.3 shows a residual plot that finds a sinusoidal trend that will not show up in a scatterplot. It was simulated with these commands:

```
x <- rep(1:10, 4)
y <- rnorm(40, mean=x + 0.05 * sin(x), sd=0.01) # small trend
res <- lm(y ~ x)
plot(fitted(res), resid(res))
```

The residual plot is one of the four diagnostic plots produced by plot.

Assessing the residuals

The residuals are used to assess whether the error terms in the model are normally distributed. As mentioned, the residuals are correlated as they add to 0, we treat them as if they are the actual error terms in the model. For example, we use either a histogram or, preferably, a quantile-normal plot to investigate if a normal assumption is appropriate. For the quantile-normal plot, deviations from a straight line indicate non-normality. One of the diagnostic plots produced by plot is a quantile-normal plot of the standardized residuals. Though normality is not essential for prediction, the sampling distributions of the coefficients depend on the error terms not being too skewed or long-tailed.

In addition to normality, an assumption of the model is also that the error terms have a common variance. A residual plot can show whether this is the case. When it is, the residuals show scatter about a horizontal line. In many data sets, the variance increases for larger values of the predictor. The commands below create a simulation of this. The graph showing the effect is in the lower-left of Figure 11.3. The absence of equal variance can sometimes be addressed by transformations or weighted least squares, though we don't pursue that here.

```
x <- rep(1:10, 4)
y <- rnorm(40, mean=1 + 1/2*x, sd=x/10)
res <- lm(y ~ x)
plot(fitted(res), resid(res))
```

The scale-location plot is one of the four diagnostic plots produced by the defaults of the plot method. This graphic also shows the residuals, but in terms of the square root of the absolute value of the standardized residuals. The graph should show points scattered along the y-axis, as we scan across the x-axis, but the spread of the scattered points should not get larger or smaller.

In some data sets, there is a lack of independence in the residuals. For example, the errors may accumulate. A lag plot, where the data is plotted against previous values of the data, may be able to show this effect. For an independent sequence, the lag plot should be scattered, whereas many dependent sequences will show some pattern. This is illustrated in the lower-right plot in Figure 11.3, which was made as follows:

```
x <- rep(1:10, 4)
epsilon <- rnorm(40, mean=0, sd=1)
y <- 1 + 2*x + cumsum(epsilon) # cumsum() correlates errors
res <- lm(y ~ x)
tmp <- resid(res)
n <- length(tmp)
plot(tmp[-n], tmp[-1])          # lag plot
```

Influential points

As we observed in Chapter 3, the regression line can be greatly influenced by a single observation that is far from the trend set by the data. The difference in slopes between the regression line with all the data and the regression line with the ith point missing will mostly be small, except for influential points. The Cook's distance is based on the difference of the predicted values of y_i for a given x_i when the point (x_i, y_i) is and isn't included in the calculation of the regression coefficients. Comparing predicted amounts, as opposed to change in slope, allows the method to generalize to more than one predictor. The Cook's distance is computed by the extractor function cooks.distance.

One of the diagnostic plots produced by the default plot method for lm objects will show the Cook's distance for the data points plotted using spikes. Another way to display this information graphically is to make the size of the points in the scatterplot depend on this distance using the cex argument. This type of plot is referred to as a bubble plot and is illustrated using the emissions (UsingR) data set in Figure 11.4. The graphic is made with the following commands:

```
res <- lm(CO2 ~ perCapita, emissions)
plot(CO2 ~ perCapita, emissions,
     cex=10*sqrt(cooks.distance(res)),
     main=expression(                      # make subscript on CO2
         paste("bubble plot of ",CO[2],
             " emissions by per capita GDP")
     ))
```

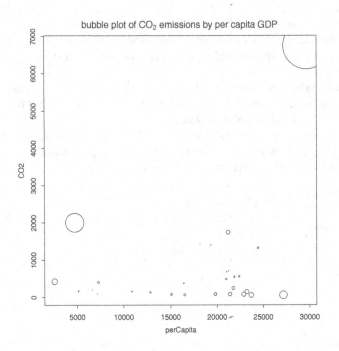

Figure 11.4: Bubble plot of CO_2 emissions by per capita GDP with area of points proportional to Cook's distance.

The square root of the distances is used, so the area of the points is proportional to Cook's distance rather than to the radius.[3]

For the maximum-heart-rate data, the four diagnostic plots produced by R with the command plot(res.mhr) are in Figure 11.5.

Prediction intervals

The value of $\hat{y}$ can be used to predict two different things: the value of a single estimate of y for a given x or the average value of many values of y for a given x. If we think of a model with replication (repeated y's for a given x, such as in Figure 11.2), then the difference is clear: one is a prediction for a given point, the other a prediction for the average of the points.

Statistical inference about the predicted value of y based on the sample is done with a *prediction interval*. As y is not a parameter, we don't call this a confidence interval. The form of the prediction interval is similar to that of a confidence interval:

---

[3]The argument to main illustrates how to use mathematical notation in the title of a graphic. See the help page ?plotmath for details.

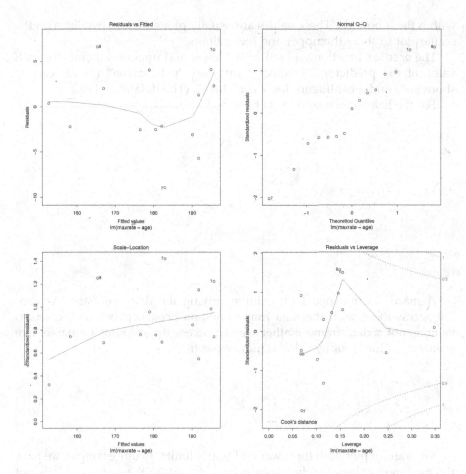

Figure 11.5: Four diagnostic plots for the maximum-heart-rate data produced by the extractor function plot.

$$\widehat{y} \pm t^* \text{SE}.$$

For the prediction interval, the standard error depends on x and is given by

$$\text{SE} = \widehat{\sigma} \sqrt{1 + \frac{1}{n} + \frac{(x - \bar{x})^2}{s_{xx}}}. \tag{11.9}$$

The value of t^* comes from the t-distribution with $n - 2$ degrees of freedom.

The prediction interval holds for all x simultaneously. Meaning, there is a $(1 - \alpha)100\%$ chance that a new data point chosen from the model will be within these bounds. These values are usually plotted using two lines on the scatterplot to show the upper and lower limits.

The predict function will return the lower and upper endpoints for each value of the predictor. We specify interval="prediction" (which can be shortened) and a confidence level with level. (The default is 0.95.)

For the heart-rate example we have:

```
pred.res <- predict(res.mhr, int = "pred")

## Warning: predictions on current data refer to _future_ responses

head(pred.res, n=3)

##      fit   lwr   upr
## 1 195.7 185.1 206.3
## 2 191.7 181.3 202.1
## 3 190.1 179.7 200.5
```

A matrix is returned with columns giving the data we want. We cannot access these with the data frame notation pred.res$lwr, as the return value is not a data frame. Rather we can access the columns by name, like pred.res[,'lwr'], or by column number, as in

```
head(pred.res[, 2])                     # the 'lwr' column

##     1     2     3     4     5     6
## 185.1 181.3 179.7 171.9 147.2 156.5
```

We want to plot both the lower and upper limits. In our example, we have the predicted values for the given values of age. As the age variable is not sorted, simply plotting will make a real mess. To remedy this, we specify the values of the age variable for which we make a prediction. We use the values sort(unique(age)), which gives just the x values in increasing order.

```
age.sort <- sort(unique(heartrate$age))
pred.res <- predict(res.mhr, newdata = data.frame(age = age.sort),
                    int="pred")
pred.res[,2]

##     1     2     3     4     5     6     7     8     9    10
## 185.1 184.3 181.3 179.7 172.7 171.9 170.3 168.7 166.3 156.5
##    11    12    13
## 154.8 147.2 141.1
```

Now we can add the prediction intervals to the scatterplot with the lines function (matlines offers a one-step alternative). The result is Figure 11.6.

```
plot(maxrate ~ age, data=heartrate)
abline(res.mhr)
lines(age.sort, pred.res[,2], lty=2) # lower curve
lines(age.sort, pred.res[,3], lty=2) # upper curve
```

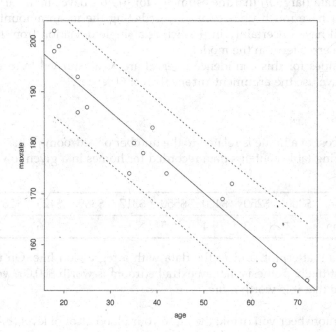

Figure 11.6: Regression line with 95% prediction intervals drawn for age versus maximum heart rate.

There is a slight curve in the lines drawn, which is hinted at in Equation 11.9. This implies that estimates near the value $(\bar{x}, \bar{y})$ have a smaller variance. This is expected: there is generally more data near this value, so the variances should be smaller.

Confidence intervals for $\mu_{y|x}$

A confidence interval for the mean value of y for a given x is given by

$$\hat{y} \pm t^* \text{SE}(\hat{y}).$$

Again, t^* is from the t-distribution with $n - 2$ degrees of freedom. The standard error used is now

$$SE(\hat{y}) = \hat{\sigma}\sqrt{\frac{1}{n} + \frac{(x - \bar{x})^2}{S_{xx}}}.$$

The standard error for the prediction interval differs by an extra term of plus 1 inside the square root. This may appear minor, but is not. If we had so much data (large n) that the estimates for the β's have small variance, we would not have much uncertainty in predicting the mean amount, but we would still have uncertainty in predicting a single deviation from the mean due to the error term in the model.

The values for this confidence interval are also returned by predict. In this case, we use the argument interval="confidence".

Problems

11.6 The cost of a home is related to the number of bedrooms it has. Suppose the following table contains data recorded for homes in a given town.

price	$300	$250	$400	$550	$317	$389	$425	$289	$389
bedrooms	3	3	4	5	4	3	6	3	4

Make a scatterplot, and fit the data with a regression line. On the same graph, test the hypothesis that an extra bedroom is worth $60,000 versus the alternative that it is worth more.

11.7 The more beer you drink, the more your blood alcohol level (BAL) rises. The following table contains a data set on beer consumption.

beers	5	2	9	8	3	7	3	5	3	5
BAL	0.10	0.03	0.19	0.12	0.04	0.095	0.07	0.06	0.02	0.05

Make a scatterplot with a regression line and 95% prediction intervals drawn. Test the hypothesis that one beer raises your BAL by 0.02% against the alternative that it raises it less. (A formula from wikipedia.org specifies a model for the mean with

$$\frac{0.906 \cdot d \cdot 1.2}{(0.49 + 0.09 \cdot 1_{a \; male}) \cdot w} - 0.017 \cdot t$$

where d is the number of drinks, w the weight in kilograms, and t the time since drinking.)

11.8 For the same blood-alcohol data as the previous exercise perform a significance test that the intercept is 0 with a two-sided alternative.

11.9 The lapse rate is the rate at which temperature drops as you increase elevation. Some hardy students were interested in checking empirically whether the lapse rate of 9.8°C/km was accurate. To investigate, they grabbed their thermometers and their Suunto® wrist altimeters and recorded the data from their hike in this table:

elevation (ft)	600	1000	1250	1600	1800	2100	2500	2900
temperature (°F)	56	54	56	50	47	49	47	45

Draw a scatterplot with regression line and investigate whether the lapse rate is 9.8°C/km. (It helps to convert to the rate of change °F per feet, which is 5.34 degrees per 1,000 feet.) Test the hypothesis that the lapse rate is 5.34 degrees per 1,000 feet against a two-sided alternative.

11.10 For the homedata (UsingR) data set, find the regression equation to predict the year-2000 value of a home from its year-1970 value. Make a prediction for an $80,000 home in 1970. Comment on the appropriateness of the regression model by investigating the residuals.

11.11 A seal population is counted over a ten-year period. The counts are reported in this table:

year	pop.	year	pop.	year	pop	year	pop
1952	724	1955	1,392	1958	1,212	1961	1,980
1953	176	1956	1,392	1959	1,672	1962	2,116
1954	920	1957	1,448	1960	2,068		

Make a scatterplot with population on the y-axis and year on the x-axis. Find the regression line. What is the predicted value for 1963? Would you use this to predict the population in 2014? Why or why not?

11.12 The deflection (UsingR) data set contains deflection measurements for various loads. Fit a linear model to Deflection as a function of load. Plot the data and the regression line. How well does the line fit? Investigate with a residual plot.

11.13 The alaska.pipeline (UsingR) data set contains measurements of defects on the Alaska pipeline that are taken first in the field and then in the laboratory. The measurements are done in six batches. Fit a linear model for the lab-defect size as modeled by the field-defect size. Find the coefficients. Discuss the appropriateness of the model.

11.14 In athletic events in which people of various ages participate, performance is sometimes related to age. Multiplying factors are used to compare

the performance of a person of a given age to another person of a different age. The data set best.times (UsingR) features world records by age and
distance in track and field.

We split the records by distance, allowing us to compare the factors for
several distances.

```
by.dist <- split(best.times, as.factor(best.times$Dist))
```

This returns a list of data frames, one for each distance. We can plot the
times in the 800-meter run:

```
plot(Time ~ age, by.dist[["800"]])
```

It is actually better to apply scale first, so that we can compare times.

Through age 70, a linear regression model seems to fit. It can be found
with

```
lm(scale(Time) ~ age, by.dist[["800"]], subset = age < 70)

##
## Call:
## lm(formula = scale(Time) ~ age, data = by.dist[["800"]], subset = age <
##      70)
##
## Coefficients:
## (Intercept)         age
##    -1.2933      0.0136
```

Using the above technique, compare the data for the 100-meter dash, the
400-meter dash, and the 10,000-meter run. Are the slopes similar?

11.15 The galton (UsingR) data set contains data collected by Francis Galton
in 1885 concerning the influence a parent's height has on a child's height. Fit
a linear model modeling a child's height by his parents'. Do a test of significance to see whether β_1 equals 1 against a two-sided alternative.

11.16 Find and plot both the prediction and the confidence intervals for the
heart-rate model: res.mhr <- lm(maxrate ~ age, data=heartrate).

11.17 The alaska.pipeline (UsingR) data set appears appropriate for a linear
model, but the assumption of equal variances does not seem appropriate. A
log-transformation of each variable does seem to have equal variances. Fit
the model

$$\log(\texttt{lab.defect}) = \beta_0 + \beta_1 \cdot \log(\texttt{field.defect}) + \epsilon.$$

Investigate the residuals and determine whether the assumption of equal variance seems appropriate.

11.18 The following commands will simulate the regression model $y_i = 1 + 2x_i + \epsilon_i$:

```
m <- 200
x <- rep(1:10, 4)
res <- replicate(m, {
  y <- rnorm(40, 1 + 2*x, 3)
  coef(lm(y ~ x))
})
plot(res[1,], res[2,])
```

Run the simulation and comment on the shape of the scatterplot. What does it say about the correlation between $\widehat{\beta}_0$ and $\widehat{\beta}_1$?

11.19 In a simple linear regression, confidence intervals for β_0 and β_1 are given separately in terms of the t-distribution as $\widehat{\beta}_i \pm t^* \mathrm{SE}(\widehat{\beta}_i)$. They can also be found *jointly*, giving a *confidence ellipse* for the parameters as a pair. This can be found easily in R with the ellipse package.[4]
If res is the result of the lm function, then plot(ellipse(res), type="l") will draw the confidence ellipse.
For the deflection (UsingR) data set, find the confidence ellipse for Deflection modeled by Load.

11.20 The linear regression model $y_i = \mu_{y|x_i} + \epsilon_i$ is flexible enough to accommodate some of the other models already encountered. The basic t-test is modeled by y ~ 1. The paired t-test becomes $y_i = \mu + x_i + \epsilon_i$ which can be modeled with y ~ offset(x). The two-sample t-test can be modeled with a predictor which is 1 for one population and 0 for the other via y ~ x.
Let's see the latter. The normtemp (UsingR) data set has normal body temperature measurements for both men and women. A two-sample t-test can be employed to perform a significance test of difference between gender, via:

```
t.test(temperature ~ factor(gender), data=normtemp)
```

Find the corresponding p-value in the output of this model:

```
lm(temperature ~ factor(gender), data=normtemp)
```

---

[4]The ellipse package is not part of the standard R installation, but it is on CRAN. You can install it with the command install.packages("ellipse").

11.3   Multiple linear regression

Multiple linear regression allows for more than one regressor to predict the value of y. Lots of possibilities exist. These regressors may be separate variables, products of separate variables, powers of the same variable, or functions of the same variable. In the next chapter, we will consider regressors that are not numeric but categorical. They all fit together in the same model, but there are additional details. We see, though, that much of the background for the simple linear regression model carries over to the multiple regression model.

Types of models

Let y be a response variable and let $x_1, x_2, \ldots, x_p$ be p variables that we will use for predictors. For each variable we have n values recorded. The multiple regression model we discuss here is

$$y_i = \beta_0 + \beta_1 x_{1i} + \cdots + \beta_p x_{pi} + \epsilon_i.$$

There are $p + 1$ parameters in the model labeled $\beta_0, \beta_1, \ldots, \beta_p$. They appear in a linear manner, just like a slope or intercept in the equation of a line. The x_i's are predictor variables, or covariates. They may be random; they may be related, such as powers of each other; or they may be correlated. As before, it is assumed that the ϵ_i values are an *i.i.d.* sample from a normal distribution with mean 0 and unknown variance σ^2. In terms of the y variable, the values y_i are an independent sample from a normal distribution with mean $\beta_0 + \beta_1 x_{1i} + \cdots + \beta_p x_{pi}$ and common variance σ^2. If the x variables are random, this is true after conditioning on their values.

Interpretation   For the simple linear regression model, the slope parameter, β_1, is easily interpreted, as one-unit change in the predictor variable will correspond to a predicted change in the mean response by β_1 units. For the multiple regression model, a similar interpretation is possible: a one-unit change in the ith predictor corresponds to a β_i-unit change in the predicted mean response *if the other predictors are held constant*. This is not always possible in practice.

● **Example 11.1: What influences a baby's birth weight?**
A child's birth weight depends on many things, among them the parents' genetic makeup, gestation period, and mother's activities during pregnancy. The babies (UsingR) data set lets us investigate some of these relationships.

This data set contains many variables to consider. We first look at the quantitative variables as predictors. These are gestation period; mother's age, height, and weight; and father's age, height, and weight.

A first linear model might incorporate all of these at once:

$$\text{wt} = \beta_0 + \beta_1 \cdot \text{gestation} + \beta_2 \cdot \text{mother's age} + \cdots + \beta_7 \cdot \text{father's weight} + \epsilon_i.$$

Why should this have a linear model? It seems intuitive that birth weight would vary monotonically with the variables, so a linear model might be a fairly good approximation. We'll want to look at some plots to make sure our model seems appropriate. ••

• **Example 11.2: Polynomial regression**
In 1609, Galileo proved mathematically that the horizontal distance traveled by an object with an initial horizontal velocity is a parabola. He based his insight on an experimental setup consisting of a ball placed at a certain height on a ramp and then released. The distance traveled was then measured. This experiment was chosen to reduce the effects of friction.[5] The data consists of two variables. Let's call them y for distance traveled and x for initial height. Galileo may have considered any of these polynomial models:

$$y_i = \beta_0 + \beta_1 x_i + \epsilon_i,$$
$$y_i = \beta_0 + \beta_1 x_i + \beta_2 x_i^2 + \epsilon_i, \quad \text{or}$$
$$y_i = \beta_0 + \beta_1 x_i + \beta_2 x_i^2 + \beta_3 x_i^3 + \epsilon_i.$$

The ϵ_i would cover error terms that are presumably independent and normally distributed. The quadratic model (the second model) is correct under perfect conditions, as Galileo demonstrated, but the data may suggest a different model if the conditions are not perfect. ••

• **Example 11.3: Predicting classroom performance**
College admissions offices are faced with the problem of predicting future performance based on a collection of measures, such as grade-point average and standardized test scores. These values may be correlated. There may also be other variables that describe why a student does well, such as type of high school attended or student's work ethic.

Initial student placement is also a big issue. If a student does not place into the right class, he may become bored and leave the school. Successful placement is key to retention. For New York City high school graduates, available at time of placement are SAT scores and Regents Exam scores. High school grade-point average may be unreliable or unavailable.

The data set stud.recs (UsingR) contains test scores and initial grades in a math class for several randomly selected students. What can we predict about the initial grade based on the standardized scores?

---

[5]This example appears in Ramsey and Schafer [49], where a schematic of the experimental apparatus is drawn.

An initial model might be to fit a linear model for grade with all the other terms included. Other restricted models might be appropriate. For example, are the verbal SAT scores useful in predicting grade performance in a future math class?   ••

Fitting the multiple regression model using `lm`

As seen previously, the method of least squares is used to estimate the parameters in the multiple regression model. We don't give formulas for computing the $\widehat{\beta}$'s but note that, since there are $p + 1$ estimated parameters, the estimate for the variance changes to

$$\widehat{\sigma}^2 = \frac{\text{RSS}}{n - (p + 1)}.$$

To find these estimates in R, again the `lm` function is used. The syntax for the model formula varies depending on the type of terms in the model. For these problems, we use + to add terms to a model, - to drop terms, and I to insulate terms so that the usual math notations apply.

For example, if x, y, and z are variables, then the following statistical models have the given R counterparts:

$$z_i = \beta_0 + \beta_1 x_i + \beta_2 y_i + \epsilon_i \qquad \text{is expressed as} \quad \text{z ~ x + y}$$
$$z_i = \beta_0 + \beta_1 x_i + \beta_2 x_i^2 + \epsilon_i \qquad \text{is expressed as} \quad \text{z ~ x + I(x\^2)}$$

Once the model is specified, the `lm` function follows this familiar format:

```
lm(formula, data=..., subset=...)
```

To illustrate with an artificial example, we simulate the relationship $z_i = \beta_0 + \beta_1 x_i + \beta_2 y_i + \epsilon_i$ and then find the estimated coefficients:

```
x <- 1:10
y <- rchisq(10,3)
z <- 1 + x + y + rnorm(10)
lm(z ~ x + y)

##
## Call:
## lm(formula = z ~ x + y)
##
## Coefficients:
## (Intercept)            x              y
##      -0.367        1.179          0.990
```

The output of lm stores much more than is seen initially (which is just the formula and the estimates for the coefficients). It is recommended that the return value be stored. Afterward, the different extractor functions can be used to view the results.

● **Example 11.4: Finding the regression estimates for baby's birth weight**
Fitting the birth-weight model is straightforward. The basic model formula is

```
wt ~ gestation + age + ht + wt1 + dage + dht + dwt
```

We've seen with this data set that the variables have some missing values that are coded not with NA but with very large values that are obvious when plotted, but not when we blindly use the functions. In particular, gestation should be less than 350 days, mother's age and height less than 99, and weight less than 999, etc. We can avoid these cases by using the subset argument as illustrated. Recall that we combine logical expressions with & for "and" and | for "or."

```
res.lm <- lm(wt ~ gestation + age + ht + wt1 + dage + dht + dwt,
 data=babies,
 subset=gestation < 350 & age < 99 & ht < 99 & wt1 < 999 &
        dage < 99 & dht < 99 & dwt < 999)
res.lm

##
## Call:
## lm(formula = wt ~ gestation + age + ht + wt1 + dage + dht + dwt,
##      data = babies, subset = gestation < 350 & age < 99 & ht <
##          99 & wt1 < 999 & dage < 99 & dht < 99 & dwt < 999)
##
## Coefficients:
## (Intercept)     gestation           age            ht           wt1
##     -105.4576        0.4625        0.1384        1.2161        0.0289
##          dage           dht           dwt
##        0.0590       -0.0663        0.0782
```

A residual plot (not shown) shows nothing too unusual:

```
plot(fitted(res.lm), resid(res.lm))
```

The diagnostic plots found with plot(res.lm) indicate that observation 261 might be a problem. Looking at babies[261,], it appears that this case is an outlier, as it has a very short gestation period. ●●

Using update **with model formulas**

When comparing models, we may be interested in adding or subtracting a term and refitting. Rather than typing in the entire model formula again, R provides a way to add or drop terms from a model and have the new model fit. This process is called updating and is done with the update function. The usage is

```
update(model.object, formula = . ~ . + new.terms - old.terms)
```

The model.object is the output of some modeling command, such as lm. The formula argument uses a . to represent the previous value. In the template above, the . to the left of the ~ indicates that the previous left side of the model formula should be reused. The right-hand-side . refers to the previous right-hand side. In the template, the +new.terms means to add term and -old.terms is used to drop terms.

• **Example 11.5: Discovery of the parabolic trajectory**
The data set galileo (UsingR) contains two variables measured by Galileo (described previously). One is the initial height and one the horizontal distance traveled.

A plot of the data illustrates why Galileo may have thought to prove that the correct shape is described by a parabola. Clearly a straight line does not fit the data well. However, with modern computers, we can investigate whether a cubic term is warranted for this data.

To do so we fit three polynomial models. The update function is used to add terms to the previous model to give the next model. To avoid a different interpretation of ^, the powers are insulated with I.

```
init.h <- c(600,700,800,950,1100,1300,1500)
h.d <- c(253, 337, 395, 451, 495, 534, 573)
res.lm   <- lm(h.d ~ init.h)
res.lm2 <- update(res.lm,  . ~ . + I(init.h^2))
res.lm3 <- update(res.lm2, . ~ . + I(init.h^3))
```

To plot these, we will use curve, but first we define a function which evaluates a polynomial given its coefficients:

```
polynomial <- Vectorize(function(x, ps) {
  n <- length(ps)
  sum(ps*x^(1:n-1))
}, "x")
```

Then we can plot as follows (Figure 11.7).

```
plot(h.d ~ init.h)
curve(polynomial(x, coef(res.lm )), add=TRUE, lty=1)
curve(polynomial(x, coef(res.lm2)), add=TRUE, lty=2)
curve(polynomial(x, coef(res.lm3)), add=TRUE, lty=3)
legend(1200, 400, legend=c("linear", "quadratic", "cubic"), lty=1:3)
```

The linear model is a poor fit, but both the quadratic and cubic fits seem
reasonable. ••

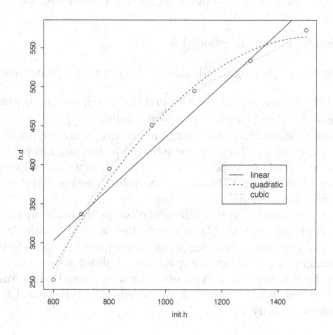

Figure 11.7: Three polynomial models fit to the Galileo data.

Interpreting the regression parameters

As mentioned, interpretation in simple regression is usually straightforward.
Changes in the predictor variable correspond to changes in the response variable in a linear manner: a unit change in the predictor corresponds to a $\hat{\beta}_1$-
unit change in the response.

However, in multiple regression this picture may not be applicable, as
we may not be able to change just a single variable. As well, when more

variables are added to a model, if the variables are correlated then the sign of the coefficients can change, leading to a different interpretation.

The language often used is that we "control" the other variables while seeking a primary predictor variable.

• Example 11.6: Does taller mean higher paid?
A University of Florida press release from October 16, 2003, read:

> "Height matters for career success..."
> The reported study, which controlled for gender, weight, and age, found that mere inches cost thousands of dollars. Each inch in height amounted to about $789 more a year in pay, the study found.

The mathematical model mentioned would be

$$\text{pay} = \beta_0 + \beta_1 \text{ height} + \beta_2 \text{ gender} + \beta_3 \text{ weight} + \beta_4 \text{ age} + \epsilon.$$

(In the next chapter we see how to interpret the term involving the categorical variable gender.) The data gives rise to the estimate $\hat{\beta}_1 = 789$. The authors interpret this to mean that each extra inch of height corresponds to a $789 increase in expected pay. So someone who is 4 inches taller, say 6 feet versus 5 feet 8 inches, would be expected to earn $3,156 more annually. ($\hat{y}$ is used to predict expected values.) The word "controlled" means that they included these variables in the model.

There are few caveats to this interpretation. First, unlike in a science experiment, where we may be able to specify the value of a variable, a person cannot simply grow an inch to see if his salary goes up. As well, it isn't realistic to imagine a person growing an inch without some change in their weight, say. So it is hard to hold all other variables equal when interpreting the coefficient. Further, as this is an observational study, causal interpretations are not necessarily valid.   ••

Statistical inferences

As in the simple linear regression case, if the model is correct, statistical inference can be made about the coefficients. In general, the estimators for a linear model are unbiased and normally distributed; from this, t-tests and confidence intervals can be constructed for the estimators, once we learn the standard errors. As before, these are output by the summary function.

• Example 11.7: Galileo, continued
For the Galileo data example, the summary of the quadratic fit contains

```
short_summary(res.lm2)

##                   Estimate Std. Error t value Pr(>|t|)
## (Intercept) -2.40e+02    6.90e+01   -3.48   0.0253 *
## init.h       1.05e+00    1.41e-01    7.48   0.0017 **
## I(init.h^2) -3.44e-04    6.68e-05   -5.15   0.0068 **
## ---
## Signif. codes:  0 '***' 0.001 '**' 0.01 '*' 0.05 '.' 0.1 ' ' 1
```

For each $\widehat{\beta}$, the standard errors are given, as is the marginal t-test, which tests for the null hypothesis that the $\widehat{\beta}$ is 0. All three have small p-values and are flagged as such with significance stars.

Finding a confidence interval for the parameters is straightforward, as the values $(\widehat{\beta}_i - \beta_i)/\mathrm{SE}(\widehat{\beta}_i)$ have a t-distribution with $n - (p+1)$ degrees of freedom if the linear model applies.

For example, a 95% confidence interval for β_1 would be

```
alpha <- 0.05
tstar <- qt(1 - alpha/2, df=4)          # n=7; p=2; df=n-(p+1)
beta1 <- 1.05
SE <- 0.141
beta1 + c(-1,1 ) * tstar * SE

## [1] 0.6585 1.4415
```

••

Model selection

Modeling is done for many reasons. One is to shine the focus on the important predictors to explain as much variation in the response as possible while avoiding the noise of unimportant factors. Doing this requires some means for determining when a predictor variable contributes sufficiently to the description of the response as to be warranted. For this we discuss a few criteria below that are easily used within R.

Before proceeding with methods to remove variables from consideration, we paraphrase some practical, general principles on building regression models for prediction provided in Section 4.6 of [25]:

- Include all input variables that might be expected to be important in predicting the response.

- Sometimes, predictors can be combined into other variables. For example, using BMI instead of both height and weight.

- For decisions on which variables to exclude:

  - If a predictor is not statistically significant and has the expected sign it is generally fine to leave it in (though the methods below will exclude it).

  - Consider removing predictors which are not statistically significant and do not have the expected sign.

  - If a predictor is statistically significant and has the expected sign, leave it in.

  - If a predictor is statistically significant and *does not* have the expected sign, then think hard about its inclusion. It might point to lurking variables, or underlying correlations with other predictors.

Partial F-test

The partial F-test is used to discriminate between two models with one being nested in the other. For example,

$$y_i = \beta_0 + \beta_1 x_{1i} + \cdots + \beta_k x_{ki} + \epsilon_i \qquad (11.10)$$
$$y_i = \beta_0 + \beta_1 x_{1i} + \cdots + \beta_k x_{ki} + \beta_{k+1} x_{(k+1)i} + \cdots + \beta_p x_{pi} + \epsilon_i.$$

The first model has $k + 1$ parameters, and the second has $p + 1$ with $p > k$ (not counting σ). Recall that the residual sum of squares, RSS, measures the variation between the data and the model. For the model with p predictors, $RSS(p)$ can be no more than $RSS(k)$ for the model with k predictors. Call the difference the *extra sum of squares*.

If the new parameters are not really important, then there should be little difference between the sums of squares when computed with or without the new parameters. If they are important, then there should be a big difference. To measure big or small, we can divide by the residual sum of squares for the full model. That is,

$$\frac{RSS(k) - RSS(p)}{RSS(p)}$$

should measure the influence of the extra parameters. If we divide the extra sum of squares by $p - k$ and the residual sum of squares by $n - (p + 1)$ (the respective degrees of freedom), then the statistic becomes

$$F = \frac{(RSS(k) - RSS(p))/(p - k)}{RSS(p)/(n - (p + 1)))} = \frac{(RSS(k) - RSS(p))/(p - k)}{\hat{\sigma}^2}. \qquad (11.11)$$

This statistic is actually a more general example of that in Equation 11.7 and has a similar sampling distribution. Under the null hypothesis that the extra β's are 0 ($\beta_{k+1} = \cdots = \beta_p = 0$), and the ϵ_i are *i.i.d.* with a Normal$(0,\sigma^2)$ distribution, F will have the F-distribution with $p - k$ and $n - (p+1)$ degrees of freedom.

This leads to the following significance test.

Partial F-test for null hypothesis of no effect

For the nested models of Equation 11.10, a significance test for the hypotheses

$$H_0 : \beta_{k+1} = \beta_{k+2} = \cdots = \beta_p = 0 \quad \text{and} \quad H_A : \text{at least one } \beta_j \neq 0 \text{ for } j > k$$

can be performed with the test statistic (11.11):

$$F = \frac{\text{extra sum of squares}/(p-k)}{\widehat{\sigma}^2}.$$

Under H_0, F has the F-distribution with $p - k$ and $n - (p+1)$ degrees of freedom. Large values of F are in the direction of the alternative. This test is called the *partial F-test*.

The anova function will perform the partial F-test. If res.lm1 and res.lm2 are the return values of two nested models, then

```
anova(res.lm1, res.lm2)
```

will perform the test and produce an analysis of variance table.

• **Example 11.8: Discovery of the parabolic trajectory revisited**
In Example 11.3 we fit the data with three polynomials, graphing each. Referring to Figure 11.7, we see that the parabola and cubic clearly fit better than the linear. But which of those two fits better? We use the partial F-test to determine whether the extra cubic term is significant.

To do this, we use the anova function on the two results res.lm2 and res.lm3. This yields

```
anova(res.lm2,res.lm3)

## Analysis of Variance Table
##
## Model 1: h.d ~ init.h + I(init.h^2)
## Model 2: h.d ~ init.h + I(init.h^2) + I(init.h^3)
##   Res.Df RSS Df Sum of Sq    F Pr(>F)
## 1      4 744
## 2      3  48  1       696 43.3 0.0072 **
```

```
## ---
## Signif. codes:  0 '***' 0.001 '**' 0.01 '*' 0.05 '.' 0.1 ' ' 1
```

The F-test is significant ($p = 0.0072$), indicating that the null hypothesis ($\beta_3 = 0$) does not describe the data well. This suggests that the underlying relationship from Galileo's data is cubic and not quadratic. Perhaps the apparatus introduced drag.   ••

The Akaike information criterion

In the partial F-test, the trade-off between adding more parameters to improve the model fit and making a more complex model appears in the $n - (p + 1)$ divisor. Another common criterion with this trade-off is Akaike's information criterion (AIC). The AIC is computed in R with the AIC extractor function. The details of the statistic involve the likelihood function, a more advanced concept, but the usage is straightforward: models with lower AICs are preferred. An advantage to the AIC is that it can be used to compare models that are not nested, a restriction of the partial F-test.

The extractor function AIC will compute the value for a given model, but the convenient stepAIC function from the MASS package will step through the submodels and do the comparisons for us.

• Example 11.9: Predicting grades based on standardized tests

The data set stud.recs (UsingR) contains five standardized test scores and a numeric value for the initial grade in a subsequent math course. The goal is to use the test-score data to predict the grade that a student will get. If the grade is predicted to be low, perhaps an easier class should be recommended.

First, we view the data using paired scatterplots

```
pairs(stud.recs)
```

The figure (not shown) indicates strong correlations among the variables.

We begin by fitting the entire model. In this case, the convenient . syntax on the right-hand side is used to indicate all the remaining variables.

```
d <- subset(stud.recs, select=-letter.grade)
res.lm <- lm(num.grade ~ ., data = d)
res.lm

##
## Call:
## lm(formula = num.grade ~ ., data = d)
##
## Coefficients:
```

```
## (Intercept)          seq.1          seq.2          seq.3          sat.v
##    -0.73953       -0.00394       -0.00272        0.01565       -0.00125
##        sat.m
##      0.00590
```

Some terms are negative, which seems odd. (Why?) Looking at the summary of the regression model we have

```
short_summary(res.lm)

##               Estimate Std. Error t value Pr(>|t|)
## (Intercept)  -0.73953    1.21128    -0.61    0.543
## seq.1        -0.00394    0.01457    -0.27    0.787
## seq.2        -0.00272    0.01503    -0.18    0.857
## seq.3         0.01565    0.00941     1.66    0.099 .
## sat.v        -0.00125    0.00163    -0.77    0.443
## sat.m         0.00590    0.00267     2.21    0.029 *
## ---
## Signif. codes:  0 '***' 0.001 '**' 0.01 '*' 0.05 '.' 0.1 ' ' 1
```

The marginal *t*-tests for whether the given parameter is 0 or not are "rejected" only for the seq.3 (for this sample of students, sequential 3 was the last high school test taken) and sat.m (the math SAT score). It is important to remember that these are tests concerning whether the value is 0 given the other predictors. They can change if correlated predictors are removed.

The stepAIC function can step through the various submodels and rank them by AIC. This gives

```
library(MASS)              # load in MASS package for stepAIC
stepAIC(res.lm, trace=0)   # trace=0 suppresses intermediate output

##
## Call:
## lm(formula = num.grade ~ seq.3 + sat.m, data = d)
##
## Coefficients:
## (Intercept)         seq.3          sat.m
##    -1.14078       0.01371        0.00479
```

The submodel with just two predictors is selected. As expected, the verbal scores on the SAT are not a useful indicator of performance. ●●

Problems

11.21 Following the example with Galileo's data, fit a fourth-degree polynomial to the galileo (UsingR) data and compare to the cubic polynomial using a partial F-test. Is the new coefficient significant?

11.22 For the data set trees, model the Volume by the Girth and Height variables. Does the model fit the data well?

11.23 The data set MLBattend (UsingR) contains attendance data for Major League Baseball for the years 1969 to 2000. Fit a linear model of attendance modeled by year, runs.scored, wins, and games.behind. Which variables are flagged as significant? Look at the diagnostic plots and comment on the validity of the model.

11.24 For the deflection (UsingR) data set, fit the quadratic model

$$\text{Deflection} = \beta_0 + \beta_1 \text{Load} + \beta_2 \text{Load}^2 + \epsilon.$$

How well does this model fit the data? Compare to the linear model.

11.25 The data set kid.weights contains age, weight, and height measurements for several children. Fit the linear model

$$\text{weight} = \beta_0 + \beta_1 \text{age} + \beta_2 \text{height} + \beta_3 \text{height}^2 + \beta_4 \text{height}^3 + \beta_5 \text{height}^4$$

Use the partial F-test to select between this model and the nested models found by using only first-, second-, and third-degree polynomials for height.

11.26 The data set fat (UsingR) contains several body measurements that can be done using a scale and a tape measure. These can be used to predict the body-fat percentage (body.fat). Measuring body fat requires a special apparatus; if our resulting model fits well, we have a low-cost alternative.

Fit the variable body.fat using each of the variables age, weight, height, BMI, neck, chest, abdomen, hip, thigh, knee, ankle, bicep, forearm, and wrist. Use the stepAIC function to select a submodel. For this submodel, what is the adjusted R^2?

11.27 The data set Cars93 (MASS) contains data on cars sold in the United States in the year 1993. Fit a regression model with MPG.city modeled by the numeric variables EngineSize, Weight, Passengers, and price. Which variables are marked as statistically significant by the marginal t-tests? Which model is selected by the AIC?

11.28 We can simulate the data to see how often the partial F-test or AIC works. For example, a single simulation can be done with the commands

```
x <- 1:10
y <- rnorm(10, 1 + 2*x + 3*x^2, 4)
require(MASS)
stepAIC(lm(y ~ x + I(x^2)), trace=0)

##
## Call:
## lm(formula = y ~ x + I(x^2))
##
## Coefficients:
## (Intercept)             x          I(x^2)
##      -0.583         3.494           2.846
```

Do a few simulations to see how often the correct model is selected.

11.29 The data set baycheck (UsingR) contains estimated populations for a variety of Bay Checkerspot butterflies near California. A common model for population dynamics is the Ricker model, for which t is time in years:

$$N_{t+1} = aN_t e^{bN_t} W_t,$$

where a and b are parameters and W_t is a lognormal multiplicative error. This can be turned into a regression model by dividing by N_t and then taking logs of both sides to give

$$\log(\frac{N_{t+1}}{N_t}) = \log(a) + bN_t + \epsilon_t.$$

Let y_t be the left-hand side. This may be written as

$$y_t = r(1 - \frac{N_t}{K}) + \epsilon_t,$$

because r can be interpreted as an unconstrained growth rate and K as a carrying capacity.

Fit the model to the baycheck data set and find values for r and K. To find y_t you can do the following:

```
d <- with(baycheck, {
  n <- length(year)
  yt <- log(Nt[-1]/Nt[-n])
  nt <- Nt[-n]
  data.frame(yt, nt)
})
```

Recall that a negative index means all but that index.

Analysis of variance

Analysis of variance, *ANOVA*, is a method of comparing means across samples based on variations from the mean. We begin by illustrating an ANOVA carried out in the traditional way, but we will see that the ANOVA model is just a special form of the linear model discussed in the previous chapter and R provides a common interface.

12.1 One-way ANOVA

A one-way analysis of variance is a generalization of the t-test for two independent samples, allowing us to compare population means for several independent samples. Suppose we have k populations of interest. From each we take a random sample. These samples are independent if the knowledge of one sample does not effect the distribution of another. Notationally, for the ith sample, let $x_{i1}, x_{i2}, \ldots, x_{in_i}$ designate the sample values.

The one-way analysis of variance applies to normally distributed populations. Suppose the mean of the ith population is μ_i and its standard deviation is σ_i. We use a σ if these are equivalent across the groups. A statistical model for the data with a common standard deviation is

$$x_{ij} = \mu_i + \epsilon_{ij},$$

where the error terms, ϵ_{ij}, are independent with Normal$(0, \sigma)$ distribution.

• Example 12.1: Number of calories consumed by month

Consider 15 subjects split at random into three groups. Each group is assigned a month. For each group we record the number of calories consumed on a randomly chosen day. Figure 12.1 shows the data. We assume that the amounts consumed are normally distributed with common variance but perhaps different means. From the figure, we see that there appears to be more clustering around the means for each month than around the grand mean or mean for all the data. Perhaps more calories are consumed in the winter?

The goal of one-way analysis of variance is to decide whether the difference in the sample means is indicative of a difference in the population means or is attributable to sampling variation. ••

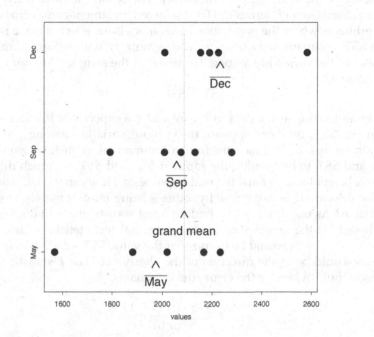

Figure 12.1: Amount of calories consumed by subjects for different months. Sample means are marked, as is the grand mean. Are the differences in the monthly means due to sampling variation or seasonal differences?

This problem is approached as a significance test. Let the hypotheses be

$$H_0 : \mu_1 = \mu_2 = \cdots = \mu_k, \quad H_A : \mu_i \neq \mu_j \text{ for at least one pair } i \text{ and } j.$$

A test statistic is formulated that compares the variations within a single group to those among the groups.

Let $\bar{x}$ be the grand mean, or mean of all the data, and $\bar{x}_i$ the mean for the ith sample. Then the total sum of squares is given by

$$\text{SST} = \sum_i \sum_j (x_{ij} - \bar{x})^2.$$

This measures the amount of variation from the center of all the data.

An analysis of variance breaks SST up into two sums:

$$\text{SST} = \sum_i \sum_j (x_{ij} - \bar{x}_i)^2 + \sum_i n_i (\bar{x}_i - \bar{x})^2. \tag{12.1}$$

The first sum is called the *error sum of squares*, or SSE. The interior sum, $\sum_j (x_{ij} - \bar{x}_i)^2$, measures the variation within the ith group. The SSE is then a measure of the within-group variability. The second term in (12.1) is called the *treatment sum of squares* (SSTr). The word treatment comes from medical experiments where the population mean models the effect of some treatment. The SSTr compares the means for each group, $\bar{x}_i$, with the grand mean, $\bar{x}$. It measures the variability among the means of the samples. We can re-express Equation 12.1 as

$$SST = SSE + SSTr.$$

From looking at the data in Figure 12.1 we expect that the SSE is smaller than the SST, as there appears to be more variation among groups than within groups. If the data came from a common mean, then we would expect SSE and SST to be roughly the same. If SSE and SST are much different, it would be evidence against the null hypothesis. How can we tell whether the differences are due to the null hypothesis being false or merely to sampling variation? As usual, we tell by finding a test statistic that can discriminate.

Based on the above observation, a natural test statistic to test whether $\mu_1 = \mu_2 = \cdots = \mu_k$ would be to consider the value SST − SSE = SSTr. "Large" values would be in the direction of the alternative. The F-statistic this comparison, but divides by the error sum of squares.

$$F = \frac{SSTr/(k-1)}{SSE/(n-k)}, \tag{12.2}$$

Large values are still consistent with a difference in the means. To get the proper scale, each term is divided by its respective degrees of freedom, yielding the mean sum of squares. The degrees of freedom for the total sum of squares are $n - 1$. For the SSE the degrees of freedom are $n - k$, so the degrees of freedom for SSTr are $k - 1$.

Under the assumption that the data is normally distributed with common mean and variance, this statistic will have a known distribution: the F-distribution with $k - 1$ and $n - k$ degrees of freedom. This is a consequence of the partial F-test discussed in Chapter 11.[1]

The one-way analysis-of-variance significance test

Suppose we have k independent, *i.i.d.* samples from populations with Normal(μ_i, σ) distributions, $i = 1, 2, \ldots, k$. A significance test of

$$H_0 : \mu_1 = \mu_2 = \cdots = \mu_k, \qquad H_A : \mu_i \neq \mu_j \text{ for at least one pair, } i \text{ and } j,$$

---

[1]This can be shown by identifying RSS(k) with the total sum of squares and RSS(p) with SSE in (11.11) and simplifying.

can be performed with test statistic

$$F = \frac{\text{SST}/(k-1)}{\text{SSE}/(n-k)}.$$

Under H_0, F has the F-distribution with $k-1$ and $n-k$ degrees of freedom. The p-value is calculated from $P(F \geq$ observed value $| H_0)$.

The R function oneway.test will perform this significance test.

• **Example 12.2: Number of calories consumed by month, continued**
The one-way test can be applied to the example on caloric intake. The two sums can be calculated directly as follows:

```
may <- c(2166, 1568, 2233, 1882, 2019)
sep <- c(2279, 2075, 2131, 2009, 1793)
dec <- c(2226, 2154, 2583, 2010, 2190)
#
xbar <- mean(c(may, sep, dec))
SST <- (15-1) * var(c(may, sep, dec))    # (n-1) * var(.) is SST
SSE <- (5-1) * var(may) + (5-1) * var(sep) + (5-1) * var(dec)
SSTr <- 5 * ((mean(may) - xbar)^2 + (mean(sep) - xbar)^2 +
             (mean(dec) - xbar)^2)
#
c(SST=SST, SSTr=SSTr, SSE=SSE)

##    SST    SSTr    SSE
## 761384 174664 586720

#
n <- 15; k <- 3
F.obs = (SSTr/(k-1)) / (SSE/(n-k))
F.obs

## [1] 1.786

pf(F.obs, df1=k-1, df2=n-k, lower.tail=FALSE)

## [1] 0.2094
```

We get a p-value that is not significant. Despite the graphical evidence, the differences can reasonably be explained by sampling variation.   ••

Using R's model formulas to specify ANOVA models

The calculations to perform an analysis of variance need not be so compli-cated, as R has functions to compute the values desired. These functions use model formulas. If x stores all the data and f is a *factor* indicating which group the data value belongs to, then

```
x ~ f
```

represents the statistical model

$$x_{ij} = \mu_i + \epsilon_{ij}.$$

The default behavior for plot of the model formula x ~ f was to make a boxplot. This is because this graphic easily allows for comparison of centers for multiple samples. The dot plot in Figure 12.1 is good for a small data set, but the boxplot is preferred for larger data sets.

Using oneway.test to perform ANOVA

The function oneway.test is used as

```
oneway.test(x ~ f, data=..., var.equal=FALSE)
```

As with the t.test function, the argument var.equal is set to TRUE if appro-priate. By default it is FALSE.

Before using oneway.test with our example of caloric intake, we put the data into the appropriate form: a data vector containing the values and a factor indicating the sample the corresponding value is from. This can be achieved with stack.

```
d <- stack(list(may=may, sep=sep, dec=dec)) # need names for list
names(d)                                     # two variables

## [1] "values" "ind"

oneway.test(values ~ ind, data=d, var.equal=TRUE)

##
##   One-way analysis of means
##
## data:  values and ind
## F = 1.786, num df = 2, denom df = 12, p-value = 0.2094
```

We get the same *p*-value as in our previous calculation, but with much less effort.

Using aov for ANOVA

The alternative aov function will also perform an analysis of variance. It returns a model object similar to 1m but has different-looking outputs for the print and summary extractor functions. These are analysis-of-variance tables that are typical of other computer software and statistics books.

Again, it is called with a model formula.

```
res <- aov(values ~ ind, data = d)
res

## Call:
##    aov(formula = values ~ ind, data = d)
##
## Terms:
##                       ind Residuals
## Sum of Squares    174664    586720
## Deg. of Freedom        2        12
##
## Residual standard error: 221.1
## Estimated effects may be unbalanced
```

The function returns the two sums of squares calculated in Example 12.1 with their degrees of freedom. The Residual standard error, $\hat{\sigma}$, is found by the square root of $RSS/(n-k)$, which in this example is

```
sqrt(586720/12)

## [1] 221.1
```

The result of aov has more information than shown, just as the result of 1m does. For example, the summary function returns

```
summary(res)

##            Df Sum Sq Mean Sq F value Pr(>F)
## ind         2 174664   87332    1.79   0.21
## Residuals  12 586720   48893
```

These are the values needed to perform the one-way test. This tabular layout is typical of an analysis of variance.

• Example 12.3: Effect of grip on cross-country skiing

Researchers at Montana State University performed a study on how various ski-pole grips affect cross-country skiing performance. There are three basic grip types: classic, modern, and integrated. Suppose 9 skiers are assigned at random to the three grip-types and for each the skier has upper-body power

Grip type	classic	integrated	modern
	168.2	166.7	160.1
	161.4	173.0	161.2
	163.2	173.3	166.8

Table 12.1: Upper-body power output (watts) by ski-pole grip type.

output measured. The data is summarized in Table 12.1. Does there appear to be a difference in power output due to grip type?

We can investigate the null hypothesis that the three grips will produce equal means with an analysis of variance. We assume that the errors are all independent and that the data is sampled from normally distributed populations with common variance but perhaps different means.

First we enter in the data. Instead of using stack, we enter in all the data at once and create a factor using gl to indicate grip type.[2]

```
UBP <- c(168.2, 161.4, 163.2, 166.7, 173.0, 173.3,
          160.1, 161.2, 166.8)
grip.type <- gl(3, 3, 9, labels=c("classic", "integrated", "modern"))
boxplot(UBP ~ grip.type, ylab="Power (watts)",
        main="Effect of cross country grip")
```

The boxplot in Figure 12.2 indicates that the integrated grip has a significant advantage. But is this due to sampling error? We use aov to carry out the analysis of variance.

```
res <- aov(UBP ~ grip.type)
summary(res)

##              Df Sum Sq Mean Sq F value Pr(>F)
## grip.type     2  116.7    58.3    4.46  0.065 .
## Residuals     6   78.4    13.1
## ---
## Signif. codes:  0 '***' 0.001 '**' 0.01 '*' 0.05 '.' 0.1 ' ' 1
```

We see that there is a small p-value that is significant at the 10% level. ••

---

[2]We could also use rep with vectorized arguments to create the factor, but gl is designed for just this task.

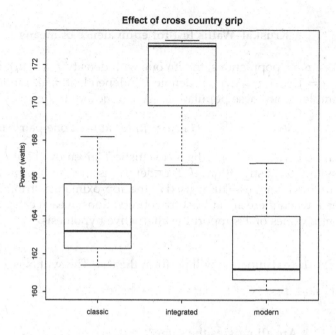

Figure 12.2: Effects of cross-country ski-pole grip on measured power output.

The nonparametric Kruskal–Wallis test

The Wilcoxon rank-sum test was discussed as a nonparametric alternative to the two-sample t-test for independent samples. Although the populations had no parametric assumption, they were assumed to have densities with a common shape but perhaps different centers.

The Kruskal–Wallis test, a nonparametric test, is analogous to the rank-sum test for comparing the population means of k independent samples.

In particular, if $f(x)$ is a density of a continuous random variable with mean 0, the assumption on the data is that x_{ij} is drawn independently of the others from a population with density $f(x - \mu_i)$. The hypotheses tested are

$$H_0 : \mu_1 = \mu_2 = \cdots = \mu_k, \qquad H_A : \mu_i \neq \mu_j \text{ for at least one pair } i \text{ and } j.$$

The test statistic involves the ranks of all the data. Let r_{ij} be the respective rank of a data point when all the data is ranked from smallest to largest, $\bar{r}_i$ be the mean of the ranks for each group, and $\bar{r}$ the grand mean. The test statistic is:

$$T = \frac{12}{n(n+1)} \sum_i n_i (\bar{r}_i - \bar{r})^2. \tag{12.3}$$

Statistical inference is based on the fact that T has an asymptotic χ^2-distribution with $k - 1$ degrees of freedom.

> **Kruskal–Wallis test for equivalence of means**
>
> Assume k populations, the ith one with density $f(x - \mu_i)$. Let $x_{ij}, i = 1, \ldots, k, j = 1, \ldots, n_i$ denote k independent, *i.i.d.* random samples from these populations. A significance test of
>
> $$H_0 : \mu_1 = \mu_2 = \cdots = \mu_k, \quad H_A : \mu_i \neq \mu_j \text{ for at least one pair } i \text{ and } j,$$
>
> can be performed with the test statistic T given by (12.3). The asymptotic distribution of T under H_0 is the χ^2-distribution with $k - 1$ degrees. This is used as the approximate distribution for T when there are at least five observations in each category. Large values of T support the alternative hypothesis.

The kruskal.test function will perform the test. The syntax is

```
kruskal.test(x ~ f, data=..., subset=...)
```

• **Example 12.4: Are all movies the same?**
The movie_data_2011 (UsingR) data set has data for movies playing domestically in the United States during 2011. An interesting question is what can be said about the predicted total gross based on the opening weekend amount? Here we look at a related question, does this ratio depend on the type of movie?

The data set is structured to have a record for each movie playing during a given week. We will need to do some data massaging to create a suitable data frame.

First we split the data by the movie variable, to get separate data frames for each:

```
x <- split(movie_data_2011, movie_data_2011$Movie)
```

Not all movies in the data base opened in 2011. We filter these out by the Previous column, as this will contain an initial NA if the movie came out in 2011.

```
x <- Filter(function(d) is.na(d$Previous[1]), x)
```

For each data frame in x, we will compute a simple summary with the following function:

```
open_to_gross <- function(d) {
  ## return Genre and open to gross ratio
  list(movie = as.character(d$Movie[1]),
```

```
        genre=as.character(d$Genre[1]), open=d$Gross[1],
        total=max(d$TotalGross))
}
l <- sapply(x, open_to_gross, simplify=FALSE)
```

This produces a list, where each component summarizes a movie. It is easier to work with this as a data frame. There are different ways to produce such, here is one where we create an initial data frame from the first component, then use Reduce to add each subsequent one:

```
d <- data.frame(l[[1]], stringsAsFactors=FALSE)
d <- Reduce(rbind, l[-1], d)
```

Not all movies are so big, to compare similar movies we only look at those with an opening weekend of $500,000 or more:

```
d <- d[d$open > 5e5,]
```

Now we compute the ratio and find a summary

```
d$ratio <- d$total / d$open
summary(d$ratio)

##    Min. 1st Qu.  Median    Mean 3rd Qu.    Max.
##    1.51    2.34    2.83    3.98    3.57   94.10
```

As should be expected, the summary shows tremendous skew. Some movies can take off and generate gross revenues week after week. We wish to look at the different genres. A simple subset of popular genres is found with:

```
genres <- c("Comedy", "Action", "Adventure", "Horror")
d1 <- subset(d, subset=genre %in% genres)
```

We can make an exploratory boxplot (Figure 12.3) through:

```
boxplot(ratio ~ genre, data=d1)
```

From the figure, it appears that perhaps not all genres have similar centers. Further the shapes are similarly skewed, so tests based on assumptions of normality would require reasonable samples sizes. Here, we perform a significance test to produce a *p*-value for comparison:

```
kruskal.test(ratio ~ factor(genre), d1)
```

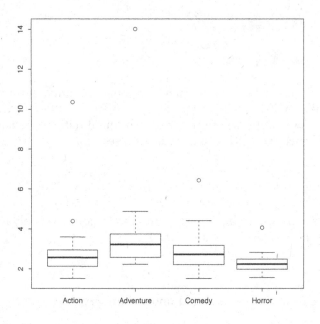

Figure 12.3: Boxplots by movie genre of the ratio of total gross earnings and opening weekend gross. Which type of movie would require more of an advertising blitz to maximize revenue?

```
##
##  Kruskal-Wallis rank sum test
##
## data:  ratio by factor(genre)
## Kruskal-Wallis chi-squared = 15.86, df = 3, p-value =
## 0.001209
```

The lack of a clear set of samples from a large population makes the interpretation of this p-value more an intuition, but the small p-value suggests a difference in this ratio amongst these genres.   ••

Problems

12.1 The morley data set contains speed-of-light measurements by Michelson and Morley. There were five experiments, each consisting of multiple runs. Perform a one-way analysis of variance to see if each of the five experiments

Lab 1	4.13	4.07	4.04	4.07	4.05
Lab 2	3.86	3.85	4.08	4.11	4.08
Lab 3	4.00	4.02	4.01	4.01	4.04
Lab 4	3.88	3.89	3.91	3.96	3.92

Table 12.2: Production of a chemical.

has the same population mean.

12.2 For the data set Cars93 (MASS) perform a one-way analysis of variance of MPG.highway for each level of DriveTrain. Does the data support the null hypothesis of equal population means?

12.3 The data set female.inc contains income data for females age 15 or over in the United States for the year 2001, broken down by race. Perform a one-way analysis of variance of income by race. Is there a difference in the mean amount earned? What is the p-value? What test did you use and why?

12.4 The data set carsafety contains car-crash data. For several makes of cars the number of drivers killed per million is recorded in Drivers.deaths. The number of drivers of other cars killed in accidents with these cars, per million, is recorded in Other.deaths. The variable type is a factor indicating the type of car.

Perform a one-way analysis of variance of the model Drivers.deaths ~ type. Is there a difference in population means? Did you assume equal variances? Normally distributed populations?

Repeat with an analysis of variance of the model Other.deaths ~ type. Is there a difference in population means?

12.5 The data set hall.fame contains statistics for several Major League Baseball players. Perform a one-way test to see whether the mean batting average, BA, is the same for Hall of Fame members (Hall.Fame.Membership) as for other players.

12.6 A manufacturer needs to outsource the production of a chemical. Before deciding on a laboratory, the manufacturer asks four laboratories to manufacture five batches each. A numeric measurement is assigned to each batch. The data is given in Table 12.2. Perform a one-way analysis of variance to see if there is a difference in the population means. Is the data appropriate for oneway.test? kruskal.test?

12.7 A manufacturer of point-of-sale merchandise tests three types of ENTER-button markings. They wish to minimize wear, as customers get annoyed when the markings on this button wear off. They construct a test of the three

test 1	63	64	95	64	60	85	
test 2	58	56	51	84	77		
test 3	85	79	59	89	80	71	43

Table 12.3: Test scores for three separate exams.

types, and conduct several trials for each. The results, in unspecified units, are recorded in the following table

Type 1	303	293	296	299	298	
Type 2	322	326	315	318	320	320
Type 3	309	327	317	315		

Is there a difference in wear time among the three types? Answer this using a one-way ANOVA.

12.8 Perform a Kruskal–Wallis test on the data in the data set PlantGrowth, where weight is modeled by the factor group. Is there a significant difference in the means?

12.9 Compare the results of a a one-way analysis of variance and the Kruskal test on the data in Table 12.3. Do they produce different p-values? Based on boxplots of the data, would you expect them to?

12.10 An instructor wishing to cut down on cheating makes three different exams and distributes them randomly to her students. After collecting the exams, she grades them. The instructor would like to know whether the three exams are equally difficult. She will decide this by investigating whether the scores have equal population means. That is, if she could give each exam to the entire class, would the means be similar? The test scores are in Table 12.3. Is there a difference in the means?

12.2   Using lm for ANOVA

The mathematics behind analysis of variance is the same as that behind linear regression. Namely, it uses least-squares estimates based on a linear model. As such, it makes sense to unify the approaches. To do so requires a new idea in the linear model.

To illustrate, we begin with an example comprising just two samples, to see how t-tests are handled with the lm function.

● **Example 12.5: ANOVA for two independent samples**
Suppose we have two independent samples from normally distributed pop-

ulations. Let $x_{11}, x_{12}, \ldots, x_{1n}$ record the first and $x_{21}, x_{22}, \ldots, x_{2n}$ the second. Assume the population means are μ_1 and μ_2 and the two samples have a common variance. We may perform a two-sided significance test of $\mu_1 = \mu_2$ with a t-test. We illustrate with simulated data:

```
mu1 <- 0; mu2 <- 1
x <- rnorm(15, mu1); y <- rnorm(15, mu2)
t.test(x, y, var.equal=TRUE)


##
##  Two Sample t-test
##
## data:  x and y
## t = -3.042, df = 28, p-value = 0.005066
## alternative hypothesis: true difference in means is not equal to 0
## 95 percent confidence interval:
##  -1.5308 -0.2987
## sample estimates:
## mean of x mean of y
##   -0.1906    0.7242
```

We see that the p-value is small, as expected.

We can approach this test differently, in a manner that generalizes to the case when there are more than two independent samples. Combine the data into a single data vector, y, and a factor keeping track of which sample, 1 or 2, the data is from. This presumes some ordering on the data after it is stored in y. For example, we can let the first n_1 values be from the first sample and the second n_2 from the last. This is what stack does. Using this order, let $1_1(i)$ be an indicator function that is 1 if the level of the factor for the ith data value is 1. Similarly, define $1_2(i)$. Then we can rewrite our model as

$$y_i = \mu_1 1_1(i) + \mu_2 1_2(i) + \epsilon_i.$$

When the data for the first sample is considered, $1_2(i) = 0$, and this model is simply $y_i = \mu_1 + \epsilon_i$. When the second sample is considered, the other dummy variable is 0, and the model considered is $y_i = \mu_2 + \epsilon_i$.

We can rewrite the model to use just the second indicator variable. We use different names for the coefficients:

$$y_i = \beta_1 + \beta_2 1_2(i) + \epsilon_i.$$

Now when the data for the first sample is considered the model is $y_i = \beta_1 + \epsilon_i$, so β_1 is still μ_1. However, when the second sample is considered, we have $y_i = \beta_1 + \beta_2 + \epsilon_i$, so $\mu_2 = \beta_1 + \beta_2$. That is, $\beta_2 = \mu_2 - \mu_1$. We say that level 1 is a reference level, as the mean of the second level is represented in reference to the first.

It turns out that statistical inference is a little more natural when we pick one of the means to serve as a reference. The resulting model looks just like a linear-regression model where x_i is $1_2(i)$. We can fit it that way and interpret the coefficients accordingly. The model is specified the same way, as with oneway.test, y ~ f, where y holds the data and f is a factor indicating which group the data is for.

To model, first we stack, then we fit with lm.

```
d <- stack(list(x=x, y=y))        # need named list.
res <- lm(values ~ ind, data=d)
short_summary(res)


##               Estimate Std. Error t value Pr(>|t|)
## (Intercept)    -0.191      0.213   -0.90   0.3778
## indy            0.915      0.301    3.04   0.0051 **
## ---
## Signif. codes:  0 '***' 0.001 '**' 0.01 '*' 0.05 '.' 0.1 ' ' 1
```

Look at the variable indy, which means the y part of ind. The marginal t-test tests the null hypothesis that $\beta_2 = 0$, which is equivalent to the test that $\mu_1 = \mu_2$. This is why the t-value of -3.042 coincides (up to a sign and rounding) with $t = -3.04$ from the output of t.test(x,y).

The F-statistic also tests the hypothesis that $\beta_2 = 0$. In this example, it is identical to the marginal t-test, as there are only two samples.

Alternatively, we can try to fit the model using two indicator functions, $y_i = \mu_1 1_1(i) + \mu_2 1_2(i) + \epsilon_i$.

This model is specified in R by dropping the implicit intercept term with a - 1 in the model formula.

```
res <- lm(values ~ ind - 1, data = d)
short_summary(res)


##       Estimate Std. Error t value Pr(>|t|)
## indx   -0.191     0.213    -0.90   0.378
## indy    0.724     0.213     3.41   0.002 **
## ---
## Signif. codes:  0 '***' 0.001 '**' 0.01 '*' 0.05 '.' 0.1 ' ' 1
```

Now the estimates have a clear interpretation in terms of the means, but the marginal t-tests are less useful, as they are testing simply whether the respective means are 0, rather than whether their difference is 0.   ••

Treatment coding for analysis of variance

The point of the above example is to use indicator variables to represent different levels of a factor in the linear model. When there are k levels, $k - 1$

indicator variables are used. For example, if the model is

$$x_{ij} = \mu_i + \epsilon_{ij}, \quad i = 1, \ldots, k, \tag{12.4}$$

then this can be fit using

$$y_i = \beta_1 + \beta_2 1_2(i) + \cdots + \beta_k 1_k(i) + \epsilon_i. \tag{12.5}$$

The mean of the reference level, μ_1, is coded by β_1, and the other β's are differences from that. That is, $\beta_i = \mu_i - \mu_1$ for $i = 2, \ldots, k$.

This method of coding is called *treatment coding* and is used by default in R with unordered factors. It is not the only type of coding, but it is the only one we will discuss.[3]

Treatment coding uses a reference level to make comparisons. This is chosen to be the first level of the factor coding the group. To change the reference level we can use the relevel function in the following manner:

```
f <- relevel(f, ref=...)
```

The argument ref specifies the level we wish to be the reference level.

• **Example 12.6: Child's birth weight and mother's smoking history**
The sampled values in the babies data set contain information on birth weight of a child and whether the mother smoked. The birth weight, wt, is coded in ounces, and smoke is a numeric value: 0 for never, 1 for smokes now, 2 for smoked until current pregnancy, 3 for smoked previously but not now, and 9 if unknown.

To perform an analysis of variance on this data set, we use subset to grab just the desired data and then work as before, only we use factor to ensure that smoking is treated as a factor. First, we see whether there appears to be a difference in the means with a boxplot (Figure 12.4).

```
library(UsingR)
d <- subset(babies, select=c("wt", "smoke"))
plot(wt ~ factor(smoke), data=d,        # notice factor() for boxplot
     main="Birth weight by smoking level")
```

Perhaps the assumption of normality isn't correct, but we ignore that. If the test is valid, it looks like level 1 (smokes now) has a smaller mean. Is this due to sampling? We fit the model as follows:

```
res <- lm(wt ~ factor(smoke), data=d)
summary(res)
```

---

[3]For more detail see ?contrasts and the section on contrasts in the manual *An Introduction to R* that accompanies R.

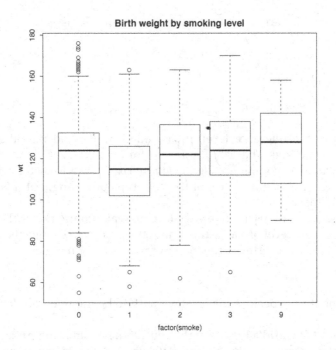

Figure 12.4: Birth weight by smoking history.

```
##
## Call:
## lm(formula = wt ~ factor(smoke), data = d)
##
## Residuals:
##    Min    1Q Median    3Q    Max
## -67.78 -11.11   0.89  11.22  53.22
##
## Coefficients:
##                Estimate Std. Error t value Pr(>|t|)
## (Intercept)     122.778      0.760  161.60  < 2e-16 ***
## factor(smoke)1   -8.668      1.107   -7.83  1.1e-14 ***
## factor(smoke)2    0.307      1.970    0.16     0.88
## factor(smoke)3    1.659      1.904    0.87     0.38
## factor(smoke)9    3.922      5.655    0.69     0.49
## ---
## Signif. codes:  0 '***' 0.001 '**' 0.01 '*' 0.05 '.' 0.1 ' ' 1
##
## Residual standard error: 17.7 on 1231 degrees of freedom
## Multiple R-squared:  0.0588,Adjusted R-squared:  0.0557
```

```
## F-statistic: 19.2 on 4 and 1231 DF,  p-value: 2.36e-15
```

The marginal *t*-tests indicate that the level 1 of the smoke factor is important, whereas the others may not contribute. That is, this is strong evidence that a mother's smoking during pregnancy decreases a baby's birth weight. The treatment coding quantifies this in terms of differences from the reference level of never smoked. The estimate, -8.668, says that the birth weight of a baby whose mother smoked during her pregnancy is predicted to be 8.688 grams less than that of a baby whose mother never smoked. ••

Comparing multiple differences

When analysis of variance is performed with lm, the output of summary displays numerous statistical tests. The *F*-test that is performed has for the null hypothesis that $\beta_2 = \beta_3 = \cdots = \beta_k = 0$ against an alternative that one or more differ from 0. That is, that one or more of the treatments has an effect compared to the reference level. The marginal *t*-tests that are performed are two-sided tests with a null hypothesis that $\beta_i = \beta_1$, one each is done for $i = 2, \ldots, k$. These test whether any of the additional treatments have a different effect from the reference one when controlled by the other variables. However, we may wish to ask other questions about the various parameters. For example, comparisons not covered by the standard output are "Do the β_2 and β_3 differ?" and "Are β_1 and β_2 half of β_3?" We show next how to handle simultaneous pairwise comparisons of the parameters, such as the first comparison.

If we know ahead of time that we are looking for a pairwise difference, then a simple *t*-test is appropriate (as in the case where we are considering just two independent samples). However, if we look at the data and then decide to test whether the second and third parameters differ, then our *t*-test is shaky. Why? Remember that any test is correct only with some probability—even if the models are correct. This means that sometimes they fail, and the more tests we perform, the more likely one or more will fail. When we look at the data, we are essentially performing lots of tests, so there is more chance of failing.

In this case, to be certain that our *t*-test has the correct significance level, we adjust it to include all the tests we can possibly consider. This adjustment can be done by hand with the simple, yet often overly conservative Bonferroni adjustment. This method uses a simple probability bound to ensure the proper significance level.

However, with R it is straightforward to perform Tukey's generally more useful and powerful "honest significant difference" test. This test covers all pairwise comparisons at one time by simultaneously constructing confidence intervals of the type

$$(\bar{y}_i - \bar{y}_j) \pm q^* \sqrt{\frac{1}{2}s^2 \left(\frac{1}{n_i} + \frac{1}{n_j} \right)}. \tag{12.6}$$

The values $\bar{y}_i$ are the sample means for the ith level and q^* is the quantile for a distribution known as the studentized range distribution. This choice of q^* means that all these confidence intervals hold simultaneously with probability $1 - \alpha$.

This procedure is implemented in the TukeyHSD function as illustrated in the next example.

• Example 12.7: Difference in takeoff times at the airport
We investigate the takeoff times for various airlines at Newark Liberty International Airport. As with other busy airports, Newark's is characterized by long delays on the runway due to requirements that plane departures be staggered. Does this effect all the airlines equally? Without suspecting that any one airline is favored, we can perform a simultaneous pairwise comparison to investigate.

First, we massage the data in ewr (UsingR) so that we have two variables: one to keep track of the time and the other a factor indicating the airline.

```
ewr.out <- subset(ewr, subset=inorout=="out", select=3:10)
out <- stack(ewr.out)
names(out) <- c("time","airline")
levels(out$airline)

## [1] "AA" "CO" "DL" "HP" "NW" "TW" "UA" "US"
```

In modeling, the reference level comes from the first level reported by the levels function. This is AA, or American Airlines.

Now plot (the boxplots in Figure 12.5) and fit the linear model as follows:

```
plot(time ~ airline, data=out)
res <- lm(time ~ airline, data=out)
short_summary(res)

##              Estimate Std. Error t value Pr(>|t|)
## (Intercept)  27.0565     0.7204   37.56  < 2e-16 ***
## airlineCO     3.8348     1.0188    3.76  0.00023 ***
## airlineDL    -2.0522     1.0188   -2.01  0.04550 *
## airlineHP     1.5261     1.0188    1.50  0.13595
## airlineNW    -4.0609     1.0188   -3.99  9.8e-05 ***
## airlineTW    -1.6522     1.0188   -1.62  0.10667
## airlineUA    -0.0391     1.0188   -0.04  0.96941
```

```
## airlineUS    -3.8304     1.0188   -3.76  0.00023 ***
## ---
## Signif. codes:  0 '***' 0.001 '**' 0.01 '*' 0.05 '.' 0.1 ' ' 1
```

The boxplots show many differences. Are these differences statistically significant? For now we treat the data as a collection of independent samples (rather than the monthly averages of varying sizes it is) and proceed using the TukeyHSD function.

The TukeyHSD function is a method of aov objects, not lm objects. Some redoing of past work is needed, though the commands are the similar:

```
## trim airlines for display
out.trimmed <- subset(out, subset=airline %in% c("CO", "AA", "NW"))
res.aov.trimmed <- aov(time ~ airline, data=out.trimmed)
TukeyHSD(res.aov.trimmed)

##   Tukey multiple comparisons of means
##     95% family-wise confidence level
##
## Fit: aov(formula = time ~ airline, data = out.trimmed)
##
## $airline
##          diff     lwr     upr p adj
## CO-AA   3.835   1.456   6.213 7e-04
## NW-AA  -4.061  -6.440  -1.682 3e-04
## NW-CO  -7.896 -10.274  -5.517 0e+00

##
res.aov <- aov(time ~ airline, data=out)
plot(TukeyHSD(res.aov), las=2)
```

The output of TukeyHSD is best viewed with the plot of the confidence intervals (Figure 12.5). This is created by calling plot on the output. The argument las=2 turns the tick-mark labels perpendicular to the axes.

Recall the duality between confidence intervals and tests of hypothesis discussed in Chapter 9. For a given confidence level and sample, if the confidence interval excludes a population parameter, then the two-sided significance test of the same parameter will be rejected. Applying this to the Newark airport example, we see several statistically significant differences at the $\alpha = .05$ level, the first few being CO-AA and NW-AA (just visible on the graph shown). ••

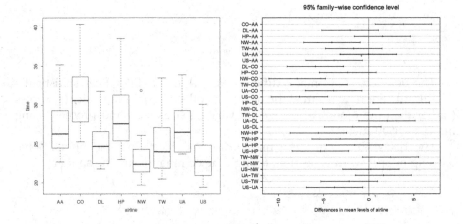

Figure 12.5: Boxplots and plots of confidence intervals given by the Tukey procedure for time it takes to takeoff at Newark Liberty International Airport by airline.

Problems

12.11 The data set MLBAttend (UsingR) contains attendance data for Major League Baseball between the years 1969 and 2000. Use lm to perform a t-test on attendance for the two levels of league. Is the difference in mean attendance significant? Compare your results to those provided by t.test.

12.12 The Traffic (MASS) data set contains data on road deaths in Sweden during 1961 and 1962. An investigation into the effect of an enforced speed limit on the number of traffic fatalities was conducted. The y variable contains the number of deaths for a given day, the year variable is the year of the data, and limit is a factor indicating when the speed limit was enforced.

Use lm to perform a t-test to investigate whether the year has an effect on the number of deaths. Repeat to test whether the variable limit has an effect.

12.13 For the data in Exercise 12.4, perform the one-way ANOVA using lm. Compare to the results of oneway.test.

12.14 For the mtcars data set, perform a one-way analysis of variance of the response variable mpg modeled by cyl, the number of cylinders. Use factor, as cyl is stored as a numeric variable.

12.15 For the mtcars data set, perform a one-way analysis of variance of the response variable mpg modeled by am, which is 0 for automatic and 1 for manual. Use factor, as am is stored as a numeric variable.

12.16 The data set npdb (UsingR) contains malpractice award information. The variable amount contains the amount of a settlement, and the variable year contains the year of the award. We wish to investigate whether the dollar amount awarded was steady during the years 2000, 2001, and 2002.

1. Make boxplots of amount broken up by year. Why is the data not suitable for a one-way analysis of variance?

2. Make boxplots of log(amount) broken up by year. Is this data suitable for a one-way analysis of variance?

3. Perform an analysis of variance of log(amount) by factor(year) for the years 2000, 2001, and 2002. Is the null hypothesis of no difference in mean award amount reasonable given this data?

12.17 Perform the Tukey procedure on the data set morley after modeling Speed by expt. Which differences are significant? Do they include all the ones flagged by the marginal *t*-tests returned by lm on the same model?

12.18 The carsafety (UsingR) data set shows a difference in means through an analysis of variance when the variable Other.deaths is modeled by type. Perform the Tukey HSD method to see what pairwise differences are flagged at a 95% confidence level. What do you conclude?

12.19 The InsectSprays data set contains a variable count, which counts the number of insects and a factor spray, which indicates the treatment given.

First perform an analysis of variance to see whether the treatments make a difference. If so, perform the Tukey HSD procedure to see which pairwise treatments differ.

12.3 ANCOVA

An analysis of covariance (ANCOVA) is the term given to models where both categorical and numeric variables are used as predictors. Performing an ANCOVA in R is also done through lm.

• **Example 12.8: Birth weight by mother's weight and smoking history**
In Example 12.2 we performed an analysis of variance of a baby's birth weight modeled by whether the mother smoked. In this example, we also regress on the numeric measurement of the mother's weight. First we make a plot, marking the points with different characters depending on the value of smoke. As smoke is stored as a numeric variable, the different plot symbols for those numbers are used.

```
babes <- subset(babies, subset = wt1 < 800)
plot(wt ~ wt1, data = babes, pch=as.numeric(smoke))
```

The plotted data in Figure 12.6 indicates a possible linear relationship. The analysis of covariance model, fit next, is essentially the model

$$\text{birth weight} = \beta_1 + \beta_2 \text{ mom's weight} + \beta_3 1_{\text{mom smokes now}}$$

This model is a parallel-lines model. For those mothers who don't smoke, the intercept is given by β_1; for those who do, the intercept is $\beta_1 + \beta_3$. The slope is given by β_2. The actual model we fit is different, as there are four levels to the smoke variable, so there would be three indicator variables, each indicating a difference in the intercept.

In R, we fit the model as follows, using factor to coerce smoke to be a factor:

```
res <- lm(wt ~ wt1 + factor(smoke), data = babes)
summary(res)

##
## Call:
## lm(formula = wt ~ wt1 + factor(smoke), data = babes)
##
## Residuals:
##     Min      1Q Median     3Q    Max
## -68.93 -10.90   0.44  11.01  52.68
##
## Coefficients:
##                 Estimate Std. Error t value Pr(>|t|)
## (Intercept)     107.0674     3.2642   32.80  < 2e-16 ***
## wt1               0.1204     0.0245    4.93 9.6e-07 ***
## factor(smoke)1   -8.3971     1.1246   -7.47 1.6e-13 ***
## factor(smoke)2    0.7944     1.9974    0.40    0.69
## factor(smoke)3    1.2550     1.9112    0.66    0.51
## factor(smoke)9    2.8683     5.6452    0.51    0.61
## ---
## Signif. codes:  0 '***' 0.001 '**' 0.01 '*' 0.05 '.' 0.1 ' ' 1
##
## Residual standard error: 17.7 on 1194 degrees of freedom
## Multiple R-squared:  0.0775,Adjusted R-squared:  0.0736
## F-statistic: 20.1 on 5 and 1194 DF,  p-value: <2e-16
```

We read this output the same way we read the output of any linear regression. For each coefficient, the marginal t-test of $\beta_i = 0$ against a two-sided alternative is performed. Three variables are flagged as highly significant. The third one for the variable factor(smoke)1 says that the value of this coefficient, -8.3971, is statistically different from 0. This value is an estimate of the

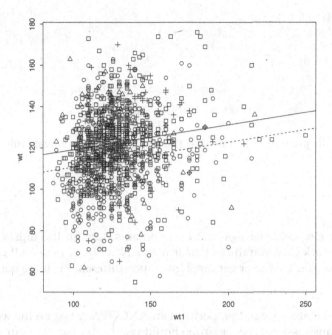

Figure 12.6: Parallel-lines model showing reference slope and slope for smokers (dashed line).

difference between the intercept for the data of nonsmoking mothers (level 0) and the data of mothers who answered "smokes now" (level 1).

We plot the data with two different but parallel regression lines in Figure 12.6.

```
plot(wt ~ wt1, pch=as.numeric(smoke), data=babes)
abline(107.0674, 0.1204)
abline(107.0674 - 8.3971, 0.1204, lty=2)
```

The last line of the output of summary(res) shows that the F-test is rejected. This is a test of whether all the coefficients except the intercept are 0. A better test would be to see whether the additional smoke variable is significant once we control for the mother's weight. This is done using anova to compare the two models.

```
res.1 <- lm(wt ~ wt1, data = babes)
anova(res.1, res)

## Analysis of Variance Table
##
```

```
## Model 1: wt ~ wt1
## Model 2: wt ~ wt1 + factor(smoke)
##   Res.Df    RSS Df Sum of Sq    F Pr(>F)
## 1   1198 394572
## 2   1194 372847  4     21725 17.4 7e-14 ***
## ---
## Signif. codes:  0 '***' 0.001 '**' 0.01 '*' 0.05 '.' 0.1 ' ' 1
```

The small p-value indicates that the additional term is warranted.   ••

Problems

12.20 The nym.2002 (UsingR) data set contains data on the finishers of the 2002 New York City Marathon. Perform an ANCOVA of time on the numeric variable age and the factor gender. How much difference is there between the genders?

12.21 For the mtcars data set, perform an ANCOVA of mpg on the weight, wt, and the transmission type, am. You should use factor(am) in your model to ensure that this variable is treated as a factor. Is the transmission type significant?

12.22 Perform an ANCOVA for the babies (UsingR) data set modeling birth weight (wt) by gestation (gestation), mother's weight (wt1), mother's height (ht), and mother's smoking status (smoke).

12.23 From the kid.weights (UsingR) data set, the body mass index (BMI) can be computed by dividing the weight by the height squared in metric units. The following will add a BMI variable:

```
kid.weights$BMI <- with(kid.weights, (weight/2.2)/(height*2.54/100)^2)
```

Model the BMI by the age and gender variables. This is a parallel-lines model. Which variables are significant? Use the partial F-test to find the preferred model. Does this agree with the output of stepAIC?

12.24 The cfb (UsingR) data set contains information on consumer expenses. In particular, INCOME contains income figures, EDUC is the number of years of education, and AGE is the age of the participant. Perform an ANCOVA modeling log(INCOME + 1) by AGE and EDUC. You will need to force EDUC to be a factor. Are both variables significant?

12.25 The data set normtemp (UsingR) contains body temperature and heart rate (hr) for 65 randomly chosen males and 65 randomly chosen females (marked by gender with 1 for males and 2 for females). Perform an AN-COVA modeling temperature by heart rate with gender treated as a factor.

12.4 Two-way ANOVA

Like the regression model, the ANOVA method generalizes to more than one predictor variable. Two-way ANOVA is the case when there are two. Higher order ANOVA models are possible, but we focus on the two-way case in the following.

- **Example 12.9: The perfect snack**

It seems that many Americans always have room in their diets for snack food. Designing the perfect snack food is a science that the major snack-food makers have perfected, but they are unlikely to share their secrets. Instead, people need to reverse engineer. For example, an infographic in the *New York Times*[4] shows an analysis by Steven A. Witherly of what makes the Dorito so appealing. One major key is to produce an item with "vanishing caloric density," which means the food tricks the brain into thinking the calories have disappeared and hence satiation is delayed. Fat-laden snacks have this effect and it is known that the percentage of fat is important. Additionally, combinations of sweet, salty, and umami (the latter provided by the additive MSG or garlic) are used to provide "flavor burst."

Without the benefit of research, an experimental sort could try various combinations to see which is preferred. For example, suppose a new snack is made which has the magical 50% fat content, what would be a good amount of salt and a good amount of sugar to use?   ••

The addition of an extra factor in the two-way model introduces the possibility of *interactions* between the two factors. In the absence of interactions, the two-way additive model for the response can be expressed as:

$$x_{ijk} = \alpha_i + \beta_j + \epsilon_{ijk},$$

where i indicates the level of the first factor, j indicates the level of the second factor, and k allows for more than one observation for the combination of level i and level j. When there is more than one, the data is said to be replicated.

The coefficients above are termed the *main effects* and have a direct interpretation. For example, the difference $\beta_j - \beta_{j'}$ is interpreted as the difference in the predicted mean response from changing from level j' to level j for the second factor assuming the first factor doesn't change.

Interactions are modeled with terms that take into account both i and j simultaneously:

---

[4]http://www.nytimes.com/interactive/2013/10/01/dining/nacho-graphic.html.

$$x_{ijk} = \alpha_i + \beta_j + \gamma_{ij} + \epsilon_{ijk}.$$

The interaction term, γ_{ij}, makes interpretation of the main effects difficult by themselves, as the change in mean response predicted by the model when one factor is held steady and another manipulated depends on the level of the factor being held steady.

Interaction plots

Though we will soon see a formal statistical test for the possible presence of an interaction, it is often best to first look graphically for interactions. For this, the *interaction plot* is widely used.

For a two-way analysis of variance there are three variables: a numeric response and two categorical predictors. To squeeze all three onto one graphic, one of the factors is selected as the trace factor. Different lines will be drawn for each level of this factor. Fix a level, for now, of the trace factor. For each level of the main factor, the mean of the data where both levels occur is plotted as a point. These points are then connected with a line segment. Repeat for the other levels of the trace factor. If the line segments for each level of the trace factor are roughly parallel—indicating a similar response regardless of the level of the trace factor—then no interaction is indicated. If the lines differ dramatically, then an interaction is indicated.

This graphic is made with the function interaction.plot. The template is

```
interaction.plot(f, trace.factor, response, legend=TRUE)
```

The response variable is the third positional argument, the first specifies the main factor, and the second is the trace.factor. By default, a legend will be drawn indicating the levels of the trace factor.

In the snacks (UsingR) data set, there is data on snack food taken from a large survey performed by the USDA. Is there an interaction between sugary amount and saturated fat (a bad fat) when using both to model the energy content (calories) of each snack in the survey?

Figure 12.7 shows the interaction plot. To some broad degree the three lines are parallel, though clearly they are not precisely parallel. A formal test will indicate if such deviations are consistent with an assumption of random variation. The interaction suggested by graph is interpreted as for medium sugary foods, the effect of going from low fat to medium fat is more pronounced than for high-sugar foods, where the effect is pronounced when comparing medium fat to high fat.

The plot was generated with the following commands:

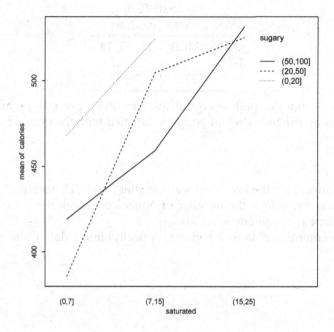

Figure 12.7: Interaction plot where snack food calories are modeled by amount of saturated fat and the sugar amount (the trace factor). An interaction is suggested, though the appearance of being not-parallel may be due to sampling variation.

```
snacks$sugary <- cut(snacks$sugar, c(0, 20, 50, 100))
snacks$saturated <- cut(snacks$fat_sat, c(0, 7, 15, 25))
with(snacks, interaction.plot(saturated, sugary, response=calories))
```

Fitting a two-way ANOVA

Before analyzing this model, we incorporate it into our linear-model picture using dummy variables. We follow the same coding (treatment coding) in terms of indicators as the one-way case. First, the data should be in long format with one variable (say x) holding the response values and two factors (say f1 and f2) to indicate which level of the two predictor variables correspond to a given response. For each factor, a reference level is chosen.

For example, Table 12.4 is simulated data inspired by a cross-sectional study on whether the family food environment exerts influences on the young children's eating. The actual study [37] used a fifty-six question survey, our data has a dramatically reduced set of factors: "satisfied with how

TV	Satisfied	
	Yes	No
0–1	0, 0	16, 15, 14
1–5	11	26, 27
5+	23, 23	36, 43

Table 12.4: Amount of high-energy drink consumed per wek by amount of television watched and whether parent is satisfied with the number of shared evening meals.

often my family eats the evening meal together" and "TV viewing" amount. The response variable is the number of ounces per week the $5 - 6$ year-old child consumes of high-energy drinks.

We can enter this data (somewhat cryptically) into a data frame with:

```
d <- data.frame(
  x =c(0, 0, 16, 15, 14, 11, 26, 27, 23, 23, 36, 43),
  satisfied=c("yes", "no")[c(1,1,2,2,2,1,2,2,1,1,2,2)],
  tv=rep(c("0-1", "1-5", "5+"), c(5,3,4))
  )
```

The basic representation of the ANOVA model uses a reference level (the first by default) and includes a term of $1_{leveli}(x)$ for each of the other levels of the first factor and corresponding ones for the second. In the above then there are two indicator values for the tv variable and one for the satisfied variable. This means there are four estimated coefficients in the additive model, taking into account the grand mean (or intercept term).

In the model with an interaction there are an additional $2 = (m - 1) \cdot (n - 1)$ coefficients for the interaction terms, each one multiplying a term of the type $1_{leveli}(x) \cdot 1_{levelj}(x)$.

In this example, as there is replication, there are enough degrees of freedom to estimate the terms and the error. For data without replication, one can estimate the interactions, but cannot perform any inference on the estimates.

Fitting the two-way ANOVA model in R involves specifying the model using the formula notation. The additive model can be specified with +, the additive model with interaction can be specified with * (or tv+satisfied + tv:satisfied, where : is just the interaction). For our data above, we have:

```
res.add <- lm(x ~ tv + satisfied, data=d)
res.int <- lm(x ~ tv * satisfied, data=d)
```

The various coefficients are estimated. Looking at the summary of the model with interaction we see:

```
summary(res.int)
```

```
##
## Call:
## lm(formula = x ~ tv * satisfied, data = d)
##
## Residuals:
##    Min    1Q Median    3Q    Max
## -3.500 -0.125  0.000  0.125  3.500
##
## Coefficients:
##                     Estimate Std. Error t value Pr(>|t|)
## (Intercept)            15.00       1.23   12.25 1.8e-05 ***
## tv1-5                  11.50       1.94    5.94 0.00102 **
## tv5+                   24.50       1.94   12.65 1.5e-05 ***
## satisfiedyes          -15.00       1.94   -7.75 0.00024 ***
## tv1-5:satisfiedyes     -0.50       3.24   -0.15 0.88243
## tv5+:satisfiedyes      -1.50       2.87   -0.52 0.62022
## ---
## Signif. codes:  0 '***' 0.001 '**' 0.01 '*' 0.05 '.' 0.1 ' ' 1
##
## Residual standard error: 2.12 on 6 degrees of freedom
## Multiple R-squared:  0.985,Adjusted R-squared:  0.973
## F-statistic: 80.7 on 5 and 6 DF,  p-value: 2.03e-05
```

The term tv1-5:satisfiedyes is an interaction term. Both are not marked with significance stars. This begs the question, is an interaction present? The interaction plot (not shown) indicates this is the case:

```
with(d, interaction.plot( tv, satisfied, x))
```

As the two models above are nested models, the partial F-test can be used, as implemented by anova:

```
anova(res.int, res.add)
```

```
## Analysis of Variance Table
##
## Model 1: x ~ tv * satisfied
## Model 2: x ~ tv + satisfied
##   Res.Df  RSS Df Sum of Sq    F Pr(>F)
## 1      6 27.0
## 2      8 28.2 -2     -1.24 0.14   0.87
```

The null hypothesis is that all coefficients related to the interaction are 0, and the large p-value gives no reason to doubt this.

It makes sense to look at the additive model. We can interpret the main effects in terms of the estimated coefficients:

```
short_summary(res.add)

##                  Estimate Std. Error t value Pr(>|t|)
## (Intercept)        15.26        0.95   16.05 2.3e-07 ***
## tv1-5              11.29        1.37    8.22 3.6e-05 ***
## tv5+               23.81        1.26   18.82 6.6e-08 ***
## satisfiedyes      -15.64        1.11  -14.09 6.2e-07 ***
## ---
## Signif. codes:  0 '***' 0.001 '**' 0.01 '*' 0.05 '.' 0.1 ' ' 1
```

Here we see all coefficients are significant. Of more interest is interpreting the coefficients. The grand mean is approximately 15. This is the predicted mean for the reference levels which with our encoding are $0 - 1$ hours of TV and not satisfied with shared evening meals. The coefficient for satisfiedyes has a negative sign, as the difference in predicted mean consumption for those satisfied with their shared meals is 15 ounces *less* per week.

• **Example 12.10: Snack food**
Returning to the snack food data, snacks, we look to see if the interaction suggested in Figure 12.7 is statistically significant.

```
res.int <- lm(calories ~ saturated * sugary, snacks)
res.add <- lm(calories ~ saturated + sugary, snacks)
anova(res.int, res.add)

## Analysis of Variance Table
##
## Model 1: calories ~ saturated * sugary
## Model 2: calories ~ saturated + sugary
##   Res.Df    RSS Df Sum of Sq    F Pr(>F)
## 1    103 253984
## 2    106 274339 -3    -20355 2.75  0.046 *
## ---
## Signif. codes:  0 '***' 0.001 '**' 0.01 '*' 0.05 '.' 0.1 ' ' 1
```

The small p-value is just significant at the $\alpha = 0.05$ level. The interaction makes the coefficients found for the main effects harder to interpret, but overall, they support that intuition that more fat and/or more sugar tends to be found in products with higher calories.   ••

Blocking variables

A blocking variable is one which is believed to influence the response variable, but how it does so is not the main question of interest. Sometimes, this may be called a *nuisance variable*. We have seen the use of a blocking variable already with the paired *t*-test. For example, the data set (shoes (MASS)) contains data on shoe wear. As the variability on shoe wear depends a lot on the person, a simple examination of shoe wear between two different sole types can be masked due to other factors. For this data, the experimental design accounted for the variation due to individuals by assigning at random one sole type to a foot of one person, and left the other foot to the second sole type. This allowed the comparison of shoe wear by looking at the differences between paired feet:

```
library(MASS)
out <- t.test(shoes$A, shoes$B, paired=TRUE)
out$p.value

## [1] 0.008539
```

The same *p*-value is returned by considering this as a two-way ANOVA model, where the second variable records the person:

```
d <- stack(shoes)
names(d) <- c("value", "sole_type")
d$person <- gl(10, 1, 20, labels=paste("P",1:10, sep=""))
xtabs(value ~ sole_type + person, data=d)

##           person
## sole_type  P1   P2   P3   P4   P5   P6   P7   P8   P9  P10
##         A 13.2  8.2 10.9 14.3 10.7  6.6  9.5 10.8  8.8 13.3
##         B 14.0  8.8 11.2 14.2 11.8  6.4  9.8 11.3  9.3 13.6

res <- aov(value ~ sole_type + person, data=d)
summary(res)

##             Df Sum Sq Mean Sq F value  Pr(>F)
## sole_type    1    0.8    0.84   11.2  0.0085 **
## person       9  110.5   12.28  163.8 6.9e-09 ***
## Residuals    9    0.7    0.07
## ---
## Signif. codes:  0 '***' 0.001 '**' 0.01 '*' 0.05 '.' 0.1 ' ' 1
```

The question of interest here is there a difference in wear between the different sole types. The small *p*-value might lead one to simply say "yes." The person variable is not of interest, we just want to account for the variability

due to different people. If we had not, then the data would have been found
to be not statistically significant:

```
summary(aov(value ~ sole_type, data=d))

##                Df Sum Sq Mean Sq F value Pr(>F)
## sole_type       1    0.8    0.84    0.14   0.72
## Residuals      18  111.2    6.18
```

• **Example 12.11: Blocking**
A student experimenter wishes to see if soft drink consumption leads to poor
health measures. They understand that there are many possible confounding
variables involved: prior consumption habits, current health measures, gender, ... As such, they limit their cohort to male candidates with a BMI less
than 25 that currently drink soft drinks rarely or never.

They are also aware of some studies that indicate there may be ethnic
differences. In their recruitment processes, it was felt this would be impolitic
to screen by this. As such, they include this variable as a blocking variable.

The student selects three treatments: 24 ounces per day of additional bot-
tled water consumption, 24 ounces per day of additional consumption of soft
drinks, 24 ounces per day of additional consumption of vitamin-enriched
drinks. The latter two having similar caloric content, though the last one
promises health benefits.

There were 12 persons recruited from 4 ethnicities. For each ethnic group,
the assignment of treatment to case was randomized. The measure was percentage change in BMI during a two-week test.

The collected data is entered into R with the following commands:

```
d <- data.frame(
  percent_change = c(1.3, -0.7,  0.3, -1.4, 0.5 ,0.9 ,1.6, 1.2,
    0.3, 1.8, 1.7, 1.3),
  ethnicity = gl(4, 1, 12, labels=paste("Ethnic", 1:4)),
  trt = gl(3, 4, 12, labels=c("Ctrl", "Soda", "Vitamin"))
  )
xtabs(percent_change ~ trt + ethnicity, data=d)

##           ethnicity
## trt        Ethnic 1 Ethnic 2 Ethnic 3 Ethnic 4
##   Ctrl          1.3     -0.7      0.3     -1.4
##   Soda          0.5      0.9      1.6      1.2
##   Vitamin       0.3      1.8      1.7      1.3
```

There is insufficient data to consider a model with interactions, as there is no replication.[5]

The additive model is fit with

```
res <- aov(percent_change ~ trt + ethnicity, data=d)
coef(res)[1:3]                          # only intercept, trt

## (Intercept)      trtSoda  trtVitamin
##     -0.1583       .1.1750      1.4000
```

The coefficients for ethnicity are ignored (and not shown) but the main effects for the treatments indicate at first glance that both the soda group and vitamin group had an increase in their BMI, with the vitamin group being the largest. In contrast, the control group appears to have dropped a bit. The question of whether these differences are statistically significant is answered with this command:

```
summary(res)

##             Df Sum Sq Mean Sq F value Pr(>F)
## trt          2   4.52   2.261    2.63   0.15
## ethnicity    3   1.07   0.358    0.42   0.75
## Residuals    6   5.15   0.859
```

The p-value of 0.15 indicates that controlling for ethic differences, the differences in treatment groups for this small trial are not statistically significant. Looking at the output of plot(TukeyHSD(res, "trt")) will demonstrate the same, with all intervals straddling 0.   ••

Problems

12.26 A politician's campaign manager is interested in the effects of television and Internet advertising. She surveys 18 people and records changes in likability after a small advertising campaign. Additionally, she records the amount of exposure her subjects have to the ad campaigns. The data is in Table 12.5.

Use an analysis of variance to investigate the following questions:

1. Is there any indication that web advertising alone is effective?

2. After controlling for television exposure, is there any indication that web advertising is effective?

---

[5]The model with replication has $1 + (3 - 1) + (4 - 1) + (3 - 1)(4 - 1) = 12$ parameters to estimate with only 12 data points to do so. As such, there is insufficient data to estimate the error terms.

TV ad exposure (viewings)		0			1–2			3+		
Web exposure	N	−1	−4	0	−1	4	1	6	2	7
	Y	1	2	2	7	5	2	3	6	1

Table 12.5: Change in likability of politician.

12.27 The grip (UsingR) data set contains more data than is used in Example 12.1. The data is from four skiers instead of one. You can view the data in a convenient manner with the command

```
ftable(xtabs(UBP ~ person + replicate + grip.type, data=grip))
```

Perform a two-way analysis of variance on the data. Check first to see whether there are any interactions, then see whether the difference in skier or grip has an effect.

12.28 In the data set mtcars the variables mpg, cyl, and am indicate the miles per gallon, the number of cylinders, and the type of transmission, respectively. Perform a two-way ANOVA modeling mpg by the cyl and am, each treated as categorical data.

Is there an indication of an interaction? Do both the number of cylinders and the type of transmission make a difference?

12.29 The data set ToothGrowth has measurements of tooth growth (len) of guinea pigs for different dosages of Vitamin C (dose) and two different delivery methods (supp).

Perform a two-way analysis of variance of tooth growth modeled by dosage and delivery method. First, fit the full model including interactions and use the F-test to compare this with the additive model.

12.30 The data set OrchardSprays contains measurements on the effectiveness of various sprays on repelling honeybees. The variable decrease measures effectiveness of the spray, treatment records the type of treatment, and rowpos records the row in the field the measurement comes from.

Make an interaction plot of the mean of decrease with treatment as a trace factor. Then fit the additive analysis-of-variance model and the model with interaction. Compare the two models using anova. Is the interaction model suggested by the results of the modeling?

12.31 What does R output when there is not enough data to estimate the parameters and an error term? We check with a simple example. In checking a comment that everything is better with butter on it, a student asks four people to rate the four combinations of bread and corn with and without butter. The data collected is:

```
d <- data.frame(
  rating=c(8, 6, 8, 4),
  food=gl(2, 2, 4, labels=c("bread", "corn")),
  butter=gl(2, 1, 4, labels=c("yes", "no")))
xtabs(rating ~ butter + food, d)

##          food
## butter bread corn
##     yes     8    8
##      no     6    4

with(d, interaction.plot(butter, food, rating)) # not shown
```

Look at the summary of the multiplicative model

```
summary(lm(rating ~ butter * food, d))
```

Do the coefficients get estimated? What is the issue then?

13

Extensions of the linear model

The linear-regression ideas are building blocks for many other statistical models. The R project's archive (CRAN, http://cran.r-project.org) warehouses over 5,000 add-on packages to R, many of which implement extensions to the linear-regression model covered in the last two chapters. In this chapter, we look at two extensions: logistic-regression models and nonlinear models. Our goal is to illustrate that most of the techniques used for linear models carry over to these (and other) models.

The logistic-regression model covers the situation where the response variable is a binary variable. Logistic regression, which is a particular case of a generalized linear model, arises in several areas, including, for example, analyzing survey data. The nonlinear models we use a function to describe the mean response that is not linear in the parameters.

13.1 Logistic regression

A binary variable is one that can have only two values, "success" or "failure," often coded as 1 or 0. In the ANOVA model we saw that we can use binary variables as predictors in a linear-regression model by using factors. But what if we want to use a binary variable as a response variable?

• Example 13.1: Spam
Junk e-mail, or spam, is a real nuisance, but it must make some business sense, as the Internet is flooded with it. Let's look at the situation from the spammer's perspective.

The spammer's problem is that very few people will open spam. How to entice someone to do so? Is it worth the expense of buying an e-mail list that includes names? Does the subject line make a difference? Imagine a test is done in which 5,000 e-mails are sent out in four different ways. The subject heading on some includes a first name, on some an offer, on some both, and on some neither. The number that are opened by the recipient is measured by an embedded image in the e-mail body that can be tracked via a web server.

If Table 13.1 contains sample data, what can we say about the importance of including a name or an offer in the subject heading? ••

		Offer in subject	
		yes	no
First name	yes	20 of 1,250	15 of 1,250
in subject	no	17 of 1,250	8 of 1,250

Table 13.1: Number of spam e-mails opened.

For simplicity, assume that we have two variables, x and y, where y is a binary variable coded as a 0 or 1. For example, 1 could mean a spam message was opened. If we try to model the response with $y_i = \beta_0 + \epsilon_i$ or $y_i = \beta_0 + \beta_1 x_i + \epsilon_i$, then, as y_i is either 0 or 1, the ϵ_i can't be an *i.i.d.* sample from a normal population. Consequently, the linear model won't apply. As having only two answers puts a severe restriction on the error term, instead the probability of success is modeled.

Let $\pi_i = P(y_i = 1)$. Then π_i is in the range 0 to 1. We might try to fit the model $\pi_i = \beta_0 + \beta_1 x_i + \epsilon_i$, but again the range on the left side is limited, whereas that on the right isn't. Even if we restrict our values of the x_i, the variation of the ϵ_i can lead to probabilities outside of $[0,1]$.

Let's change tack. For a binary random variable, the probability is also an expected value. That is, after conditioning on the value of x_i, we have $E(y_i|x_i) = \pi_i$. In the simple linear model we called this $\mu_{y|x}$, and we had the model $y_i = \mu_{y|x} + \epsilon_i$. Interpreting this differently will let us continue. We mentioned that the assumption on the error can be viewed two ways. Either assuming the error terms, the ϵ_i values, are a random sample from a mean 0 normally distributed population, or, equivalently that each data point y_i is randomly selected from a Normal$(\mu_{y|x}, \sigma)$ distribution independently of the others. Thus, we have the following ingredients in simple linear regression:

- The predictors enter in a linear manner through $\beta_0 + \beta_1 x_1$.

- The distribution of each y_i is determined by the mean, $\mu_{y|x}$, and some scale parameter σ.

- There is a relationship between the mean and the linear predictors ($\mu_{y|x} = \beta_0 + \beta_1 x_1$).

The last point needs to be changed to continue with the binary regression model. Let $\eta = \beta_0 + \beta_1 x_1$. Then the change we make is to assume that η can be transformed to give the mean by some function $m()$ via $\mu_{y|x} = m(\eta)$, which can be inverted to yield back $\eta = m^{-1}(\mu_{y|x})$. The function $m()$ is called a *link function*, as it links the predictor with the mean.

The logistic function $m(x) = e^x/(1 + e^x)$ is often used (see Figure 13.1), and the corresponding model is called *logistic regression*. For this, we have

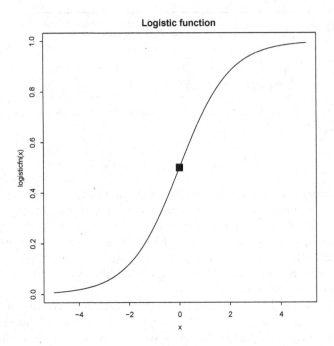

Figure 13.1: Graph of logistic function, $m(x) = e^x/(1 + e^x)$. The inflection point is marked with a square.

$$\pi_i = m(\beta_0 + \beta_1 x_i) = \frac{e^{\beta_0 + \beta_1 x_i}}{1 + e^{\beta_0 + \beta_1 x_i}}.$$

The logistic function turns values between $-\infty$ and ∞ into values between 0 and 1, so the numbers specifying the probabilities will be between 0 and 1. When $m()$ is inverted we have

$$\log(\frac{\pi_i}{1 - \pi_i}) = \beta_0 + \beta_1 x_i. \qquad (13.1)$$

This log term is called the *log-odds ratio*. The odds associated to some probability are $p/(1 - p)$, which is evident if we understand that an event having odds a to b means that in $a + b$ *i.i.d.* trials we expect a wins. Thus the probability of success should be $a/(a + b)$. Reversing, if the probability of success is $a/(a + b)$, then the ratio becomes $(a/(a + b))/(1 - a/(a + b))$ or a/b, which is the ratio of the odds. The expression $e^{\log\text{-odds ratio}}$ will give the odds.

To finish the model, we need to specify the distribution of y_i. It is Bernoulli with success probability π_i, or more compactly, Bernoulli$(m(\beta_0 + \beta_1 x_i))$.[1]

Generalized linear models

Logistic regression is an example of a *generalized linear model*. The key ingredients are as above: a response variable y and some predictor variables $x_1, x_2, \ldots, x_p$. The predictors enter into the model via a single linear function:

$$\eta = \beta_0 + \beta_1 x_1 + \cdots + \beta_p x_p.$$

The mean of y given the x values is related to η by an invertible link function $m()$ as $\mu = m(\eta)$ or $m^{-1}(\mu) = \eta$. Finally, the distribution of y is given in terms of its mean and, perhaps, a scale parameter such as σ.

Thus, the model is specified by the coefficients β_i, a link function $m()$, and a probability distribution that may have an additional scale parameter.

Fitting the model using `glm`

Generalized linear models are fit in R using the `glm` function. Its usage is similar to that of `lm`, except that we need to specify the probability distribution and the link function. A template for usage is

```
res <- glm(formula, family=..., data=...)
```

The formula is specified as though it were a linear model. The argument `family` allows us to specify the distribution and the link. Details are in the help page `?family` and in the section "Generalized linear models" in the manual *An Introduction to R* accompanying R. We will use only two: the one for logistic regression and one to compare the results with simple linear regression.

For logistic regression the argument is specified by `family="binomial"`, as the default link function is what we want. For comparison to simple linear regression, the link function is just an identity, and the family is specified as `family="gaussian"`.[2]

As an illustration, let's compare using `glm` and `lm` to analyze a linear model. We will use simulated data so we already "know" the answer.

● **Example 13.2: Comparing `glm` and `lm`**
We first simulate data from the model that y_i has a Normal$(x_{1i} + 2x_{2i}, \sigma)$ distribution and fit using `lm`:

---

[1]This should be compared to the description of the simple regression model through Normal$(\beta_0 + \beta_a x_i, \sigma)$.

[2]Gaussian is a mathematical term named for Carl Gauss that describes the normal distribution.

```
x1 <- rep(1:10, 2)
x2 <- rchisq(20, df=2)
y <- rnorm(20, mean=x1 + 2*x2, sd=2)
res.lm <- lm(y ~ x1 + x2)
short_summary(res.lm)

##               Estimate Std. Error t value Pr(>|t|)
## (Intercept)    -0.490      1.074   -0.46     0.65
## x1              1.065      0.146    7.30  1.2e-06 ***
## x2              1.990      0.199   10.00  1.6e-08 ***
## ---
## Signif. codes:  0 '***' 0.001 '**' 0.01 '*' 0.05 '.' 0.1 ' ' 1
```

Both the coefficients for x1 and x2 are flagged as significantly different from 0 in the marginal t-tests.

The above can all be done using glm. The only difference in the function call is that the modeling involves specifying the family argument. We show all the output below.

```
res.glm <- glm(y ~ x1 + x2, family="gaussian")
summary(res.glm)

##
## Call:
## glm(formula = y ~ x1 + x2, family = "gaussian")
##
## Deviance Residuals:
##     Min      1Q  Median      3Q     Max
##   -3.29   -1.65    0.24    1.23    2.30
##
## Coefficients:
##               Estimate Std. Error t value Pr(>|t|)
## (Intercept)    -0.490      1.074   -0.46     0.65
## x1              1.065      0.146    7.30  1.2e-06 ***
## x2              1.990      0.199   10.00  1.6e-08 ***
## ---
## Signif. codes:  0 '***' 0.001 '**' 0.01 '*' 0.05 '.' 0.1 ' ' 1
##
## (Dispersion parameter for gaussian family taken to be 3.295)
##
##     Null deviance: 465.965  on 19  degrees of freedom
## Residual deviance:  56.014  on 17  degrees of freedom
## AIC: 85.36
##
## Number of Fisher Scoring iterations: 2
```

The same coefficients are found. This is not surprising, but technically a different method is used. For each coefficient, a two-sided significance test is done with null hypothesis that the value is 0. For this model, the results are identical, as with lm. No information about the F-statistic is given, as the theory does not apply here in general. Rather, the AIC is given. Recall that this could be used for model selection with models having lower values being preferred. ••

Now we fit a logistic model.

• **Example 13.3: Premature babies**

It is well known that risk factors associated with premature births include smoking and maternal malnutrition. Do we find this to be the case with the data in the babies (UsingR) data set?

We'll need to manipulate the data first. First we extract just the variables of interest, using the subset argument to eliminate the missing values.

```
babies.prem = subset(babies,
  subset= gestation < 999 & wt1 < 999 & ht < 99 & smoke < 9,
  select=c("gestation","smoke","wt1","ht"))
```

A birth is considered premature if the gestation period is less than 37 full weeks.

```
babies.prem$preemie = with(babies.prem, as.numeric(gestation < 7*37))
table(babies.prem$preemie)

##
##    0    1
## 1079   96
```

For glm with binomial models the response variable can be numeric, as just defined, or a factor (the first level is "failure," the others are "success").

We will use the body mass index (BMI) as a measure of malnutrition, though expect that this might be a poor proxy. The BMI is the weight in kilograms divided by the height in meters squared. If there is some dependence, we will investigate further.

```
babies.prem$BMI = with(babies.prem, (wt1/2.2) / (ht*2.54/100)^2)
hist(babies.prem$BMI)                    # looks okay
```

We can now model the variable preemie by the levels of smoke and the variable BMI. This is similar to an ANCOVA, except that the response variable is binary.

```
res <- glm(preemie ~ factor(smoke) + BMI, family=binomial,
           data=babies.prem)
short_summary(res)

##              Estimate Std. Error z value Pr(>|z|)
## (Intercept)   -3.4246     0.7116   -4.81  1.5e-06 ***
## factor(smoke)1  0.1935     0.2357    0.82     0.41
## factor(smoke)2  0.3137     0.3890    0.81     0.42
## factor(smoke)3  0.1011     0.4050    0.25     0.80
## BMI             0.0401     0.0304    1.32     0.19
## ---
## Signif. codes:  0 '***' 0.001 '**' 0.01 '*' 0.05 '.' 0.1 ' ' 1
```

None of the variables are flagged as significant. This indicates that the model with no effects is, perhaps, preferred. (The sampling distribution under the null hypothesis is different from the previous example, so the column gets marked with "z value" as opposed to "t value.") We check which model is preferred by the AIC using stepAIC from the MASS package.

```
library(MASS)
stepAIC(res, trace=0)

##
## Call:  glm(formula = preemie ~ 1, family = binomial, data = babies.pre
##
## Coefficients:
## (Intercept)
##        -2.42
##
## Degrees of Freedom: 1174 Total (i.e. Null);  1174 Residual
## Null Deviance:      665
## Residual Deviance: 665  AIC: 667
```

The model of constant mean is chosen by this criteria, indicating that these risk factors do not show up in this data set.   ••

• **Example 13.4: The spam data**
Let's apply logistic regression to the data on spam in Table 13.1. Set y_i to be 1 if the e-mail is opened, and 0 otherwise. Likewise, let x_{1i} be 1 if the e-mail has a name in the subject, and x_{2i} be 1 if the e-mail has an offer in the subject. Then we want to model y_i by x_{1i} and x_{2i}. To use logistic regression, we first turn the summarized data into 5,000 samples.

```
first.name <- gl(2, 2500, 5000, labels=c("yes", "no"))
offer      <- gl(2, 1250, 5000, labels=c("yes", "no"))
opened <- c(rep(1:0, c(20, 1250-20)), rep(1:0, c(15, 1250-15)),
            rep(1:0, c(17, 1250-17)), rep(1:0, c( 8, 1250-8)))
xtabs(opened ~ first.name + offer)

##              offer
## first.name yes no
##       yes   20 15
##        no   17  8
```

We remark that the value of opened could have been defined a bit more quickly using a function and sapply:

```
f <- function(x) rep(1:0, c(x, 1250-x))
opened <- c(sapply(c(20, 15, 17, 8), f))
```

Now to fit the logistic regression model.

```
res.glm <- glm(opened ~ first.name + offer, family="binomial")
short_summary(res.glm)

##              Estimate Std. Error z value Pr(>|z|)
## (Intercept)    -4.042      0.199  -20.32  <2e-16 ***
## first.nameno   -0.341      0.263   -1.29   0.196
## offerno        -0.481      0.267   -1.80   0.072 .
## ---
## Signif. codes:  0 '***' 0.001 '**' 0.01 '*' 0.05 '.' 0.1 ' ' 1
```

Although only the intercept is flagged as significant at the 0.05 level, suppose the estimates are correct. How can we interpret them? The coding is such that when no first name or offer is included, the log-odds ratio is $-4.864 = -4.0419 - 0.3407 - 0.4813$. When the first name is included but not the offer, the log-odds ratio is $-4.5232 = -4.0419 - 0.4813$. When both are included, it's -4.0419. Let o_0 be the odds ratio when neither a name nor an offer is included:

$$o_0 = \text{odds ratio} = \frac{\pi}{1-\pi} = e^{-4.864}.$$

If we include the first name, the odds ratio goes up to $e^{-4.864+0.341} = o_0 \cdot e^{0.341}$, which is an additional factor of $e^{0.341} = 1.406$. So, if the original odds were 2 to 100, they go up to $2(1.406)$ to 100. ●●

Avoiding replication   In the previous example the data was replicated to
produce variables first.name, offer, and opened with 5,000 values, so that
all the recorded data was present. The interface for glm conveniently allows
for tabulated data when the binomial family is used.

A two-column matrix is used, with its first column recording the number
of successes and its second column the number of failures. In our example,
we can construct this matrix using cbind as follows:

```
opened <- c(8,15,17,20)
opened.mat <- cbind(opened=opened, not.opened=1250 - opened)
opened.mat

##       opened not.opened
## [1,]       8       1242
## [2,]      15       1235
## [3,]      17       1233
## [4,]      20       1230
```

The predictor variables match the levels for the rows. For example, for
the values of 8 and 15 for opened, offer was 0 and first.name was 0 then 1.
Continuing gives these values:

```
offer <- c(0, 0, 1, 1)
first.name <- c(0, 1, 0, 1)
```

Finally, the model is fit as before, using opened.mat in place of opened:

```
glm(opened.mat ~ first.name + offer, family="binomial")

##
## Call:  glm(formula = opened.mat ~ first.name + offer, family = "binomia
##
## Coefficients:
## (Intercept)    first.name         offer
##      -4.864         0.341         0.481
##
## Degrees of Freedom: 3 Total (i.e. Null);  1 Residual
## Null Deviance:      5.77
## Residual Deviance: 0.736  AIC: 24.7
```

(The reference levels are different than previously, so the coefficients don't
match without some work.)

13.2   Nonlinear models

The linear model is called "linear" because of the way the coefficients β_i enter
into the formula for the mean. These coefficients simply multiply some term.

A nonlinear model allows for more complicated relationships. For example, an exponential model might have the response modeled as

$$y_i = \beta_0 e^{-\beta_1 x_i} + \epsilon_i.$$

Here, $\mu_{y|x} = \beta_0 e^{-\beta_1 x}$ is not linear in the parameters due to the position of β_1. Variations on the exponential model are

$$y_i = \beta_0 x_i e^{-\beta_1 x_i} + \epsilon_i \quad \text{and} \quad y_i = \beta_0 (e^{-\beta_1 x_i}(1 - \beta_2) + \beta_2) + \epsilon_i.$$

The exponential model, with $\beta_1 > 0$, may be used when the response variable decays as the predictor increases. The second model has a growth-then-decay phase, and the third a decay, not to 0 but to some threshold amount $\beta_0 \cdot \beta_2$.

In general, a single-response, nonlinear model can be written as follows:

$$y_i = f(x_i | \beta_0, \beta_1, \ldots, \beta_r) + \epsilon_i.$$

We have $r + 1$ parameters and only one predictor with an additive error. More general models could have more predictors and other types of errors, such as multiplicative.

The possibilities seem endless but in fact are constrained by the problem we are modeling. When using nonlinear models we typically have some idea of which types of models are appropriate for the data and then fit just those. If the model has *i.i.d.* errors that are normally distributed, then using the method of least squares allows us to find parameter estimates and use AIC to compare models.

Fitting nonlinear models with nls

Nonlinear models can be fit in R using nls. The nls function computes nonlinear least squares. Its usage appears similar to that of lm, but there are differences. A basic template is

```
res <- nls(formula, data=..., start=c(...), trace=FALSE)
```

The formula notation used to specify nonlinear models is interpreted differently than for lm. The formula again looks like response ~ mean, but the mean is specified using ordinary math notations. For example, the exponential model for the mean could be written with y ~ N * exp(-r*(t-t0)), where N, r, and t0 are parameters. It is often convenient to use a function to return the mean, such as y ~ f(x, beta0, beta1, ...). That is, a function that specifies the parameter values by name.

Starting points The method of nonlinear least squares uses an algorithm that usually needs to start with parameter values that are close to the actual ones. The argument start=c(...) is where we put the initial guesses for the parameters. This can be a vector or list using named values, such as start=c(beta0=1, beta1=2). Finally, the optional argument trace=TRUE can be used if we want to see what is happening during the execution of the algorithm. This can be useful information if the algorithm does not converge. By default it is FALSE.

The initial parameter guesses are often found by doing some experimental plots. These can be done quickly using the curve function with the argument add=TRUE, as illustrated in the examples. When we model with a given function, it helps to have a general understanding of how the parameters change the graph of the function. For example, the parameters in the exponential model, written $f(t|N,r,t_0) = Ne^{-r(t-t_0)}$, may be interpreted by t_0 being the place where we want time to begin counting, N the initial amount at this time, and r the rate of decay. For this model, the mean of the data decays by $1/e$, or roughly $1/3$ in $1/r$ units of time.

Some models have self-starting functions programmed for them, which means specifying a starting point is unnecessary. The help page for selfStart lists the pre-defined ones.

• Example 13.5: Yellowfin tuna catch rate
The data set yellowfin (UsingR) contains data on the average number of yellowfin tuna caught per 100 hooks in the tropical Indian Ocean for various years. This data comes from a paper by Myers and Worm (see ?yellowfin) that uses such numbers to estimate the decline of fish stocks (biomass) since the advent of large-scale commercial fishing. The authors fit the exponential decay model with some threshold to the data.

We can repeat the analysis using R. First, we plot (Figure 13.2).

```
plot(count ~ year, data=yellowfin)
```

A scatterplot is made, as the data frame contains two numeric variables. The count variable does seem to decline exponentially to some threshold. We try to fit the model

$$Y = N\left(e^{-r(t-1952)}(1-d) + d\right) + \epsilon.$$

(Instead of β_i we give the parameters letter names.)
To fit this in R, we define a function for the mean

```
f <- function(t, N, r, d) N*(exp(-r*(t-1952)))*(1-d) + d)
```

We need to find some good starting points for nls. The value of $N = 7$ seems about right, as this is the starting value when $t = 1952$. The value r is a decay rate. It can be estimated by how long it takes for the data to decay by roughly $1/3$. We guess about 10, so we start with $r = 1/10$. Finally, d is the percent of decay, which seems to be $.6/6 = .10$.

We plot the function with these values to see how well they fit.

```
curve(f(x, N=6, r=1/10, d=0.1), add=TRUE)
```

The fit is good (the solid line in Figure 13.2), so we expect nls to converge with these starting values.

```
res.yf <- nls(count ~ f(year, N, r, d), start=c(N=6, r=1/10, d=0.1),
  data=yellowfin)
res.yf

## Nonlinear regression model
##    model: count ~ f(year, N, r, d)
##     data: yellowfin
##      N        r        d
## 6.0202 0.0938 0.0536
##    residual sum-of-squares: 15.5
##
## Number of iterations to convergence: 8
## Achieved convergence tolerance: 3.06e-06
```

The numbers below the coefficients are the estimates. Using these, we add the estimated line using curve again. This time it is drawn with dashes, and it visually seems to fit all the data a little better.

```
curve(f(x, N=6.02, r=0.0939, d=0.0539), add=TRUE, lty=2, lwd=2)
legend(1980, 6, legend=c("exploratory", "exponential"), lty=1:2)
```

The estimate for d has only 5.36% of the initial amount remaining. ••

Using predict to plot the prediction line The output of nls has many of the same extractor functions as lm. In particular, the predict function can be used to make predictions for the model. You can use this in place of curve to draw the predicted line for the mean response. For example, to draw the line for the yellowfin tuna data, we create a range of values for the year variable, and then call predict with a named data frame.

```
tmp <- 1952:2000
lines(tmp, predict(res.yf, data.frame(year = tmp)))
```

• **Example 13.6: Sea urchin growth**
The urchin.growth (UsingR) data set contains growth data of reared sea urchins over time. Typical growth starts at 0 and progresses to some limiting size. Some models for growth include logistic growth

$$g(t|y_\infty, k, t_0) = y_\infty \cdot (1 + e^{-k(t-t_0)})^{-1}$$

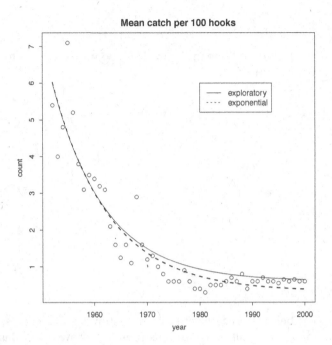

Figure 13.2: Mean catch per 100 hooks of yellowfin tuna in the tropical Indian Ocean. An exponential decay model with threshold is given by the dashed line.

and a Richards growth model

$$f(t|y_\infty, k, t_0) = y_\infty \cdot (1 - e^{-k(t-t_0)})^m.$$

The logistic-growth function is identical to that used in logistic regression, although it is written differently. Our goal here is to fit both of these models to the data, assuming *i.i.d.*, additive error terms, and decide between the two based on AIC. As the Richards model has more parameters, its fit should be much better to be considered a superior model for the data.

We follow the same outline as the previous example: define functions, find initial guesses by plotting some candidates, and then use nlm to get the estimates.

We define two functions and plot the jittered scatterplot (Figure 13.3).

```
logistic <- function(t, Y, k, t0, m) Y * (1 + exp(-k * (t-t0)))^(-1)
richards <- function(t, Y, k, t0, m) Y * (1 - exp(-k * (t-t0)))^m

plot(jitter(size) ~ jitter(age,3), data=urchin.growth,
     xlab="age", ylab="size", main="Urchin growth by age")
```

Next, we try to fit the logistic model. The parameters can be interpreted from the scatterplot of the data. The value of Y corresponds to the maximum growth of the urchins, which appears to be around 60. The value of t0 is where the inflection point of the graph occurs. The inflection point is when the curve stops growing faster. A guess is that it happens around 2 for the data. Finally, k is a growth rate around this point. It should correspond to roughly 1 over the time it takes to grow one-third again after the value at t_0. We guess 1 from the data. With these guesses, we do an exploratory graph with curve (not shown but looks okay).

```
curve(logistic(x, Y=60, k=1, t0=2), add=TRUE)
```

We fit the model with nls

```
res.logistic <- nls(size ~ logistic(age, Y, k, t0),
                start = c(Y=60, k=1, t0=2),
                data = urchin.growth)
res.logistic

## Nonlinear regression model
##    model: size ~ logistic(age, Y, k, t0)
##     data: urchin.growth
##     Y     k    t0
## 53.90  1.39  1.96
##   residual sum-of-squares: 7299
##
## Number of iterations to convergence: 5
## Achieved convergence tolerance: 6.89e-06
```

```
curve(logistic(x, Y=53.903, k=1.393, t0=1.958), add=TRUE)
```

Finally, so we can compare, we find the AIC:

```
AIC(res.logistic)

## [1] 1561
```

Next, we fit the Richards model. First, we try to use the same values, to see if that will work (not shown).

```
curve(richards(x, Y=53.903, k=1.393, t0=1.958, m = 1), add=TRUE)
legend(4, 20, legend=c("logistic growth", "Richards"), lty=1:2)
```

It is not a great fit, but we try these as starting points for the algorithm anyway:

```
res.logistic <- nls(size ~ logistic(age,Y,k,t0,m),
                    data  = urchin.growth,
                    start = c(Y=53, k=1.393, t0=1.958, m=1))
```

```
## Error: singular gradient matrix at initial parameter estimates
```

This is one of the error messages that can occur when the initial guess isn't good or the model doesn't fit well.

Using a little hindsight, we think that the problem might be t_0 and k. For this model, a few exploratory graphs indicate that we should have $t \geq t_0$ for a growth model, as the graphs decay until t_0. So, we should start with $t_0 < 0$. As well, we slow the rate of growth.

```
res.richards <- nls(size ~ richards(age, Y, k, t0, m),
                data  = urchin.growth,
                start = c(Y=53, k=0.5, t0=0, m=1))
res.richards
```

```
## Nonlinear regression model
##    model: size ~ richards(age, Y, k, t0, m)
##     data: urchin.growth
##      Y       k       t0       m
## 57.265  0.784  -0.859  6.064
##   residual sum-of-squares: 6922
##
## Number of iterations to convergence: 9
## Achieved convergence tolerance: 1.83e-06
```

```
curve(richards(x, Y=57.26, k=0.78, t0=-0.8587, m = 6.0636),
      add=TRUE, lty=2)
```

Now we have convergence. The residual sum-of-squares, 6,922, is less than the 7,922 for the logistic model. This is a good thing, but if we add parameters this is often the case.[3] We compare models here with AIC.

```
AIC(res.richards)
```

```
## [1] 1550
```

This is a reduction from the other model. As such, we would select the Richards model as a better fit by this criteria.   ••

---

[3]We do not have nested models, for which this would always be the case.

Urchin growth by age

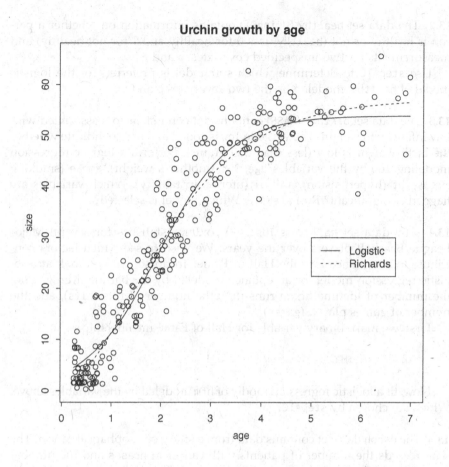

Figure 13.3: Sea urchin growth data, with logistic model fit in solid and Richards model fit in dashed line.

Problems

13.1 The data set tastesgreat (UsingR) is data from a taste test for *New Goo*, a fictional sports-enhancement product. Perform a logistic regression to investigate whether the two covariates, age and gender, have a significant effect on the enjoyment variable, enjoyed.

13.2 The data set healthy (UsingR) contains information on whether a person is healthy or not (healthy uses 0 for healthy and 1 for not healthy) and measurements for two unspecified covariates, p and g.

Use stepAIC to determine which submodel is preferred for the logistic model of healthy, modeled by the two covariates p and g.

13.3 The data set birthwt (MASS) contains data on risk factors associated with low infant birth weight. The variable low is coded as 0 or 1 to indicate whether the birth weight is low (less than 250 grams). Perform a logistic regression modeling low by the variables age, lwt (mother's weight), smoke (smoking status), ht (hypertension), and ui (uterine irritability). Which variables are flagged as significant? Run stepAIC. Which model is selected?

13.4 The data set hall.fame (UsingR) contains statistics for several Major League Baseball players over the years. We wish to see which factors contribute to acceptance into the Hall of Fame. To do so, we will look at a logistic regression model for acceptance modeled by the batting average (BA), the number of lifetime home runs (HR), the number of hits (hits), and the number of games played (games).

First, we make binary variable for Hall of Fame membership.

```
hfm = hall.fame$Hall.Fame.Membership != "not a member"
```

Now, fit a logistic regression model of hfm modeled by the variables above. Which are chosen by stepAIC?

13.5 The esoph data set contains data from a study on esophageal cancer. The data records the number of patients with cancer in ncases and the number of patients in the control group with ncontrols. The higher the ratio of these two variables, the worse the cancer risk. Three factors are recorded: the age of the patient (agegp), alcohol consumption (alcgp), and tobacco consumption (tobgp).

We can fit an age-adjusted model of the effects of alcohol and tobacco consumption with an interaction as follows:

```
res.full <- glm(cbind(ncases, ncontrols) ~ agegp + tobgp * alcgp,
                data = esoph, family = binomial())
```

A model without interaction is fit with

```
res.add  <- glm(cbind(ncases, ncontrols) ~ agegp + tobgp + alcgp,
                data = esoph, family = binomial())
```

Use AIC to compare the two models to determine whether an interaction term between alcohol and tobacco is hinted at by the data.

13.6 The data set Orange contains circumference measurements for several trees (Tree) based on their age. Use a logistic growth model to fit the data for tree 1. What are the estimates?

13.7 The data set ChickWeight contains measurements of weight and age (Time) for several different chicks (coded with Chick). For chick number 1, fit a logistic model for weight modeled by Time. What are the coefficients?

13.8 The data set wtloss (MASS) contains weight measurements of an obese patient recorded during a weight-rehabilitation program. The variable Weight records the patient's weight in kilograms, and the variable Days records the number of days since the start of the program. A linear model is not a good model for the data, as it becomes increasingly harder to lose the same amount of weight each week. A more realistic goal is to lose a certain percentage of weight each week. Fit the nonlinear model

$$\text{Weight} = a + b2^{-\text{Days}/c}.$$

The estimated value of c would be the time it takes to lose b times half the excess weight.

What is the estimated weight for the patient if he stays on this program for the long run? Suppose the model held for 365 days. How much would the patient be expected to weigh?

13.9 The reddrum (UsingR) data set contains length-at-age data for the red drum fish. Try to fit both the models

$$l = b_0(1 - e^{-k(t-t_0)}) \quad \text{and} \quad l = (b_0 + b_1 t)(1 - e^{-k(t-t_0)}).$$

(These are the von Bertalanffy and "linear" von Bertalanffy curves.) Use the AIC to determine which is preferred.

Good starting values for the "linear" curve are 32, 1/4, 1/2, and 0.

13.10 The data set midsize (UsingR) contains values of three popular mid-size cars for the years 1990 to 2004. The 2004 price is the new-car price, the others values of used cars. For each car, fit the exponential model with decay. Compare the decay rates to see which car depreciates faster. (Use the variable year=2004-Year and the model for the mean $\mu_{y|x} = Ne^{-rt}$.)

Programming

This appendix touches briefly on programming in R. Though the basics can be learned quickly, there are many little details to learn for mastery. For that, we defer to the much more authoritative resources including [8] by a primary architect of R; [6]; [40], a more recent addition; and [60] for a very promising work in progress. Our goal here is to cover some of the basics, leaving the grittier details to the already very good alternatives.

A.1 Functions

Programming at its core involves the writing and organization of functions. We cover here the basics of functions and the basics of how R chooses a function to call.

The basic template for defining a new function in R is

```
function_name <- function(argument_list...) {
  function_body
}
```

At first glance, this gives at least three things to discuss—the name, the arguments, and the body. We begin with some R-specific conventions related to naming a function.

Function names

First, as previously discussed, a function object created by function need not have a name to be useful. In Chapter 4 we saw that using anonymous functions for arguments to functions like sapply (in small doses) can lead to more readable code. In this section we discuss valid names for R functions and some naming conventions employed that allow functions to be called in R-specific manners.

Syntactically valid names When functions are named, there are a few conventions that are R-specific. The R documentation (?quote) discusses *syntactic* *names* consisting of a sequence of letters, digits, the period (".") and the underscore, with the convention that they must not start with a digit nor underscore, nor with a period followed by a digit. For example, mean, mean.default are syntactic names, as is .mean (a coding convention is to use a leading period for private functions), but not _mean.

Non-syntactically valid names   However, *non*-syntactic function names are possible. To define such a function requires the quoting of the function name. The convention is to use back quotes, but most times standard quotes work well. Quoting is always an option, but required for using two special naming conventions discussed next.

Infix notation   In R functions are typically called by name followed by parentheses (e.g., sd(x)). However, there are other function calls that happen, that at first glance may not seem like function calls at all. A common example is when we add two numbers: 2 + 2. This gets carried out by calling the function name +:

```
'+'

## function (e1, e2)  .Primitive("+")
```

The above command actually wraps the function + in backquotes so that its value will be displayed. We see that the "plus" function has two arguments and then does something with a "primitive" function.[1] So, though rarely done, it is possible to to use the "plus" function in this manner:

```
'+'(1, 2)

## [1] 3
```

This style where the function name is first, then the arguments, is termed *prefix notation*. The more conventional style for binary operations is called *infix notation*.[2]

Infix notation is certainly desirable for addition, and other familiar mathematical symbols. R allows the user to define other functions that can be used in this manner, following a naming convention: the name is wrapped in percent symbols, as in %text%.

Base R uses this notation to define an infix version for match:

```
'%in%'

## function (x, table)
## match(x, table, nomatch = 0L) > 0L
```

---

[1] In base R, many functions are dispatched to "primitive functions" written in C.

[2] There is also *postfix notation*, which was adopted by some calculator manufacturers.

```
## <bytecode: 0x1009cb380>
## <environment: namespace:base>
```

Which allows for a more natural looking query of existence in a container:

```
evens <- seq(0, 20, by=2)
2 %in% evens
```

```
## [1] TRUE
```

A sometimes useful trick in programming is to have a lookup return a default value if something is "falsy" (this term comes from JavaScript conventions). For R, falsy might include a value is FALSE, NA, NULL, NaN, or "". Here is how this could be coded:

```
'%||%' <- function(val, default) {
  if(is.null(val) || is.na(val) || (is.character(val) && val == "")
     || (is.numeric(val) && is.nan(val)))
    default
  else
    val
}
```

The first bit does many checks. We use the shortcut operators discussed in the following section. These allow the construction of logical checks that might fail if NA values were involved, as any NA values are caught first.

To see this function work, imagine we have the following list and want to query it, while imagining it is such a big list you can't check just by looking:

```
l <- list(a=1, b="two", c="III")
l$a %||% "a"                          # there so returns l$a
```

```
## [1] 1
```

```
l$bee %||% "b"                        # not there, so returns "b"
```

```
## [1] "b"
```

Here is another useful function to allow expressions to be read left to right. We will use it in the sequel when we want the focus to be on the first function call:

```
'%|%' <- function(x, FUN) FUN(x)
3 %|% sin %|% cos                     # finds cos(sin(3))
```

```
## [1] 0.9901
```

Replacement functions In computer programming, *pure function* is a term from functional programming that, roughly, refers to functions that do not have any noticeable side effects after evaluation, and for which the return value is only dependent on the values of the inputs.

R is designed for statistical programming where tradition has it that modifications to data should be done only deliberately by the user. For the most part, this makes a functional programming style consistent with most of R's idioms. For example, in the design of R, function arguments are copied on modification.[3] Further, in the body of a function, a programmer needs to go out of their way to modify an object outside of the function.

Of course, users want to modify data. For example, in data cleaning, it might be desirable to convert coded values to NA. To illustrate, in a previous example we discussed the cost of a hip-replacement procedure at various hospitals. Suppose the hospitals that couldn't report were coded with a −1, a clearly impossible value. These missing values could be converted to R's NA convention as follows:

```
hip_cost <- c(10500, 45000, 74100, -1, 83500, 86000, 38200, -1,
              44300, 12500, 55700, 43900, 71900, -1, 62000)
hip_cost[hip_cost == -1] <- NA
```

This last command does modify a data set, and the user initiates it by putting the indexing on the left-hand side of an assignment.[4] But how does R know to modify values?

Perhaps, it is simpler to consider the function to change names of an object:

```
x <- c("a"=1, "b"=2)
names(x) <- c("A", "B")
```

This last expression has a side effect—it modifies the names attribute of x. In the section "Subset assignment" of the R Language Manual, it is explained that the above is essentially translated into the following code:

```
tmp <- x                              # a temporary copy of x
x <- "names<-"(tmp, value=c("A", "B"))  # call a different function
rm(tmp)                               # clean up
```

This utilizes a copy of x called tmp (actually it uses '*tmp*', suitably quoted). More importantly, what seems like a usage of the names function actually employs a different function, names<-. More subtly, the use of the

---

[3]They are sent as references to save memory, but when modified a copy is created to avoid modifying the object that was passed in.

[4]Sure it could be the right-hand side if for some unusual reason the -> operator was used for assignment. Chambers [8] says this operator is a historic leftover from the days before inline editing.

named argument value places a restriction on the argument names for a re-placement function: conventionally the first is x and the last must be value. (Named arguments are discussed in the next section.)

In the definition of names<- it is important that the function returns the value that x should become, due to the assignment in the second line above.

Summarizing, these are the necessary rules: The name ends in <-, x is customarily the first argument, value is the name of the last argument, and the function should return the modified value of x. Following these few rules, a replacement operator is fairly simple to define. Here is a function to recode values:

```
"re_code<-" <- function(x, old, value) {
  x[x == old] <- value

  x
}

hip_cost <- c(10500, 45000, 74100, -1, 83500, 86000, 38200, -1,
              44300, 12500, 55700, 43900, 71900, -1, 62000)
re_code(hip_cost, -1) <- NA          # re_code(x, old_value) <- value
hip_cost                             # it has been modified

## [1] 10500 45000 74100     NA 83500 86000 38200     NA 44300 12500
## [11] 55700 43900 71900     NA 62000
```

Arguments

The arguments of a function allow a user to parameterize a function call, making single function calls able to do a wider variety of tasks. A simple example might be to write a function to compute points on a line:

```
f <- function(x, m, b) m * x + b
```

This takes three arguments: an x to find a matching y for, and two other arguments to specify the slope and intercept parameters. By using additional arguments, one function can handle many cases.

The args function will return a description of the arguments of a function.[5] For example, here is the argument list for the sapply function:

```
args(sapply)

## function (X, FUN, ..., simplify = TRUE, USE.NAMES = TRUE)
## NULL
```

---

[5]For programming purposes, formals gives a more useful description.

From the output we can see a few features that make programming in R a bit more convenient:

- Functions are first-class objects and can be passed as arguments to other functions (the FUN argument).

- Function arguments can be given default values (e.g., simplify=TRUE) which makes interactive usage more convenient.

- There is the ability to pass in a variable number of arguments (the special argument ...).

First, sapply is typically called as follows:

```
sapply(mtcars, mean)                        # mean of each variable

##      mpg      cyl    disp       hp     drat       wt     qsec
##  20.0906   6.1875 230.7219 146.6875   3.5966   3.2172  17.8487
##       vs       am     gear     carb
##   0.4375   0.4062   3.6875   2.8125
```

From the above output of args, it is clear that mtcars matches the X argument and mean the FUN argument. This is done by position in the argument list.

Had we wanted to avoid the simplification step (returning a list instead of an array), we could have called the function with:

```
sapply(mtcars[,1:2], mean, simplify=FALSE) # two columns only

## $mpg
## [1] 20.09
##
## $cyl
## [1] 6.188
```

The argument simplify=FALSE would match by name, not position. Using keyword arguments is achieved by placing the construct name= before the argument value.[6] The name part need not be a total match of the argument name, as R will attempt to resolve it by partial matching. If the actual argument can be uniquely identified—and the argument is before any ... in the definition—then the partial name will match.

The value for USE.NAMES is taken from the default value specified by the programmer of sapply. Default values are specified when defining the function with the pattern name=default, similar to how a value is specified by name when a function is called.

---

[6]This is not name<-default, which would assign the value to the name and then match this argument by position. This is one area where a distinction is made between the use of = and <-, as in this context = is not used for assignment.

The basic matching algorithm employed by R is: exact name match, partial matching by name prefix (though never after ..., when present), positional match.

Variadic arguments   The ... in the arguments of sapply is an R convention to handle a variable number of arguments. This allows us to call sapply as follows:

```
sapply(mtcars, mean, na.rm=TRUE)

##     mpg     cyl    disp      hp    drat      wt    qsec
## 20.0906  6.1875 230.7219 146.6875  3.5966  3.2172 17.8487
##      vs      am    gear    carb
##  0.4375  0.4062  3.6875  2.8125
```

The argument na.rm=TRUE is not for sapply, but is passed to the mean function when it is called for each variable in mtcars. The following code would be a simplified version of sapply:

```
simple_sapply <- function(X, FUN, ...) {
  out <- list()
  for( i in 1:ncol(X)) {
    out[[i]] <- FUN(X[[i]], ...)
  }
  simplify2array(out)
}
simple_sapply(mtcars, mean, na.rm=TRUE)

## [1]  20.0906   6.1875 230.7219 146.6875   3.5966   3.2172
## [7]  17.8487   0.4375   0.4062   3.6875   2.8125
```

The command FUN(X[[i]], ...) would pass along any additional named arguments given to simple_sapply not matched by exact name or position to the function passed in through fun. In the example, this is na.rm=TRUE.[7]

One can work with the ... example directly. It can be converted into a list with named components through list(...). What can't be done is to strip out some of the arguments before passing it along, other tricks are used. (This would be useful if passing the values to two different functions would be desired, but only part of the values to one, and part to another.)

• Example A.1: Our own histogram function
The default hist function in R is a bit lacking. First, as a histogram reminds us of the underlying density, an argument can be made that the default histogram should be on the density scale. That is, it should be normalized so

---

[7]Partial matching by name is not a good habit, and with a good editor—like RStudio's—unnecessary.

the area is 1. With that convention, it is helpful to automatically estimate the density and add it to the graphic. Finally, the default choice of bins can be improved. Our choice below follows the truehist function of the MASS.

We walk through some of the above to get what we want. Here is a first attempt:

```
ourhist <- function(x) {
  hist(x, breaks="Scott", probability=TRUE)
  lines(density(x))
}
```

If we were to try it out, say ourhist(rivers), we would see that tt works fine. But what if we wanted to use a different rule for breaks=? It would be nice to be able to override our settings. One way would be to define a breaks= argument:

```
ourhist <- function(x, breaks) {
  hist(x, breaks=breaks, probability=TRUE)
  lines(density(x))
}
```

Though we could call ourhist(rivers, breaks="Scott"), our typical usage, ourhist(rivers), would yield an error with a complaint Argument "breaks" is missing, with no default. We remedy that by setting a default value.

```
ourhist <- function(x, breaks="Scott") {
  hist(x, breaks=breaks, probability=TRUE)
  lines(density(x))
}
```

Both ourhist(rivers) and ourhist(rivers, breaks="Sturges") will now work. The two commands show a difference in the number of bins, the second using the "Sturges" rule to compute the number instead of the default.

Still, the histogram drawn looks bland. Let's add some color to it—the color purple. The hist function—like other plotting functions—has a col argument to set the color of the boxes. We can make the color be purple by default with this modification:

```
ourhist <- function(x, breaks="Scott", col="purple") {
  hist(x, breaks=breaks, probability=TRUE, col=col)
  lines(density(x))
}
```

Trying it out gives

```
ourhist(rivers)
ourhist(rivers, "Sturges")              # use different bin rule
ourhist(rivers, "Sturges","green")  # green before purple
ourhist(rivers, "green")                # Oops

## Error: 'arg' should be one of "sturges", "fd", "freedman-diaconis",
"scott"
```

We see that we can make changes to the defaults quite easily. The third line uses the Sturges' rule and green as the color.

However, we also see that we can make an error. Look at the last command. We tried to change the color to green but keep the default for the breaks rule. This didn't work. That is because R was expecting a breaks rule as the second argument to the function. To override the positional matching of arguments we should use named arguments in the function call:

```
ourhist(rivers, col="green")
```

Of course, partial matching will also work:

```
ourhist(rivers, c="green")
```

But not if we just get the name wrong:

```
ourhist(rivers, color="green")                        # col -- not color

## Error: unused argument (color = "green")
```

There are many other things we might want to modify about our histogram function, but mostly these are already there in the hist function. For example, changing the x-axis label. It would be nice to be able to pass along arguments to our function ourhist to the underlying hist function. R does so with the ... argument. When our function contains three periods in a row (...) in the argument and in the body of the function, all extra arguments to our function are passed along. You may notice if you read the help page for hist that it too has a ... in its argument list.

Again, modify the function, this time to

```
ourhist <- function(x,breaks="Scott",col="purple",...) {
  hist(x,breaks=breaks,probability=TRUE,col=col,...)
  lines(density(x))
}
```

Then we can do these things

```
ourhist(rivers, xlab="histogram of river lengths") # change the x label
ourhist(rivers, xlab="histogram of river lengths", col="green")
```

But, there are issues! We can't pass in probability as an argument, it will be sent twice to hist: once through the specification probability=TRUE and through ... To avoid that, R programmers sometimes define a local function to mask arguments, as is done here (where we also fix the *y*-limit to avoid truncation):

```
ourhist <- function(x, breaks="Scott", col="purple",...) {
  lhist <- function(x, breaks, col, probability, xlim, ylim, ...) {
    h <- hist(x,  breaks=breaks, plot=FALSE)
    d <- density(x)
    hist(x, breaks=breaks, col=col, probability=TRUE,
        xlim = range(c(h$breaks,  d$x)),
        ylim = range(c(h$density, d$y)), ...)
  }
  lhist(x, breaks=breaks, col=col,...)
  lines(density(x), lwd=2)
}
```

••

Body

The body of a function is comprised of one or more expressions which R steps through as it evaluates a function. Which expressions get evaluated can be controlled through various control flow constructs. Which variable bindings are in force during the evaluation of a function is the issue of scoping. As previously mentioned, the return value is the result of the last expression evaluated. This is often the last expression in the block of code but may be forced through a return call.

Control flow

The if statement is one of the basic statements in R to ensure conditional evaluation of an expression. The basic template is:

```
if (condition) {
  expression
} else {
  expression
}
```

where condition is evaluated and coerced to a logical value. If TRUE the first expression is evaluated, if FALSE the second. The evaluated condition should be coercible to a length 1 logical value, if it is a vector with length greater than 1 the first value is used and a warning is issued, and if coercion returns NA, an error is thrown.[8]

The else is optional. For "else-if" expressions, one can use another if statement in an expression. As with functions, if the expression is not a compound expression, the braces can be omitted. A few examples follow.

This code will print out if a value is "odd" or "even:"

```
x <- 2
if(x %% 2 == 0) "even" else "odd"

## [1] "even"
```

The condition there returns the remainder after division by 2. This is numeric, but is implicitly coerced to logical (only 0 is coerced to FALSE). In the above, each expression is a single statement, so the braces may be omitted.[9]

This function determines the sign of a number using a nested if statement to consider the three cases.

```
x <- 42
if(x < 0) {
  print("negative")
} else if(x > 0) {
  print("positive")
} else {
  print("zero")
}

## [1] "positive"
```

Related to if is the ifelse construct. This has the form

```
ifelse(condition, if_true, if_false)
```

This can be used as an alternative to an expression like

```
if(condition) if_true else if_false
```

but is more useful, as the condition here can be a logical vector. This can be used directly with vectorized code. Returning to the even-odd problem, we could have done:

---

[8]It is a common programming gotcha to have conditions that evaluate to a length 0 expression (when NULL is present, say) or evaluate to vectors of longer length through vectorization.

[9]This example will have problems if x is NULL or if x is longer than length 1.

```
x <- 1:8
ifelse( x %% 2 == 0, "even", "odd")

## [1] "odd"  "even" "odd"  "even" "odd"  "even" "odd"  "even"
```

Short-circuit evaluations The logical operators & and | perform element-wise "and" and "or" operations on vectors. The similar looking && and || binary operators have a different role. They are used for short-circuit logical evaluations—only evaluating an expression on the right if after evaluating the one on the left the answer is not clear. (If the left value is FALSE and && then the answer must be FALSE or if the left value is TRUE and || then the answer must be TRUE, so the right evaluation need not happen.)

We saw this just used to test for NULL and NA before testing the value to match "". (Neither NULL=="" or NA=="" will evaluate to a TRUE or FALSE.)

These operators are typically employed in the conditional expression of an if statement, as was done in a previous example.

A less typical use might be the following construct as an avoidance of nested if statements:

```
centre <- function(x, type) {
  type == "mean"    && return(mean(x))
  type == "median"  && return(median(x))
  type == "trimmed" && return(mean(x, trim=.1))
  which(max(table(x)) == table(x))       # mode like
}
centre(mtcars$mpg, type="median")

## [1] 19.2
```

The above evaluates the return statement only if the first condition is true, so it effectively uses type to switch between the choice of center.

switch The above example—along with its British spelling of center—comes from the help page for switch. This is the more R-like idiom for switching between which expression to evaluate based on the value of some variable. The above would be more naturally written as:

```
centre <- function(x, type) {
  switch(type,
         mean    = mean(x),
         median  = median(x),
         trimmed = mean(x, trim=0.1),
         which(max(table(x)) == table(x))
         )
}
```

As above, when the first argument to switch is evaluated and returns a character, then the rest of the arguments are basically combined as a list and matching is done by name (no partial matching). The last value is the catch all if no match is present.

When the expression evaluates to a number then that value is used to select a case. For example, this function could convert between the ordering of the pos argument of text from an integer to verbal description:

```r
par_convert <- function(x) {
  switch(x,
         "bottom",
         "left side",
         "top",
         "right side")
}
```

Values not in 1:4 have a NULL return value.

Repeated evaluation   In Chapter 4 for loops were discussed briefly before a lengthier discussion on alternatives often employed when working with R.

To recap, a for loop loops like:

```r
for(var in collection) {
  expression
}
```

where the collection can be a range (e.g., 1:10 or seq_along(x)), vector, or a list (such as a data frame). The expression is evaluated as many times as elements in the collection with var looping over the values of each element in the collection. The looping can be broken through the keyword break.

A simple example is this implementation of "bizz-buzz" where for multiples of three we print bizz and for multiples of five, buzz. For values where both three and five are multiples, we print bizz-buzz.

```r
res <- character(100)
for (i in 1:100) {
  out <- character(0)
  if (i %% 3 == 0) out <- c(out, "bizz")
  if (i %% 5 == 0) out <- c(out, "buzz")
  res[i] <- if(length(out) > 0) paste(out, collapse="-") else i
}
res %|% head

## [1] "1"     "2"     "bizz" "4"     "buzz" "bizz"
```

While loops Another looping construct is the while loop which will repeat an expression until a condition is met or the expression executes a break command. While loops are useful in statistical programming of iterative algorithms.

Base R uses the following to find the quantiles of the birthday distribution. This is related to an interesting question from probability:

> In a classroom of n students, what is the probability of there being two or more (non-twins) sharing the same birthday? Assume all birthdays are equally likely.

We won't answer this, it is too fun to work out by oneself. For the impatient, it is computed by pbirthday(n).

A related question, might be how large would the class need to be so that there is a 95% chance of this event occurring? This is a quantile problem, so should be given by qbirthday(0.95) (and is). Here we compute it directly by inverting pbirthday using a while loop:

```
n <- 2
while( pbirthday(n) < 0.95) n <- n + 1
n
```

```
## [1] 47
```

That is a class of 47 students has a 95% chance or better of having two students sharing a birthday.

This solves for the smallest n with pbirthday(n) >= 0.95, by increasing n if the value of pbirthday(n) is less than 0.95. If this event could never happen, then this loop would go on without termination.

Exception handling The tryCatch function is one of R's constructs for handling unusual conditions. It is useful for programming when something may not work. For example, programmatically downloading files may be problematic: the remote service may be down, the URL may have changed, ... R will often throw an error if such things occurs. The tryCatch tries an expression, and if an error is thrown the "Catch" part is called to handle the exceptional case.

Here is the basic pattern:

```
f <- "http://flaky.webresource.com"
tmp <- tempfile()
tryCatch(download.file(f, destfile=tmp),
      error = function(e) message("Download of ", f, " failed:", e))
```

A final expression can be specified to run on conclusion, but that isn't so appropriate here. The above will gracefully print out a message, instead of a script-stopping error should the resource pointed to by f not be available.

Variable scope

For simple functions, the body is just a set of expressions that get evaluated in an order from first to last unless directed otherwise by some control-flow expression. The subtleties involved have to do with the simple question: what is a variable referring to inside the body of a function? This is the issue of *scope*.

In many cases, the subtleties do not arise. For example, let's look at the sd function:

```
stats::sd

## function (x, na.rm = FALSE)
## sqrt(var(if (is.vector(x)) x else as.double(x), na.rm = na.rm))
## <bytecode: 0x10acf33c0>
## <environment: namespace:stats>
```

We can see the body involves the following functions and constructs: sqrt, var, if/else, is.vector, as.double; and the following variables: x, na.rm.

The functions referenced in sd are found by following a sequence of environments listed by the function search.

The variables, x and na.rm, are found within an environment constructed when the function is called. That is, when sd(rivers) is executed, the argument matching results in a environment that has the argument x bound to the rivers data set (matched by position and found by searching in package:datasets) and the argument na.rm bound to FALSE, as this is the default value. This evaluation environment references the environment of sd and so through some search process, all the variables and function names in the body of sd are defined. It may be complicated to think about, but this is basically what should be expected—values are found where you expect them to be.

What makes this tricky, are the following: default arguments are expressions that can depend on other variables; the arguments are not actually evaluated until needed (*lazy evaluation*); and not all variables within a function body need be passed into the function through its arguments.

Following Chambers [8], we look at the function mad:

```
mad

## function (x, center = median(x), constant = 1.4826, na.rm = FALSE,
##        low = FALSE, high = FALSE)
## {
##        if (na.rm)
##            x <- x[!is.na(x)]
##        n <- length(x)
##        constant * if ((low || high) && n%%2 == 0) {
```

```
##          if (low && high)
##              stop("'low' and 'high' cannot be both TRUE")
##          n2 <- n%/%2 + as.integer(high)
##          sort(abs(x - center), partial = n2)[n2]
##      }
##      else median(abs(x - center))
## }
## <bytecode: 0x10eb0cd50>
## <environment: namespace:stats>
```

The arguments are:

```
formals(mad)  %|% function(x) head(x, n=4)

## $x
##
##
## $center
## median(x)
##
## $constant
## [1] 1.483
##
## $na.rm
## [1] FALSE
```

Here the formal arguments contain an argument with no default (x), arguments with specific defaults (e.g., constant=1.4826), and an argument with
a reference to another argument center=median(x). The latter may cause
confusion—how can it know x when the function is defined, as it isn't known
until the function is actually called?

When R creates the environment for evaluation within a function, the arguments are not evaluated immediately. Instead a promise is created with the
unevaluated expression. When the variable is needed, the promise is evaluated: promises corresponding to actual arguments are evaluated in the environment the function is called from, but promises corresponding to default
arguments are evaluated in the environment created for the call. So, when
mad calls center it looks for the binding of x within the environment of the
call. The value of center is called after x has possibly been stripped of NA values. This change would be reflected in the call to center, as this is the binding
for x at the time the promise is evaluated.

So to understand how a function is evaluated, it is important to know
about its enclosing environment, as this can determine some values. This
environment is returned by the environment function.[10]

---

[10] As such, functions in R are not "pure functions," in that they aren't necessarily determined
solely by their inputs.

Closures

What happens when a function returns a function? This returned function would have its enclosing environment be the environment made in the call to the outer function. For example, this simple function:

```
expand <- function(rate=2) function(x) x * rate
```

returns a function that by default will double its argument, but if a different value is passed in for rate, will have a different factor:

```
tripler <- expand(3)
tripler(4)

## [1] 12
```

Here rate in the tripler function is looked for in the environment made for the function call. It isn't found, so the search extends to the enclosing environment—the environment made for the call to expand. Here rate comes not from a default argument, but is passed in so is resolved in the enclosing environment of expand. Somewhat complicated to think about, but functions that return functions can be very powerful.

We've seen the Vectorize function which takes a function and "vectorizes" it, returning a new, related function which will apply itself to each element of the container, in the example below the variables in a data frame:

```
m <- mtcars[, 1:3]
median(m)                          # fails

## Error: need numeric data

vmedian <- Vectorize(median)
vmedian(m)                         # works now

##    mpg   cyl  disp
## 19.2   6.0 196.3
```

• **Example A.2: Function composition**
Mathematically, there are two familiar notations for the composition of two functions: $f(g(x))$ and $(f \circ g)(x)$. The former mirrors typical R usage—$g(x)$ is a new value which is evaluated by the function f. The latter works at the function object level, combining functions to make a new function $f \circ g$ which is then evaluated at x. To make an R function that composes two functions (of a single variable) can be done as follows:

```
compose <- function(f, g) {
  function(...) f(g(...))
}
```

We can add an infix operator as follows:

```
'%.%' <- function(f, g) compose(f, g)
```

To call this we have:

```
h <- sqrt %.% var                        # another sd function
h(1:10)

## [1] 3.028
```

(Due to evaluation precedence, (sqrt %.% var) (1:10) would also work, but not without the first pair of parentheses.) ••

A.2 Generic functions

R employs the concept of a *generic function* that allows different functions to be dispatched depending on the manner in which the function is called. This is a very useful feature. For example, it allows many different types of plots to be made from the same function call, plot(x): if x is numeric, a simple plot against the index is produced; if a factor, a bar chart after tabulation is produced; if a fitted model object, a collection of diagnostic plots is produced; or if a formula, a scatterplot or boxplot is produced depending on the model and variables employed. With generic functions, a function name reflects a general idea, the various implementations the specific details. This is a very useful level of abstraction.

R has a few different ways to implement this behavior, which is more formally called *polymorphism*. We describe the S3 implementation. As well, there is a newer style called S4 and a different style called Reference Classes.

S3 methods

S3 methods are part of the base R language. The basic idea is simple:

- A function is registered as a generic function.

- When a generic function is called, there is a means to resolve which implementation (method) of the generic function should be dispatched.

We begin with the notion of dispatch. An R object is an object with a class attribute.[11] A class attribute is a character vector listing the classes from which an object inherits. These may be explicitly defined or implicitly. Some common classes for data objects are integer for integer data (such data is also numeric) and double for floating point numbers (which are also numeric). A matrix is of class "matrix" and of class given by the underlying data. A data frame is of class data.frame and also class list.

The class function will list the class of an object. The following lists many at once:

```
c(class(1L), class(1.0), class(matrix(1)), class(data.frame(1)),
  class(list(1)))

## [1] "integer"    "numeric"    "matrix"     "data.frame"
## [5] "list"
```

There are also "is" functions for checking if an object is of a certain class. For example:

```
is.integer(1)

## [1] FALSE

is.integer(1L)                          # 1L forces an integer

## [1] TRUE

is.double(1.0)

## [1] TRUE
```

There is an alternate form for this is(x, "class") (though not exactly):

```
is(mtcars, "data.frame")

## [1] TRUE

is(1.0, "double")                       # not exactly the same

## [1] FALSE
```

Coercion from one class to another can be implicit:

```
x <- 1
class(x)
```

---

[11]The ?UseMethod help page has much detail on this.

```
## [1] "numeric"

x <- c(1, "two")
class(x)

## [1] "character"
```

Or can be explicit through an "as" function:

```
as.character(1)

## [1] "1"
```

The class attribute can also be manipulated through the class<- method. For example, here we add a class to a numeric value:

```
x <- 1
class(x) <- c("Newclass", class(x))
class(x)

## [1] "Newclass" "numeric"
```

Class-based dispatch S3 dispatch happens through the class of the first argument of a generic function. Suppose an R object x has classes "classA" and then "classB". Then a call generic(x) will search first for a function generic.classA, if not found it will continue to look for generic.classB, if that fails, then a search for generic.default is made.[12] For a successful search that function will be called with the given arguments.

Knowing this allows one to create new methods for dispatch.[13] For example, in Chapter 3 we mentioned that the UsingR package, following an idea in the mosaic package, extends the mean function to have a formula interface.

To do this is not hard. The actual function definition was mentioned in Chapter 4. What we didn't know then is that we just need to name the function mean.formula, as formula is the name of the class for formula objects:

```
mean.formula <- function(formula, data, ..., na.action=na.omit) {
  out <- aggregate(formula, data, mean, ..., na.action=na.action)
  xtabs(formula, out)
}
```

---

[12]For this reason, it is best to not use . in the name of a function, unless S3 dispatch is intended.

[13]S3 is a form of object-oriented programming, in that functions are dispatched based on the class of an object. In many object-oriented languages, a function is bound to a class and termed a method of the class. This isn't exactly what happens in R, as generic functions are not bound to a class. Therefore, the term "method" for functions that are dispatched to can be seen as a slight abuse of the vocabulary of object-oriented programming.

We can check how this works by taking group means of miles per gallon by the cyl and am variables:

```
mean(mpg ~ am + cyl, data=mtcars)

##      cyl
## am       4      6      8
##    0 22.90 19.12 15.05
##    1 28.07 20.57 15.40
```

We get the expected output. To be clear, we don't call mean.formula (though we could), rather we let R find it through dispatch on the first argument, the formula mpg ~ am + cyl.

Registering a generic function   To create a new S3 generic function requires a call to UseMethod. To illustrate, we create a generic function size for giving a quick sense of the size of a data object, this being a mixture of length and dim for rectangular data.

To define the generic function is done simply as follows:

```
size <- function(x, ...) UseMethod("size")
```

· A default method is optional, but a good idea. In this catchall we just use the length of the object, which is suitable for data vectors:

```
size.default <- function(x) length(x)
```

For a rectangular array, we want the number of rows and columns, not the length. For a matrix, the length would be the product of the two values, not the pair. To adjust that for our size method we have:

```
size.matrix <- function(x) dim(x)
```

For a list, we can recursively define the size method:

```
size.list <- function(x) sapply(x, size, simplify=FALSE)
```

The sapply function will try to simplify to an array, which is likely not what we want, so we use the argument simplify=FALSE.

Recall a data frame has a list and a matrix interface. Internally it is created from a list of variables, but for purposes of S3 dispatch, it is not treated as a list. So like a matrix, without a separate method, the size method would dispatch to size.default, which would return the number of columns of a data frame.

To get the number of rows and columns could be done with dim, as with a matrix, but here we sneak in a call to NextMethod:

```
size.data.frame <- function(x, ...) c(nrow(x), NextMethod())
```

The NextMethod function will continue the search for a matching size method. In this case, ending up at size.default and returning the number of columns. Though unnecessary here, this functionality allows a subclass to modify a method of a superclass without needing to recopy the entire method definition of the superclass.[14]

To see this work, we have:

```
size(list(numeric   = rnorm(10),
          matrix    = matrix(rnorm(10), nrow=2),
          dataframe = data.frame(x=rnorm(10), y=rnorm(10))))

## $numeric
## [1] 10
##
## $matrix
## [1] 2 5
##
## $dataframe
## [1] 10  2
```

S4 classes and methods

The ease and power of programming with generic functions makes S3 an enticing style to use. For many uses, it is an appropriate choice. For larger projects, or projects that will be shared amongst many different users, a more flexible and robust system is available, this being the S4 style. We don't try and discuss S4 programming here, instead we direct the curious reader to the book by Chambers [8].

Reference classes

Reference classes are a relatively new edition to R. The style of reference classes more closely follows that of other object-oriented programming paradigms: classes are more rigorously defined and methods are attached to classes and their subclasses. In addition, the term "reference" refers to the fact that values may be accessed by reference—and not by copy—so values may be updated within a function call. Some refer to these objects as *mutable*, or able to be modified. Reference classes are well suited to programming graphical user interfaces, for example the shiny packages employs them in its design.

---

[14]A *subclass* is a class that inherits from a *superclass*. In S3, this simply means that in the output of class, the subclass comes first.

We will discuss the main details through an example. The manual page ?ReferenceClasses has much more detail. In the following we define a class for polygons and specializations to hold rectangles, squares, and triangles.

New classes and fields   The creation of a class is done through setRefClass. Unlike S3 programming, a class can have defined fields with required types. In S3 programming, anything can be made a class of some type simply by re-defining the class attributes—nothing enforces a class to have any structure. This makes it very easy in complicated programs for misassumptions about objects to be made.

For reference classes (like the S4 system they are written in), fields are specified by name and optionally class. In our example below, field names and types are defined through the fields argument. We recommend the style of specifying the field names and types with a list, though one can just use a character vector to specify the names, The lone field of our Polygon class is pts, a matrix with two columns holding the x and y coordinate pairs. The specification of the class for pts ensures us that only a matrix can be present for that value. This rigor is a good thing, as when we use this subsequently we can be assured of the type of object without needing to do any type checking.

Methods   The methods argument is a *named* list allowing the definition of one or more methods for the class. In reference classes, a method is a property of the class that holds a function (not a data object like pts) which has some extra features for evaluation attached to it.

Some methods have reserved names with special meanings. For example, we specify an initialize function below. This method, when defined, is called when a new instance is created. In our definition, we pass along values we don't use to any superclass through the callSuper(...) call. The default function just takes any named arguments and tries to set them as property values. In our method we wish to change the more convenient x and y spec-ification of a polygon into a more convenient-to-program matrix with two columns. We replicate the first point to ensure a closed polygon and save the user from having to do so.

The following logic is employed. A check is made to test if a value for pts is passed in. If so, then the pts fields is initialized. The use of the double arrow assignment is important. Inside a reference class method, assignment with <- or = is *local* to the body of the method. In this case, we want to modify the pts property of the class—not a pts object in the environment defined for the function call, hence the use of the double arrow which will search and find the appropriate environment to bind pts.[15] An alternate to assignment in this way, is the initFields methods which is used for field initialization.

---

[15] R will produce a warning when a named field is not assigned through the double arrow assignment.

```
Polygon <- setRefClass("Polygon",
                        fields=list(
                          pts="matrix"
                          ),
                        methods=list(
                          initialize = function(pts, ...) {
                          if(!missing(pts))
                            pts <<- pts
                          callSuper(...)
                        }
                      ))
```

The pts field must be a matrix, but not all matrices specify valid polygons. Best to do some validation before a user gets too far. Reference classes are primarily an extension of S4 classes. In S4 programming, the method validObject is used to check an object's validity. Below, we add a validObject methods which throws an error if the polygon entered is not convex. The check of this is from a mathematical characterization whose details are not important here.

```
Polygon <- setRefClass("Polygon",
        fields=list(
          pts="matrix"
          ),
        methods=list(
          initialize = function(pts, ...) {
            if(!missing(pts)) {
              initFields(pts=pts)
              validObject()
            }
            callSuper(...)
          },
          validObject=function() {
            is_convex() || stop("Not a convex polygon")
          },
          is_convex=function() {
            "Is this a convex polygon?"
            if (nrow(pts) < 3)
              stop("Our polygons have 3 or more vertices")

            q <- rbind(pts, pts[1:2]) # needed for last check
            cp <- function(x, y)  x[1]*y[2] - x[2]*y[1]
            signs <- sapply(1:(nrow(q)-1),
                        function(i) cp(q[i,], q[i+1,]))
            all(signs >= 0) || all(signs <= 0)
```

```
    }
))
```

The above will throw an error if a user tries to specify a non-convex poly-gon. Generally speaking, defensive programming against invalid user input is a good thing.

At this point, we have a new class, Polygon, and a constructor with the same name, as we assigned it that way. To create a *new* instance of the class, we call the classes' new method (which always exists, and in this case will consult the just-defined `initialize` method).

Calling a reference class method is done using R's object-oriented notation for calling a method:

```
theta <- seq(0, 2*pi, length=6)
theta <- theta[-length(theta)]              # drop last one
pts <- cbind(cos(theta), sin(theta))

p <- Polygon$new(pts=pts)                    # a pentagon
```

We assume—like the base `polygon` function—that we don't need to repeat the first and last vertex. This is much more convenient in general, but in this example has us trim off the last value of `theta`.

The last line shows the style to call a class method `class$class_method`. Another class method is `help`, which when called will list the other available methods.

To see that we need to specify a convex polynomial, we try to set a bow tie with a crossing pair of lines:

```
pts <- cbind( c(0,1,1,0), c(0,1,0,1) )
Polygon$new(pts=pts)

## Error: Not a convex polygon
```

Adding methods to a class can be done simply by adding more *named components* to the `methods` list above, or through the `methods` class method, as illustrated here where an area method is added. This uses a formula from http://www.mathopenref.com/coordpolygonarea.html for the area of a poly-gon specified by its vertices.

```
Polygon$methods(
    area = function() {
        "Compute area of a polygon, cf.
        http://www.mathopenref.com/coordpolygonarea.html"
        q <- rbind(pts, pts[1,])
        m <- nrow(q)
```

```
    i <- 1:m; j <- c(2:m,1)
    sum(q[i,1] * q[j,2] - q[j,1] * q[i, 2]) /2
})
```

The first string documents the method. This documentation is available through:

```
Polygon$help("area")

## Call:
## $area()
##
##
## Compute area of a polygon, cf.
##      http://www.mathopenref.com/coordpolygonarea.html
```

More importantly, is the calling of a method on an instance of a class. The syntax is instance$method_name, where the instance is sometimes known as the receiver. The following calls the area method on the instance p of the Polygon class:

```
p$area()

## [1] 2.378
```

Within the body of the method, the pts object is found in the fields, and the functions like sum are found through a regular search. Methods can also be used without reference to the class, and will be searched for before searching in the usual placed.

Continuing, we can add a method to compute the perimeter. We do this by finding the lengths between each successive pair of points.

```
Polygon$methods(
    perimeter = function() {
        "Compute perimeter by adding length of pieces"
        q <- rbind(pts, pts[1,])
        m <- nrow(q)
        lens <- sapply(1:m, function(i)
            stats::dist(q[if(i < m) i:(i+1) else c(1,m),])[1])
        sum(lens)
    }
)
```

We stop to see that this all works. The following creates a regular *n*-gon that is inscribed in the unit circle and compares its perimeter to the circumference of a circle:

```
ngon <- function(n) {
  theta <- seq(0, 2*pi, length=n+1)
  theta <- theta[-(n+1)]                    # Don't add closing one
  pts <- cbind(cos(theta), sin(theta))
  Polygon$new(pts)
}
p <- ngon(17)
p$perimeter() / (2 * pi)

## [1] 0.9943
```

Next we add in a few methods for convenience.

The dist method shows how we can override a binding, in this case the dist function from the stats package, and still use the original function by referencing its package. (The dist function computes distances of a matrix, our dist method is a convenience function for the distance between two points.) If dist is called within a reference class method, this method will be used which subsequently calls the dist function from stats after some data mangling.

```
Polygon$methods(
  dist=function(x,y) stats::dist(cbind(x,y))[1]
)
```

Now we do a visualization method, a simple plot. Again, we need to reference a plot function in the base graphics package. As polygons are vertices connected by edges, we don't allow the user to modify the type argument. The rotate method employs a mathematical trick to rotate coordinates (matrix multiplication). The last two lines are different. The use of the double assignment shows that this method will modify the value of pts. Reference classes are mutable, and calling this function will update the points without any assignment.

The last line returns the magic .self object which refers to the class instance when the method is called. This allows the chaining of calls, a style that is popular in JavaScript programming, though perhaps in this application causes more confusion than it is worth.

```
Polygon$methods(
  plot=function(type, xlab="", ylab="", ...) {
    graphics::plot(pts, type="l", xlab=xlab, ylab=ylab, ...)
  },
  rotate=function(alpha) {
    a <- rbind(
      c( cos(alpha), sin(alpha)),
      c(-sin(alpha), cos(alpha))
```

```
      )
    pts <<- pts %*% a
    invisible(.self)
  }
  )
```

Here we show how the plotting may proceed. The asp argument for plot is passed on via the ... argument and just sets the aspect ratio to be 1-1. The second bit of code, shows how we can chain method calls when we return the reference .self.

```
p$plot(asp=1)
## rotate by pi/10 and then plot repeatedly
for(i in 1:10) {
  Sys.sleep(.05)
  p$rotate(pi/10)$plot(xlim=c(-1,1), ylim=c(-1,1), asp=1)
}
```

Subclasses Now we define specialized subclasses of our Polygon class. First a rectangle class. A rectangle is a convex polygon with four vertices, and whose diagonals are of equal length. This latter part is codified in the is_rectangle method, convexity is checked in the callSuper call.

Rectangles have a much simpler pair of formulas for the area and perimeter, assuming we can find their width and height. We do this through get_wh and store these values as additional properties. The contains argument to setRefClass indicates that this new class is a subclass of Polygon and so inherits its methods and fields. To these we add the w and h fields.

```
Rectangle <- setRefClass("Rectangle",
          contains="Polygon",
          fields = list(
            w="numeric",
            h="numeric"
            ),
          methods=list(
            initialize=function(pts, ...) {
              if(!missing(pts)) {
                wh = get_wh(pts)
                initFields(pts=pts, w = wh[1], h=wh[2])
                validObject()
              }
              callSuper(pts, ...)
            },
            validObject=function() {
```

```
        is_rectangle() || stop("Not a rectangle")
        callSuper()
      },
      is_rectangle=function() {
        "are these points of a rectangle?"
        x <- pts[,1]; y <- pts[,2]
        ((length(x) == length(y)) && (length(y) == 4)) &&
          isTRUE(all.equal(dist(x[c(1,3)], y[c(1,3)]),
                           dist(x[c(2,4)], y[c(2,4)])))
      },
      get_wh=function(pts) {
        "get width, height from points"
        x <- pts[,1]; y <- pts[,2]
        c(dist(x[1:2], y[1:2]), dist(x[2:3], y[2:3]))
      },
      area=function() w * h,
      perimeter = function() 2*w + 2*h
    ))
```

By redefining area for this class, instances of it will use this definition, not that inherited from Polygon.

Here we provide a constructor for this class, making it easier for the user:

```
rectangle <- function(x,y) Rectangle$new(pts=cbind(x,y))
```

The rectangle class requires four points. Here, our first attempt fails, as we try to specify the replicated point to close the polygon:

```
r <- rectangle(c(0,2,2,0,0), c(0,0,1,1,0)) # fails! (5 vertices)

## Error: Not a rectangle
```

Rather, we just specify the coordinates in some clockwise or counter-clockwise order:

```
r <- rectangle(c(0,2,2,0), c(0,0,1,1))
c(r$w, r$h)

## [1] 2 1

r$area()

## [1] 2
```

The validity checks are defined with callSuper. This passes on the call to the superclass so all the inherited checks are in place. This specification will fail as the polygon is not convex, not because the diagonals are a different length:

```
rectangle(c(0,1,1,0), c(0,1,0,1) )          # a bow tie

## Error: Not a convex polygon
```

A square is a rectangle with four equal-length sides. This further subclass shows how to subclass `Rectangle`:

```
Square <- setRefClass("Square",
            contains="Rectangle",
            methods=list(
                initialize=function(pts, ...) {
                  if(!missing(pts)) {
                    initFields(pts=pts)
                    validObject()
                  }
                  callSuper(pts, ...)
                },
                validObject=function() {
                  is_square() || stop("Not a square")
                  callSuper()
                },
                is_square=function() {
                  wh <- get_wh(pts)
                  isTRUE(all.equal(wh[1], wh[2]))
                }
                ))
```

Again, a simplified constructor

```
square <- function(x, y) Square$new(pts=cbind(x,y))
```

Here are some examples, the first fails as we try to specify a rectangle:

```
s <- square(c(0,2,2,0), c(0,0,1,1))         # fails

## Error: Not a square

s <- square(c(0,1,1,0), c(0,0,1,1))
s$perimeter()                               # uses that of Rectangle

## [1] 4
```

A triangle is not a specialization of a rectangle, so our class here is a subclass of the `Polygon` class. The only additional method is one to check that the polygon specifies a triangle, in this case defined by just 3 vertices (though this allows for the case of degenerate triangles).

```
Triangle <- setRefClass("Triangle",
            contains="Polygon",
            methods=list(
                initialize=function(pts, ...) {
                    if(!missing(pts)) {
                        initFields(pts=pts)
                        validObject()
                    }
                    callSuper(pts, ...)
                },
                validObject=function() {
                    is_triangle() || stop("Not a triangle")
                    callSuper()
                },
                is_triangle = function() {
                    x <- pts[,1]; y <- pts[,2]
                    length(x) == length(y) && length(y) == 3
                }
            ))

triangle <- function(x, y) Triangle$new(pts=cbind(x,y))
```

This shows the area method is inherited from the Polygon superclass.

```
t <- triangle(c(0,2,1), c(0,0,1))
t$area()

## [1] 1
```

Bibliography

[1] Alan Agresti. *An Introduction to Categorical Data Analysis*. Wiley-Interscience, 2nd edition, 2007.

[2] Alan Agresti and Barbara Finlay. *Statistical Methods for the Social Sciences*. Pearson, 4th edition, 2008.

[3] Jim Albert. *Bayesian Computation with R*. Springer, 2nd edition, 2009.

[4] Jim Albert. *LearnEDA: Functions for Learning Exploratory Data Analysis*, 2012. R package version 1.2.

[5] Jeanine Baker and Samara McPhedran. Gun Laws and Sudden Death: Did the Australian Firearms Legislation of 1996 Make a Difference? *British Journal of Criminology*, 2007. doi:10.1093/bjc/azl084.

[6] W. John Braun and Duncan J. Murdoch. *A First Course in Statistical Programming with R*. Cambridge University Press, 2008.

[7] Angelo Canty and B. D. Ripley. *boot: Bootstrap R (S-Plus) Functions*, 2013. R package version 1.3-9.

[8] John M. Chambers. *Software for Data Analysis*. Springer, 2008.

[9] Winston Chang. *R Graphics Cookbook*. O'Reilly Media, 2013.

[10] William S. Cleveland. *Visualizing Data*. Hobart Press, 1993.

[11] George Cobb. The Introductory Statistics Course: A Ptolemaic Curriculum? *Technology Innovations in Statistics Education*, 2007.

[12] Richard Cotton. *Learning R: A Step-by-Step Function Guide to Data Analysis*. O'Reilly Media, 2013.

[13] Michael Crawley. *The R Book*. Wiley, 2nd edition, 2012.

[14] Michael J. Crawley. *Statistics: An Introduction using R*. Wiley, 2005.

[15] Peter Dalgaard. *Introductory Statistics with R*. Springer, second edition, 2008.

[16] A. C. Davison and D. V. Hinkley. *Bootstrap Methods and Their Applications.* Cambridge University Press, Cambridge, 1997. ISBN 0-521-57391-2.

[17] Meyer Dwass. Modified Randomization Tests for Nonparametric Hypotheses. *The Annals of Mathematical Statistics*, 28(1):181–187, 1957.

[18] Bradley Efron. Bootstrap methods: Another look at the jackknife. *Ann. Statist.*, 7(1):1–29, 1979. http://projecteuclid.org/euclid.aos/1176344552.

[19] Michael D. Ernst. Permutation Methods: A Basis for Exact Inference. *Statistical Science*, 19(4):676–685, 2004. DOI 10.1214/088342304000000396.

[20] Julian J. Faraway. *Linear Models with R.* Chapman & Hall/CRC, 2004.

[21] Julian J. Faraway. *Extending the Linear Model with R: Generalized Linear, Mixed Effects and Nonparametric Regression Models.* Chapman & Hall/CRC, 2005.

[22] John Fox. *Applied Regression Analysis and Generalized Linear Models.* SAGE Publications, second edition, 2008.

[23] David Freedman, Robert Pisani, and Roger Purves. *Statistics.* W. W. Norton & Company, fourth edition, 2007.

[24] Michael Friendly. *HistData: Data sets from the history of statistics and data visualization,* 2012. R package version 0.7-0.

[25] Andrew Gelman and Jennifer Hill. *Data Analysis Using Regression and Multilevel/Hierarchical Models.* Cambridge University Press, 2006.

[26] William Gosset. The probable error of the mean. *Biometrika*, 6:1–25, 1908.

[27] Garrett Grolemund and Hadley Wickham. Dates and Times Made Easy with lubridate. *Journal of Statistical Software*, 40(3):1–25, 2011.

[28] Deborah Lynn Guber. Getting What You Pay For: The Debate Over Equity in Public School Expenditures. *Journal of Statistics Education*, 7(2), 1999. http://www.amstat.org/publications/jse/secure/v7n2/datasets.guber.cfm.

[29] Frank E. Harrell. *Regression Modeling Strategies: With Applications to Linear Models, Logistic Regression, and Survival Analysis.* Springer, 2010.

[30] David Hemenway. How to find nothing. *Journal of Public Health Policy*, 30(3):260–268, 2009.

[31] Torsten Hothorn, Kurt Hornik, Mark A. van de Wiel, and Achim Zeileis. A Lego System for Conditional Inference. *The American Statistician*, 60(3):257–263, 2006.

[32] Torsten Hothorn, Kurt Hornik, Mark A. van de Wiel, and Achim Zeileis. Implementing a Class of Permutation Tests: The coin Package. *Journal of Statistical Software*, 28(8):1–23, 2008.

[33] Rosenthal J.A., Lu X., and Cram P. Availability of consumer prices from U.S. hospitals for a common surgical procedure. *JAMA Internal Medicine*, pages 1–6, 2013.

[34] Frank E. Harrell Jr., with contributions from Charles Dupont, and many others. *Hmisc: Harrell Miscellaneous*, 2013. R package version 3.10-1.1.

[35] Matthew Kadey. Nutrition. trailrunnermag.com, April 2013.

[36] Daniel T. Kaplan. *Statistical Modeling: A Fresh Approach*. CreateSpace Independent Publishing Platform, 2009.

[37] Campbell K.J., Crawford D.A., and Ball K. Family food environment and dietary behaviors likely to promote fatness in 5–6 year-old children. *International Journal of Obesity*, 30(8):1272–80, 2006. http://www.ncbi.nlm.nih.gov/pubmed/16491108.

[38] Roger Koenker. *quantreg: Quantile Regression*, 2013. R package version 4.98.

[39] Robin H. Lock, Patti Frazer Lock, Kari Lock Morgan, Eric F. Lock, and Dennis F. Lock. *Statistics: Unlocking the Power of Data*. Wiley, 2012.

[40] Norman Matloff. *The Art of R Programming: A Tour of Statistical Software Design*. No Starch Press, 2011.

[41] David Meyer, Achim Zeileis, and Kurt Hornik. *vcd: Visualizing Categorical Data*, 2012. R package version 1.2-13.

[42] David S. Moore and George P. McCabe. *Introduction to the Practice of Statistics*. W.H. Freeman & Co., fifth edition, 2006.

[43] Jad Mouawad and Christopher Drew. Airline Industry at Its Safest Since the Dawn of the Jet Age. *The New York Times*, February 12, 2013. p. A1.

[44] Paul Murrell. *R Graphics*. CRC Press, second edition, 2011.

[45] Christine Neill and Andrew Leigh. Weak Tests and Strong Conclusions: A Re-Analysis of Gun Deaths and the Australian Firearms Buyback. CEPR Discussion Papers 555, Centre for Economic Policy Research, Research School of Economics, Australian National University, June 2007.

[46] Joe Nocera. The 'Die Hard' Quandary. *The New York Times*, February 12, 2013. p. A31.

[47] Karl Pearson. On Lines and Planes of Closest Fit to Systems of Points in Space. *Philisophical Magazine*, 2:559–572, 1901. http://stat.smmu.edu.cn/history/pearson1901.pdf.

[48] José Pinheiro and Douglas Bates. *Mixed-Effects Models in S and S-PLUS*. Springer, 2000.

[49] Fred Ramsey and Daniel Schafer. *The Statistical Sleuth: A Course in Methods of Data Analysis*. Cengage Learning, 2002.

[50] Peter Reuter and Jenny Mouzos. *Evaluating Gun Policy: Effects on Crime and Violence*, chapter Australia: A Massive Buyback of Low-Risk Guns, pages 121–156. Brookings Institution, 2003.

[51] Elisabeth Rosenthal. The Price for a Hip Replacement? Many Hospitals Are Stumped, Research Shows. *The New York Times*, February 12, 2013. p. A21.

[52] Deepayan Sarkar. *Lattice: Multivariate Data Visualization with R*. Springer, 2008.

[53] Andrew Ross Sorkin. A database of names and how they connect. *The New York Times*, February 12, 2013.

[54] Stanley Smith Stevens. On the Theory of Scales of Measurement. *Science*, 103:677–680, 1946.

[55] Stephen M. Stigler. *The History of Statistics*. Belknap Press of Harvard University Press, 1986.

[56] Paul Teetor. *R Cookbook*. O'Reilly Media, 2011.

[57] W. N. Venables and B. D. Ripley. *Modern Applied Statistics with S*. Springer, New York, fourth edition, 2002. ISBN 0-387-95457-0.

[58] Amanda Wachsmuth, Leland Wilkinson, and Gerard E. Dallal. Galton's Bend: A Previously Undiscovered Nonlinearity in Galton's Family Stature Regression Data. *The American Statistician*, 57(3):190–192, August 2003.

[59] M. P. Wand. Data-Based Choice of Histogram Bin Width. *Statistical Computing and Graphics*, 1997.

[60] Hadley Wickham. Advanced R development: making reusable code. https://github.com/hadley/devtools/wiki.

[61] Hadley Wickham. *ggplot2: Elegant Graphics for Data Analysis*. Springer New York, 2009.

[62] Hadley Wickham. The Split-Apply-Combine Strategy for Data Analysis. *Journal of Statistical Software*, 40(1), 2011.

[63] Peter Wolf and Uni Bielefeld. *aplpack: Another Plot PACKage: stem.leaf, bagplot, faces, spin3R, and some slider functions*, 2012. R package version 1.2.7.

[64] Alain Zuur, Elena N. Ieno, and Erik Meesters. *A Beginner's Guide to R*. Springer, 2009.

Index

Printed in the United States
by Baker & Taylor Publisher Services